Geothermal Reservoir Engineering

Second Edition

Geothermal Reservoir Engineering

Second Edition

Malcolm A. Grant
Paul F. Bixley

AMSTERDAM • BOSTON • HEIDELBERG • LONDON • NEW YORK • OXFORD
PARIS • SAN DIEGO • SAN FRANCISCO • SINGAPORE • SYDNEY • TOKYO

ELSEVIER

Academic Press is an imprint of Elsevier

Academic Press is an imprint of Elsevier
30 Corporate Drive, Suite 400, Burlington, MA 01803, USA
525 B Street, Suite 1800, San Diego, CA 92101-4495, USA
The Boulevard, Langford Lane, Kidlington, Oxford, OX51GB, UK

Second edition 2011

Library of Congress Cataloging-in-Publication Data
Grant, Malcolm A.
  Geothermal reservoir engineering. – 2nd ed. / Malcolm A. Grant, Paul F. Bixley.
    p. cm.
  Includes bibliographical references and index.
  ISBN 978-0-12-810375-3 (paperback)
  1. Geothermal engineering. I. Bixley, Paul F. II. Title.
  TJ280.7.G7 2011
  621.44–dc22
                              2010051008

British Library Cataloguing in Publication Data
A catalog record for this book is available from the British Library

For information on all Academic Press publications
visit our web site at books.elsevier.com

Printed   and   bound   in   the   USA

11 12 13   10 9 8 7 6 5 4 3 2 1

ISBN: 978-0-12-810375-3

Working together to grow
libraries in developing countries

www.elsevier.com | www.bookaid.org | www.sabre.org

ELSEVIER   BOOK AID
            International   Sabre Foundation

Contents

Foreword

When the first edition of *Geothermal Reservoir Engineering* appeared in the early 1980s, it was first of its kind. The field of geothermal reservoir engineering appeared to be still something of a novelty at that time, since the professional title of "reservoir engineer" had only just begun to be used in the geothermal industry, although it had been common in the petroleum industry for decades. The migration of petroleum reservoir engineering principles into the geothermal field was perhaps begun in the late 1960s by the late Professor Henry J. Ramey, Jr., and gathered steam (so to speak) in the 1970s as U.S. oil companies, spearheaded by UNOCAL, joined the geothermal industry. Nonetheless, the principles of reservoir engineering had been applied in geothermal development from the beginning, without use of the title. As geothermal development expanded rapidly in the 1970s, spawned by the two oil crises, practitioners of reservoir engineering were drawn from among the ranks of geoscientists and engineers throughout the geothermal community. In fact, this confluence of geoscience and engineering is one of the characteristics that distinguishes geothermal reservoir engineering from petroleum reservoir engineering. It is not an accident that the authors of this book come from very different backgrounds: Malcolm Grant, originally an applied mathematician, and Paul Bixley, a geologist. Despite their disparate technical origins (or perhaps because of it), the two can be recognized clearly as among the most experienced geothermal reservoir engineering experts in the world. Geothermal reservoirs themselves have a degree of dissimilarity from one to the next, which made the development of procedures and practices problematic at times. Applying basic principles, and at the same time recognizing the uniqueness of each reservoir, is an important aspect of the reservoir engineering of geothermal fields. That is why this book is so important: First, because its authors have firsthand experience in very many of the world's geothermal reservoirs, and, second, because the book itself includes specific case examples, adding the reality to the theory.

For nearly 30 years, we have used this book as a reference for our graduate course in geothermal reservoir engineering at Stanford University, and students of our university and many other universities and training programs have benefitted greatly from it. However, the first edition has been out of print for many years, and it has been difficult to find outside of libraries. This second edition will be welcomed back by those who mourned the disappearance of the first. More important, much has been added to the field of geothermal reservoir engineering in the past 30 years, both in theory and in new lessons learned. The return of this book in the form of a second edition is most welcome, especially since it coincides with the renaissance of geothermal reservoir development in response to the rebirth of interest in renewable energy. The geothermal community will look forward to using it for the next 30 years!

Roland N. Horne

Thomas Davies Barrow Professor of Earth Sciences, Stanford University, California

July 2010

The preparation of a second edition has highlighted just how much progress has been made in geothermal science in the last three decades. The first edition was consciously an attempt to meet a developing need, providing a text for a new discipline served only moderately well by the literature in related fields. Now there is a well-defined geothermal reservoir engineering profession, and specialist courses are available in several universities worldwide. This second edition is therefore devoted more to document current practice.

The growth of geothermal development has provided the basic information to define the processes of reservoir engineering. The discipline revolves around the flow of fluid through rock, and it is only by actual observation of thermodynamic changes with such flow that theory can be validated or refuted. By providing a body of evidence, measuring changes on large and small scales, theoretical models have become actual methods to be applied for useful purposes.

The second edition follows the first in being designed as both a text for students and a manual for working professionals. As a textbook it aims to present a complete introduction to geothermal reservoir engineering for the student who has a background in geosciences, engineering, or mathematics. All of the basic material is included, and for each significant point either an explanation with examples is provided or reference is made to readily available published material.

This edition presents the state of the art at the present time. Important concepts and issues are explained as they are presently understood. In all sections there is a strong emphasis on practicality. The techniques described are those that are used in field practice and have been found to generate useful results. Theoretical results that have not been validated are discussed only briefly, if at all.

The data and the examples presented in this book are predominantly from high-temperature fields in volcanic terrain. Many of the methods are also applicable to low-temperature fields, and some examples of these are also included.

The layout of the book corresponds roughly to the chronological development of a field, starting with some initial concepts, then well testing, and finally on to reservoir simulation and development. For the reservoir engineer or student, all of the chapters should be of interest. For engineers and scientists of other disciplines, Chapters 2 and 10 cover the concepts of reservoir engineering and their relation to field models. Chapter 4 will be of interest to geoscientists who wish to correlate their observations with reservoir conditions as inferred from downhole data.

Malcolm A. Grant
Paul F. Bixley
Auckland, New Zealand

Acknowledgments

Many people and organizations assisted in the preparation of this edition. For the generous access to field data we thank New Zealand's geothermal operators, specifically Contact Energy, Mighty River Power (through its associates Rotokawa Joint Venture and Kawerau Geothermal), Tuaropaki Power Company, Ngati Tuwharetoa Geothermal Assets, and Top Energy. We also thank Mighty River and Paul Beere for preparing most of the figures.

The first edition contained a number of numerical errors (mostly in Appendix 1), and we thank those who found them. Particular thanks go to the Geological Survey of Japan and also to Brian Barnett, Dick Glover, Graham Weir, Barbara Simpson, Mark McGuiness, Rick Allis, Yunus Razali, and Jorge Acuña.

Discussions with colleagues have proven invaluable in clarifying our thoughts, and in particular we thank Tom Powell, Sabodh Garg, Bill Cumming, Greg Ussher, and Christine Siega for their input.

Geothermal Reservoirs

1.1. INTRODUCTION

This book examines the flow of fluid underground, how geothermal reservoirs came to exist, the reservoirs' characteristics, and how they change with development. In essence, the historic flow of fluid created reservoirs, and the modification of this flow by exploitation is the basis for the science of geothermal reservoir engineering.

Geothermal resources have been used for cultural purposes and mineral extraction for the last 2000 years. The first modern "deep" drilling (>100 meters) to investigate the deeper resource commenced at Larderello in 1856, and the first power generation began almost 50 years later in 1904 (Cataldi et al., 1999). Relative to petroleum or groundwater resources, the development of geothermal resources that followed was slow. The first "real" power development, of 250 kWe at Larderello commenced in 1913, building up to more than 100 MWe by the 1940s. This was done with a steam-dominated resource. It was not until 1958 that electricity was first generated in significant amounts from a high-temperature liquid geothermal resource at Wairakei. Since that time, the development of steam- and liquid-dominated resources for power production has begun in many countries worldwide, with a total installed capacity exceeding 10,000 MWe in 2010.

In both groundwater and petroleum resources, scientific research went along with expanding exploitation. At first this research was primarily directed toward prospecting to find and extract the fluid. As the effects of fluid extraction

Geothermal Reservoir Engineering. ISBN: 978-0-12-383880-3 DOI: 10.1016/B978-0-12-383880-3.10001-0

on the underground resource became apparent, research into the behaviors of wells and underground reservoirs to understand these changes and learn about the underground resource expanded. The first text on the flow of fluids through porous media was published in 1937 (Muskat, 1937), and by the 1940s, groundwater hydrology and petroleum reservoir engineering were scientific disciplines in their own right. Since then, the continuing rapid development in exploiting these resources has led to a significant increase in field experience and research that has resulted in the development of a large and sophisticated reservoir engineering industry focused on the analysis and prediction of subsurface reservoir and well performance.

In comparison, geothermal reservoir engineering is a much smaller industry, with a correspondingly smaller professional workforce, but it too has now accumulated experience to form a specialized profession.

1.2. THE DEVELOPMENT OF GEOTHERMAL RESERVOIR ENGINEERING

Research efforts related to geothermal systems and their exploitation have followed a pattern similar to that for groundwater and petroleum reservoirs. The exception, of course, is that because of its later start, the high-precision tools and computers that were not available in the early days of groundwater and petroleum exploration have naturally played a larger role in the development of geothermal reservoir engineering. Without the sophisticated exploration methods now available to locate potential underground geothermal resources, during the early geothermal developments of the 1950s and 1960s, most exploration wells were drilled into areas defined by the discharge of steam and/or hot water and associated surface thermal activity. The initial geothermal developments were relatively small, with the result that the resource was not stressed, and a reliable understanding of reservoir behavior or for geothermal reservoir engineering was not required to predict future behavior.

This does not mean that no scientific work was carried out. Early ideas of subsurface flow associated with geothermal discharges in Iceland were put forward by Bunsen in the 1840s (Björnsson, 2005), Von Knebel (1906), and Thorkelsson (1910) (Einarsson, 1942). Conduction of heat away from an isolated magmatic intrusion was discussed at about the same time by Ingersoll and Zobel (1913). Research in this area was limited, and publications relating directly to geothermal phenomena were intermittent.

Hot springs have been used for millennia around the world for bathing, cooking, and hydrotherapy, and mineral deposits associated with the surface discharges have been exploited at least since the nineteenth century (Cataldi et al., 1999). The earliest exploitation for electrical energy was the use of geothermal steam at Larderello, Italy, starting in 1904. The progressive development of Larderello during the first half of the twentieth century gave practical experience in handling geothermal steam for power generation but produced little in the way of subsurface reservoir engineering technology. In Iceland, Einarsson (1942) developed the idea of deep circulation as the mechanism supplying surface discharges of geothermal fluid, and Bodvarsson (1951) began defining the heat transfer problems associated with geothermal exploitation. With the initiation of drilling in Wairakei, New Zealand, in the early 1950s, the first substantial amount of subsurface data from a liquid-dominated reservoir became available.

Two approaches to geothermal reservoir assessment developed. The first was to map the reservoir, collecting as much information as possible and using this to define the physical properties of the subsurface object that was being explored. The second approach was to investigate the processes that might be occurring underground in order to see what roles they might play in the reservoir being exploited. In practice these two approaches have continued together in many geothermal fields that have been developed since that time.

During the 1950s at Wairakei, the first approach led to the mapping of subsurface temperatures across the field (Banwell, 1957) and, from these maps, to inferences about the pattern of fluid flow in the reservoir. The second approach suggested the presence of thermal convection due to heat at depth and to theoretical studies of large-scale convection systems in porous media (Wooding, 1957, 1963) and the first numerical modeling (Donaldson, 1962). In a similar vein, studies at Steamboat Springs, Nevada (White, 1957), and Iceland (Bödvarsson, 1964) led to an improved understanding of how cold meteoric water may circulate to a certain depth and flow up to the surface to charge a geothermal field.

More detailed analyses of the form now prominent in geothermal reservoir engineering slowly developed. Pressure transient analyses were applied sporadically in most areas of early geothermal exploration (see, for example, de Anda et al., 1961). In the 1960s, systematic analyses were made of fields in Iceland and Kamchatka (Thorsteinsson & Eliasson 1970; Sugrobov, 1970). Later in the decade, the first attempts to apply petroleum reservoir engineering were made (Whiting & Ramey, 1969; Ramey, 1970).

By the mid-1960s, there was considerable geothermal exploration and development in progress around the world. The first power plants were commissioned in The Geysers in California, and extensive study programs were started in Mexico, Chile, Turkey, El Salvador, Japan, and various fields in the Imperial Valley of California. Data from a range of different fields were being produced and some of it published, and more problems were encountered and analyzed.

The phrase "geothermal reservoir engineering" first appeared in the 1970s, and the area emerged as a distinct discipline. During this decade, scientific effort moved away from theoretical studies of what processes might possibly be important in the reservoir to practical analyses driven by the data now becoming available and the actual problems experienced in development.

Coherent conceptual models of reservoirs were developed, consistent with both the large-scale system hosting the reservoir and the local detail determined from well testing. Field developments were normally sized on the basis of volumetric reserve estimates of some form or sometimes on the available well flow alone.

At the beginning of the 1980s, the first numerical simulation codes were developed, and a trial in which several codes were used to simulate a set of test problems demonstrated their consistency (Sorey, 1980). During the 1980s and 1990s, reservoir simulators became more capable, and increasing computing power meant that by the 1990s, it was reasonable to simulate a reservoir with enough blocks to be able to represent known geological structures and varying rock properties within the reservoir. As a result, by the 1990s, larger new developments were being sized on the basis of simulation results (see the discussion of Awibengkok in Chapter 12).

At the same time, downhole instruments were steadily improving in capability and resolution, making detailed temperature-pressure-spinner profiles possible. More important than either simulation or better instruments was the accumulation of collective experience within the profession. Although the basic concepts were all developed by the 1970s, and the fact that many recent technical papers are superficially similar to some from that time, the weight of experience means that these concepts are now being applied more rigorously and more consistently with observation. In some aspects of the geothermal reservoir engineer's work, it is now possible to refer to normal practice to define the procedures and expected results.

In the first decade of the twenty-first century, experience, simulation, and instrumentation have all continued to improve. Perhaps the most significant change has been the increasing importance of environmental impacts of development. Seismic effects have become a limiting factor on some Engineered Geothermal System (EGS) projects (Glanz, 2010). Impacts on surface springs have always been an issue in development of the associated deep resource, and this concern has prevented some developments.

Geothermal reservoir engineering is now clearly a distinct discipline. Distinctive features of geothermal reservoirs generally include the following:

1. The primary permeability is in fractured rock.
2. The reservoir extends a great distance vertically.
3. For liquid-dominated reservoirs a caprock is not essential, and usually the high-temperature reservoir has some communication with surrounding cool groundwater.
4. The vertical and lateral extent of the reservoir may not be clear.

Basic to all reservoir engineering is the observation that almost everything that happens is the result of fluid flow. The flow of fluid (water, steam, gas, or mixtures of these) through rock, fractures, or a wellbore is the unifying feature of all geothermal reservoir analysis.

1.3. DEFINITIONS

Because many different terms are used when discussing geothermal systems (or sections or groupings of such systems), a nomenclature has been selected here that is followed throughout the book. The terms have been defined to keep their meanings clear and consistent. Unfortunately, the limited number of terms commonly used makes for considerable difficulty, since many of these terms have general meanings as well as the particular meanings chosen here.

Most areas of geothermal activity are given some geographic name. Provided they are distinct and separate from neighboring areas of activity, they have been described as *geothermal fields*. The term is intended to be purely a convenient geographic description and makes no presumption about the greater geothermal *system* that has created and maintains the field activity. The many fields in the world that have double names (Mak-Ban, Karaha-Bodas, Bacon-Manito) illustrate that exploration has shown that surface activity originally thought to be associated with separate fields is later found to be part of a single, larger field.

The total subsurface hydrologic system associated with a geothermal field is here termed a *geothermal system*. This includes all parts of the flow path, from the original cold source water, its path down to a heat source, and finally its path back up to the surface.

Finally and most important, there is the *geothermal reservoir*. This is the section of the geothermal field that is so hot and permeable that it can be economically exploited for the production of fluid or heat. It is only a part of the field and only a part of the hot rock and fluid underground. Rock that is hot but impermeable is not part of the reservoir. Whether a reservoir exists depends in part on the current technology and energy prices. It is a fairly common experience to drill deeper into an existing field, proving additional reservoir volume at greater depth. In the most extreme contrast, an EGS (see Chapter 14) project aims to create a reservoir where none exists by creating permeability in hot, otherwise impermeable rock.

1.4. ORGANIZATION OF THIS BOOK

Chapter 2 covers the concepts of geothermal reservoirs. After briefly considering conductive heat flow, the main topic is convective geothermal systems. The need for water circulation to great depth is shown, along with a basic conceptual model of a field driven by an upflow of hot fluid. The dynamic nature of the natural state is stressed. The boiling point for depth (BPD) model is introduced as representing conditions in high-temperature upflow areas. Fields with lateral outflow and fields without boiling are treated as being equivalent. Vapor-dominated fields are related to their natural upflow of steam.

Exploitation can introduce the flow of additional hot and cold fluids, the formation of a free surface, and an increase in boiling in the reservoir.

Conceptual models form the basis of quantitative modeling, but some qualitative inferences can be made directly.

Chapter 3 considers some quantitative models and different approaches to simplifying real situations. The two dominant approaches are pressure-transient and lumped-parameter models. Linking them are the concepts of flow (transmission of fluid and heat) and storage (the ability of the reservoir to store fluid in response to pressure change). After discussing homogeneous porous media, possible differences introduced by fractured media are considered.

Beginning with Chapter 4, the book follows the reservoir engineering work that would be performed as part of a field development in roughly chronological order. Chapter 4 discusses the principles of interpreting measurements in geothermal wells and the sometimes complex relationship between measurements in a well and the physical state in the reservoir. The end objective of measurements on a well is a summary description of the well as a model—a specification of zones of permeability, with permeability, reservoir pressure, and temperature/enthalpy determined for each zone.

Chapter 5 covers downhole PTS instruments and their limitations. Chapter 6 covers measurements during drilling, which provide information about pressure and sometimes temperature in formations above the reservoir. Chapter 7 covers the measurements made at well completion, with the objectives being defining the model of the well, the depth and magnitude of its permeable zones, and the reservoir conditions at each zone. Measurements during warmup confirm or refine these interpretations.

Chapter 8 covers the discharge of the well, which provides the definite measurement of the well's productive capacity and the permeability of its feed zones. Methods for initiating and measuring discharge are discussed. A well may discharge various fluids at the wellhead: liquid water, dry steam, or a two-phase mixture, with the last being the most common. Single-phase flows can be measured by orifice plates or weirs, and two-phase flows are measured by separator, "lip pressure and weir," or tracer dilution. From the variation of mass flow and enthalpy with wellhead pressure, it is possible to make inferences about the reservoir fluid, well permeability, and the state of the well. The calculation of wellhead pressure under flowing conditions is described for single-phase flows, and wellbore simulations are used to model two-phase flows.

Chapter 9 is a case study of well BR2 at Ohaaki. It has a 32-year history, from its first drilling in 1966 to its abandonment in 1998. During this period it exhibited a wide variety of behaviors. Chapters 10−13 turn from properties of the individual well to reservoir-scale properties, with the end objective being the formulation of a conceptual and numerical model of the reservoir and often simplified models of some aspects of reservoir behaviors. Chapter 10 discusses the quantitative and qualitative inferences that can be made from downhole data. The mapped distribution of pressure, temperature, or chemical composition supplies information about the reservoir structure. Chapter 11 discusses reservoir simulation and the role of the reservoir engineer in the simulation, from the production of detailed well data to the review of results. It is stressed that a simulation should be calibrated against as wide a set of data as possible. Additional types of data provide more constraints on possible model structure and improve model quality.

Chapter 12 reviews the observed history of seven fields: Wairakei, The Geysers, Awibengkok (Salak), Svartsengi, Balcova-Narlidere, Palinpinon, and Mak-Ban. All have been exploited long enough to show significant changes. Wairakei shows a field being developed from little initial knowledge about geothermal to long-term operation and a sophisticated simulation. The Geysers' history shows its overdevelopment and later decline, which were mitigated by injection. Awibengkok shows successful development of a new resource. Svartsengi shows development and exploration. Balcova-Narlidere is a well-documented low-temperature field. Palinpinon is a field with initially severe injection returns that was converted into a long-term operation. Mak-Ban is a very high-quality resource successfully developed without major problems. The unexploited Patuha field, a distinctive hybrid between vapor-dominated and liquid dominated, is also summarised. Chapter 13 covers field management with simple lumped-parameter models or decline models, deposition, tracer testing, and injection management.

Chapter 14 covers well stimulation and EGS. The first stimulation method, which is often overlooked, is thermal stimulation by injecting cold water and is the most cost-effective. Acid stimulation provides a means of mitigating deposition problems. Well stimulation of deep sedimentary aquifers is now practiced in many places. Finally, a "true EGS" project is outlined: the creation of permeability and a reservoir in hot rock with little permeability, and the results to date are reviewed. Four appendices review the pressure transient theory, gas corrections for output testing, the equations of state and flow in porous media, and provide a glossary of field names referred to in the book.

1.5. REFERENCES AND UNITS

The geothermal literature is now so extensive that a comprehensive survey has not been attempted. Examples have been chosen and field experience quoted where it has given a useful illustration, but these examples are simply those that were appropriate and conveniently available. Similarly the References are the works that have been cited in the text, but they are not meant to be a comprehensive coverage of the literature. The text does not pretend to be a representative survey of the current literature. Where possible, references have been used that are readily available on the Web and, for preference, those that are available at no cost.

Equations are written in SI units. In some cases, where it is conventional to use other units—for example, lip pressure or orifice plate calculations—SI units are not used, and these equations are marked with an asterisk (*). In text and figures, pressures are usually given in bar (1 bar $= 10^5$ Pa) and wellhead pressures in bars gauge. Pressures are absolute unless gauge is specified. Casing and wellbore sizes are conventionally given in inches, and inches are used in the text, but calculations use meters. The SI unit for permeability is the meter squared. A more convenient unit is 10^{-12} m^2, which is defined to be 1 darcy (1 d). (The definition of the *darcy* uses pressure in atmospheres, with the result that 1 d $= 0.987 \times 10^{-12}$ m^2, but it is more convenient to redefine it as exactly 10^{-12} m^2. In any event, no geothermal permeability is measured with an accuracy of 1%.) The fractional unit md $= 10^{-3}$ d is also used. Note that the unit of temperature is the Kelvin, denoted by K, so, for example, thermal conductivity has the dimensions of W/m.K, not W/m.°C. However, temperatures are measured as temperatures Celsius (°C) to differentiate from absolute temperatures: 0°C $= 273.15$ K. The terms *water* and *steam* are used to refer to the liquid and vapor phases of water substance. Where gas content is significant, *liquid* and *vapor* are used for the liquid and vapor phases of the water-gas mixture, and *water* and *steam* refer to that part of the respective phase that is water substance.

Concepts of Geothermal Systems

2.1. INTRODUCTION

This chapter discusses fluid distribution, pressure, and temperatures that can occur in geothermal reservoirs. These concepts provide a background for the following chapters, which describe the testing of wells and reservoirs, the construction of a conceptual model of the resource, and then a quantitative model.

Much of the following discussion relates to unexploited fields in which the thermodynamic state of the field is determined by the natural processes of heat and fluid transport. This is because the results of testing done early in a new field development are the most challenging to interpret and because the natural state of the field influences its subsequent response to exploitation. As a geothermal field is developed and information from more wells, along with the system response to production and injection, becomes available, the conceptual model and the numerical simulation model can be further refined.

Anyone who has ever watched a geyser in action or a hot pool bubbling and wondered where the water and heat came from has some sort of mental picture of this small part of a geothermal system. The accuracy of such a mental picture depends to a great extent on the available data and on the individual's interpretation of that data based on preconceived ideas and past experience. Each of

Geothermal Reservoir Engineering. ISBN: 978-0-12-383880-3 DOI: 10.1016/B978-0-12-383880-3.10002-2

these mental models is thus unique to the individual concerned and is closely linked with that person's background and experience.

In the scientific arena, such mental models, which are based on a range of data from various disciplines, together with experience in related research, form the basis for the development of *conceptual models* that should bring together all the available information into a single coherent model. Of course, these conceptual models will vary from one expert to another, depending on the individual's background and the weight given to specific data. However, since a model should be consistent with all the observable aspects of the system, they should all give very similar basic results. These models will vary among individuals, depending on their need. For example, the fine detail that is essential in a model of flow around a particular well will become insignificant as the scale increases to include the complete geothermal resource.

2.2. CONDUCTIVE SYSTEMS

2.2.1. The Thermal Regime of the Earth

Over almost the entire surface of the Earth is a flux of heat through the crust upward to the ground surface. This heat is transported to the surface by conduction through the crustal rocks. The average geothermal gradient in the shallowest part of the crust is around 30°C/km. Since the heat flux varies from place to place over the Earth's surface, and the thermal conductivity varies with different strata, conductive gradients of up to 60°C/km can be encountered. Thus, higher temperatures are encountered when drilling or mining deep into the crust, and temperatures more than 100°C are often found in deep oil or gas wells. One means of prospecting for geothermal reservoirs that is not evident from surface discharge of geothermal fluids ("blind systems") is identifying areas of anomalous heat flow by measuring temperature gradient in wells— either shallow "temperature gradient" wells or existing deep oil, gas, or groundwater wells. A region of potential geothermal resource may be associated with high heat flow. The near surface temperature gradients can be extrapolated through impermeable strata to obtain deep temperatures. Such extrapolation cannot be continued through permeable strata because the presence of permeability means that convection will determine the temperature distribution (Benoit, 1978; Salveson & Cooper, 1979).

2.2.2. Warm Groundwater Basins

One source of water at temperatures above mean surface is from aquifers that are so deep that their temperature is raised by the normal geothermal gradient. The mechanism heating the water in such systems is then simply the vertical conduction of heat through the crust. The fluid flow in the aquifer must be sufficiently slow that there is time for the water to be heated by this conductive heat flow. The general reduction of permeability with depth implies that successful production from greater than a few kilometers requires anomalously high permeability, and the chance of such anomaly decreases with increasing depth.

In some large groundwater basins, a fraction of the heated water circulates back up to the surface where there is suitable permeability and structure. Otherwise the heated fluids may be confined within a particular stratum.

2.2.3. Deep Sedimentary Aquifers

Deep sedimentary aquifers heated by the normal thermal gradient are found in many continental environments. These aquifers are usually not part of a currently active circulation system. Figure 2.1 shows such a system. This simple two-well system in an aquifer is very common in groundwater or petroleum engineering. The only difference is that the fluid is warm. Figure 2.1 in its simplicity makes a sharp contrast to the figures that follow, which show active geothermal systems. A typical example of this type of system is a carbonate/sandstone aquifer that was developed using production-injection well pairs for district heating in the Paris Basin. Other examples are found in most larger-scale basin areas where elevated temperatures are encountered around the world.

2.2.4. Warm Springs and Fracture and Fault Systems

Many warm springs are found along major fault and fracture lineations throughout the world, suggesting that these major fault systems provide the channels for the flows of warm water that feed the springs. Such channels provide the means for circulation of cold meteoric water to depths where it is heated by the normal geothermal temperature gradient and then returned to the surface to form warm springs. These are a form of convective system, with convection along the plane of the fault being heated by conductive heat transfer into the fault zone. The driving force for the circulation is the density difference between the cool descending water column and the hotter rising column. This mechanism differs from a full convective systems discussed later in this chapter in that it is confined to a narrow fault plane with no extensive reservoir and is

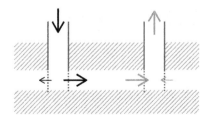

FIGURE 2.1 Injection well (dark) and production well (gray) in sedimentary aquifer.

located in an area with a normal geothermal gradient. An example of a fault-controlled spring system is the hot spring area around Banff, Canada.

2.2.5. Geopressured Systems

Geopressured geothermal reservoirs are closely analogous to geopressured oil and gas reservoirs. Fluid caught in a permeable stratigraphic trap may, by crustal motion over millions of years, be raised to lithostatic pressure. Such reservoirs are generally relatively deep, at least 2 km, so that the geothermal gradient ensures reservoir temperatures over 100°C. A number of such reservoirs have been found in petroleum exploration.

Where these reservoirs are found associated with petroleum, the water is generally saturated with methane, and the methane may be a more important energy source than the heat in the water. Reservoir engineering of such a system is more like a petroleum reservoir than a hydrothermal reservoir or groundwater aquifer. Experiments were conducted on existing wells, originally drilled for petroleum exploration, in the 1980s in the U.S. Gulf Coast, including power generation using an experimental "hybrid power system," but the results were discouraging, with operational problems due to high salinity and high CO_2 content. No further studies have been done since that time (Griggs, 2005). EGS exploration in the Cooper Basin (see Chapter 14) has revealed similar abnormally high pressures in the ancient granite terrain that is overlaid by about 4 km of more recent sediments. Successful development of the deep geothermal resource would operate at similar high pressures.

2.2.6. Hot, Dry Rock or Engineered Geothermal Systems

In some locations, rock of low permeability that has been heated to useful temperatures can be found. The heat source may be from volcanism or an abnormally high geothermal gradient, or there may be impermeable rock on the flanks of a hydrothermal system. Compared with the other systems, these do not have sufficient intrinsic permeability, but they do contain heat.

Exploitation of such a system depends on creating permeability by controlled fracturing such that fluid can be circulated through the rock and heat can be extracted. The fracturing creates a reservoir that did not previously exist. This subject is discussed further in Chapter 14.

2.3. CONVECTIVE SYSTEMS: LIQUID DOMINATED

2.3.1. Introduction: The Dominance of Convection

Hydrothermal convective systems are geothermal systems with high temperatures and usually with surface activity. At the present time, all major geothermal power stations operate on such systems.

In contrast to conductive systems, it is the flow of hot fluid through the system that determines the temperature and fluid distribution. The natural state of the geothermal reservoir in a convective system is thus dynamic, and knowledge of the natural fluid flow is needed to understand how this natural state was formed. Surface features such as geysers, springs, fumaroles, cold gas vents, and mudpools may be associated with this type of reservoir, and they are the end points for some part of the natural thermal flow.

Such natural flows play a dominant role in establishing the state of the fluid within the reservoir, and an understanding of them provides information about reservoir parameters such as vertical permeability, which cannot be determined by other means. Because new flow patterns created by exploitation will usually overwhelm the natural flows in the reservoir, it is important that appropriate information and data be collected early in the exploration program of a new resource.

In low-temperature systems the reservoir fluid is always liquid water, while in higher-temperature systems, steam can also be present. All geothermal reservoirs located to date can be divided into two types—liquid dominated and vapor dominated—depending on whether liquid or steam is the mobile phase. A few reservoirs have separate regions of both types. The majority of reservoirs are liquid dominated and have a vertical pressure distribution that is close to hydrostatic. In vapor-dominated reservoirs the vertical pressure distribution is close to steam-static. In each case the dominant mobile phase, either liquid or steam, controls the pressure distribution, although the other phase may be present in significant amounts. The remainder of this section considers only liquid-dominated reservoirs. Vapor-dominated reservoirs are discussed later in this chapter.

2.3.2. Deep Circulation and Magmatic Heat

Conductive geothermal systems do not require a great deal of heat at depth and can occur anywhere in the world. High-temperature convective systems demand some additional heat above the normal conductive gradient.

One of the earliest conceptualizations of a geothermal system to be built on a detailed analysis of technical evidence was by Bunsen in the 1840s (Björnsson, 2005), who showed that the water discharged by springs in Iceland was meteoric water. For his description of the hot spring system in west Iceland, Einarsson (1942) visualized something akin to a deep groundwater basin. His geothermal flux had to be higher than normal to produce the higher-temperature spring discharge, and his aquifers were factures and fissures in the otherwise impermeable basalts. Variations of Einarsson's model are still accepted for some hot springs in Iceland. For the more intensely active areas of central Iceland, Bödvarsson (1964) allowed deeper circulation and called on a magmatic source for the heat energy.

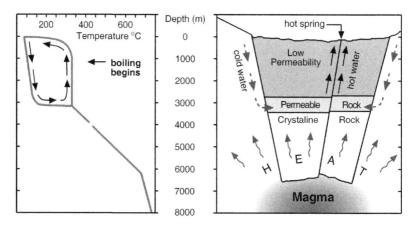

FIGURE 2.2 Model of the large-scale circulation of fluid in the natural state of a geothermal system. *Source: White, D.E., 1967 "Some principles of geyser activity, mainly from Steamboat Springs, Nevada" Am. J. Sci. 265, pp641–684.*

A model similar to Bödvarsson's was produced by White (1967, 1968) for the system associated with Steamboat Springs, where isotopic evidence indicated that about 95% of the water discharged in the springs was of meteoric origin. White produced the model shown in Figure 2.2, where water originating at the ground surface percolates downward through faults and fissures or structures in otherwise impermeable rock to considerable depth.

In Figure 2.2, a circulation depth of 3 km is indicated. White suggested a possible range of 2 to 6 km. A significant number of geothermal fields have now been drilled beyond 3 km without finding a bottom to the system, so the fluid circulation depths would appear to be larger rather than smaller. In White's model the water is heated at depth, probably by close contact with some magmatic body, to the high temperatures encountered in reservoirs associated with such systems. The buoyancy imbalance between the hot and cold columns drives this fluid back up to the surface through other permeable channels.

These systems, which require large amounts of heat compared to the normal crustal heat flux, are generally found in regions of relatively recent volcanism. This accounts for the large number of geothermal fields associated with volcanic arc and crustal rifting. The fractures or flow paths for the water circulation appear to be associated with structures such as regional rift zones or calderas. The total amount of heat transported out of convective geothermal systems over their lifetimes is large—so large, in fact, that not only must circulating water make close contact, but this magma itself must be convecting or replenished by mechanisms such as crustal spreading. Some geothermal fields are long-lived. Browne (1979) imputes a lifetime of hundreds of thousands of years for Kawerau. At Coso a lifespan of 300,000 years has been suggested, at least intermittently (Adams et al., 2000). Silberman and colleagues (1979) suggest that Steamboat Springs may have existed for 3 million years, and Villa and Puxeddu (1994) suggest that Larderello may be as much as 4 million years old. In contrast, some fields have much shorter lifetimes. The Salton Sea geothermal field has an estimated lifetime of 3,000 to 20,000 years (Kasameyer et al., 1984; Heizler & Harrison, 1991).

Long lifetimes of geothermal systems cannot be sustained by a single emplacement of magma. Simulations of single magmatic intrusions in aquifers show that the thermal disturbance lasts only about 10,000 years (Cathles, 1977; Norton & Knight, 1977). For long-lived fields, even if the magma was extensive, many cubic kilometers would be required to supply the cumulative heat discharged (White, 1968; Lachenbruch et al., 1976). Larderello would have required 32,000 km^3 of magma to maintain activity over 4 million years (Villa & Puxeddu, 2004). Similarly Banwell (1957) estimated that over its lifetime Wairakei required at least 10,000 km^3 of magma for its heat supply. For both Wairakei and Larderello the magma volume is too large to be stored under the field. This suggests that the magma source itself must be convecting, so molten rock remains near the zone where heat is exchanged between the magma and the fluid in the geothermal system.

2.3.3. Exploitation and System Circulation

The deep circulation feature of geothermal systems implies that in general the changes induced by exploitation will not greatly affect the natural upflow from depth. Consider the case of a field with a natural flow of 100 kg/s and a base temperature of about 300°C. Assuming a depth of 5 km to the base of the geothermal system, this flow is driven by the pressure difference due to the fluid density between the hot and cold columns, which in this case is about 100 bar. An additional pressure difference of 25 bar due to drawdown in the reservoir would increase the natural flow by 25%, or 25 kg/s. This is not a significant contribution in relation to the typical production flows of about 1000 kg/s.

Thus, the natural state of the reservoir is dynamic, and the fluid distribution is controlled by a dynamic balance of mass and heat flow. Once exploitation occurs, fluid flow to and from wells is generally much greater than the natural flow. This may create a significant flow from parts of the reservoir beyond the depth or areal extent of the wells. With large-scale production and reinjection, the primary induced flow is from the injection wells to the production wells, and induced flow changes outside this area are significantly smaller.

2.3.4. The Vertical Upflow Model and Boiling Point for Depth Models

Having considered the processes in the geothermal system as a whole, this section focuses on the smaller, relatively shallow part of the system that contains the reservoir: the area where fluid rises within an exploitable depth from the surface. The simplest case is when the upflow rises vertically from greater depth to ground surface. The upflow at great depth consists of water or supercritical fluid. As this fluid ascends, the pressure decreases, and at some point, depending on the temperature and fluid chemistry, the ascending fluid forms two phases, gas and liquid, which both rise to the surface. The upflow continues toward the surface as gas and liquid. From conservation of mass and energy it is possible to estimate the form of this upflow. The following assumes that the fluid at depth is liquid and boils when it reaches its saturation pressure, as illustrated in Figure 2.2.

For most purposes, conduction can be ignored as a means of heat transport. For an example, before development, the Wairakei field discharged 400 MW thermal to the ground surface over an area of about 11 km^2—a heat flux of 40 W/m^2. The upflow at depth was most likely confined to a smaller area of 2−3 km^2, and in this case the convective heat flux in the deep recharge area would have been about 180 W/m^2. The original temperature at 400 m was about 250°C, giving a gradient of 250/400 = 0.6 K/m. Assuming a conductivity of 2 W/m.K, the conductive heat flux would have been about 1 W/m^2, two orders of magnitude less than the conductive heat flow.

Assuming that the pressure at the boiling level depends on the temperature in the upflowing liquid zone, as boiling commences, saturation conditions must apply. Below the boiling level, the temperature distribution is given by

$$T = T_b \tag{2.1}$$

where T_b is the constant "base temperature." Above the boiling level, temperature is given by the saturation relation

$$T = T_{sat}(P) \tag{2.2}$$

The pressure gradient at any depth is equal to the local hydrostatic gradient plus the dynamic gradient caused by the upflow. In most cases the latter is less than 10% of the static gradient (Donaldson et al., 1981), and it may be much less. In a field where the mass flux density (upflow per unit area) is low, the excess dynamic gradient is correspondingly small. This excess gradient is present only in the area of upflow, and toward the margins where there is lateral flow, the pressure gradient will be close to hydrostatic. Ignoring the dynamic gradient, the BPD approximation is obtained from

$$\frac{dP}{dz} = \rho_w g \tag{2.3}$$

The BPD pressure profile is that of a static column of water whose temperature is everywhere at saturation for the local pressure. The BPD approximation implies that steam saturation is close to residual (see Appendix 2). The BPD approximation thus not only approximates the pressure and temperature in the reservoir, but it also specifies something about the reservoir fluid: that little mobile steam is present in the boiling zone. Figure 2.3a shows the fluid

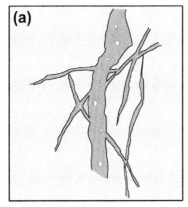 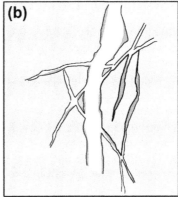

FIGURE 2.3   Distribution of liquid (dark shading) and steam (no shading) in liquid-dominated and vapor-dominated reservoirs. Matrix denoted by pale shading.

distribution in the reservoir fracture network (the higher-permeability paths). Figure 2.3b shows the complementary situation for a vapor-dominated reservoir. In the liquid-dominated reservoir in Figure 2.3a, the fractures are occupied by water, with occasional steam bubbles. The matrix is water-saturated. In the vapor-dominated reservoir in Figure 2.3b, the fractures are occupied by steam, with a film of water clinging to the fracture walls and in dead-end spaces, and the matrix is fully or partly saturated with water.

As a model, BPD—that is, a static fluid column everywhere at boiling point (until the constant-temperature liquid-water section is reached)—is for many purposes a good approximation of the initial state of the upflowing core of the reservoir. Note that this is only an approximation; pressures and temperatures can be higher or lower, and it is incorrect to regard BPD as any sort of theoretical maximum temperature. The BPD profile is naturally of less value where flows are relevant and actual reservoir pressure gradients are necessary, as in comparing pressures between different wells to determine a lateral pressure gradient.

In reality, the rising fluid is cooled by dilution and conduction as much as it would be by boiling, which reduces the amount and extent of boiling. The BPD model is still valid as long as there is some boiling. Typically, boiling conditions will be found in the core of the upflow, with cooler conditions toward the margins.

In all high-temperature geothermal systems, noncondensable gases (NCGs) are present in the reservoir fluids. The presence of these gases together with dissolved salts changes the saturation relation for the reservoir fluid from that for pure water, with the effect that the pressure at which the liquid phase first boils is greater than that for pure water; in other words, boiling starts deeper. A modified boiling curve can be computed by adding conservation of gas to conservation of mass and energy.

2.3.5. Systems with Lateral Outflow

The assumption that the natural flow is entirely vertical is an idealization. Structural control by permeability variation and topographic effects will usually impose some degree of lateral flow. The BPD approximation requires a component of upflow, since boiling conditions can only be maintained if there is some continued upflow of fluid. Most two-phase geothermal fields have associated nonboiling lateral flows away from the boiling upflow region. If the natural flow is horizontal or turns downward, boiling ceases and liquid reservoir conditions are encountered.

Thus, the BPD profile may be applicable only in the upflow region of this type of reservoir. The lateral outflows from this region will be liquid water (although there will usually be some shallow boiling in places along the top of the outflow tongue). Exploration wells will be expected to encounter different regimes within the reservoir, depending on their location. In the upflow region, boiling conditions should be expected, and in outflow regions, high-temperature liquid conditions with temperature inversions (declining temperatures with depth) can be expected. Usually the outflow region of a field is initially explored, and with further drilling, the flow is traced toward the high-temperature source area. For example, at Ahuachapan, El Tatio, Wairakei, Yangbajan, and Tiwi, the outflow region was initially investigated. Some examples follow.

Tongonan

The Tongonan field is located on the island of Leyte in the Philippines. It is a large field with an installed capacity of 703 MW in the greater Tongonan area, which includes the adjacent Mahanagdong field. An overview of the development history is given by Gonzalez and colleagues (2005). Figure 2.4 shows a conceptual cross section.

Tongonan is a liquid-dominated field with a base temperature of over 320°C and chloride content of up to 11,000 ppm. Before development, steam-heated

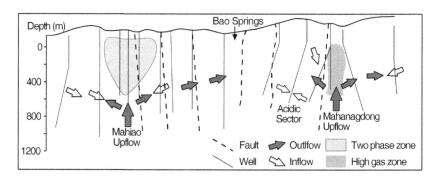

FIGURE 2.4 Conceptual model of Greater Tongonan. *Source: Seastres et al., 1996.* © *Geothermal Resources Council*

waters existed near the surface. The Tongonan upflow rises under Mahiao toward one end of the reservoir. Here, steam and water rise, creating a limited region of two-phase conditions in the natural state. Water flowed laterally away from the upflow area to ultimately discharge at the Bao Springs and possibly elsewhere. This combination of upflow and horizontal outflow is very common. The surface activity in a field of this type is a guide to reservoir fluid distribution. The principal springs are at Bao, but the highest temperatures are beneath the steam-heated activity at Mahiao, and it is such steam-heated activity that indicates the area of upflow. There is an impermeable region separating Tongonan proper from the separate upflow at Mahanagdong. Under production, large pressure drawdown has occurred in the Tongonan area, and the upper part of the reservoir here has become vapor dominated.

Dixie Valley

Dixie Valley is a high-temperature field in the Basin and Range province of the western part of the United States. As is typical of many Basin and Range fields, it is associated with permeability developed along a fault zone that provides the primary source of deep recharge. The reservoir is on the fault zone rather than an outflow from it. Figure 2.5 shows a cross section through Dixie Valley. The reservoir is a region around the fault zone.

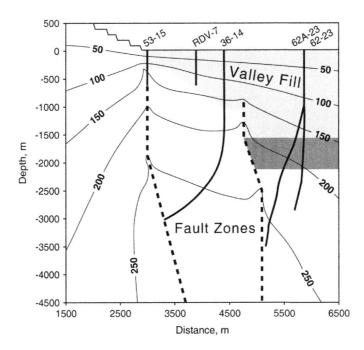

FIGURE 2.5 Dixie Valley thermal cross section. *Source: Blackwell et al., 2000. © Geothermal Resources Council*

2.3.6. Inferences from Pressure Distribution

The discussion of systems with outflow introduced the effects of permeability contrasts on the pattern of the natural flow. The presence or absence of permeable or impermeable features is of great importance in assessing the possible effects of exploitation. The temperature profiles measured in exploration wells, together with geological information, provide a guide to the permeability structure of the reservoir, but pressure distribution (when this is available) gives more definite and direct information about how different parts of the reservoir are interconnected. Often there is little or no information about reservoir pressures above the casing depths during the early stages of field development. Such information is vital to predicting the response of the reservoir to development, and in particular to evaluating the potential interconnection between the high-temperature resource and surrounding cool aquifers. Stage testing of exploratory wells, or drilling specific wells to gain this information once the deeper reservoir has been identified, can provide this essential data.

In petroleum or groundwater resources, a pressure difference between two reservoirs or aquifers at different depths that is significantly different from hydrostatic normally implies that the two reservoirs are not connected by permeable structures. This inference is not always valid for geothermal reservoirs and can be difficult to evaluate due to the variation of the vertical pressure gradient with temperature. Because the natural state of a geothermal reservoir is dynamic, its pressure distribution is as well, and pressure differentials may be caused by the natural flow, the temperature difference, or some combination of these factors.

Figure 2.6 illustrates the natural state (predevelopment) pressure distribution with depth in three New Zealand fields. The depth is measured from the highest surface elevation of discharge of chloride water. At Wairakei and Kawerau, the pressure profile is smooth and extrapolates to atmospheric at surface. No large-scale surface confining layer is apparent. Pressures deep in the reservoir are overpressured compared to surface (for a hydrostatic column allowing for temperature). This overpressure is caused by the natural upflow through the reservoir, not by any confining bed. Because of the coarse nature of the data, this conclusion is only valid on the large scale. With more details, the effects of lower-permeability capping structures can be observed.

Data from Ngawha illustrate a truly confined geothermal field, with only a very small flow of liquid from the reservoir to the surface; a larger flux of gas seeps through the confining formations. Beneath the caprock, production is found in fractured greywacke, and within this reservoir, pressure gradients are near hydrostatic for the reservoir temperature, 225–230°C. The reservoir is overpressured with respect to ground surface. In this case the overpressure is due to the confining layer rather than the dynamic pressure gradient due to vertical flow.

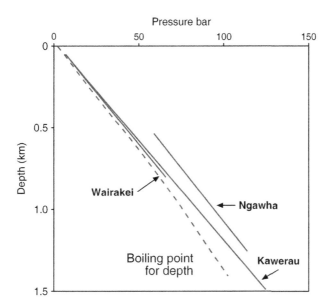

FIGURE 2.6 Pressure distribution with depth in three New Zealand geothermal fields. *Source: Grant 1981.*

2.3.7. Summary

In its natural state the basic components of the geothermal *system*, of which the reservoir is the hot, exploitable part, are: (1) an aquifer or fracture network containing hot fluid, (2) a path through which cold water can flow to recharge the *system* or an input of magmatic fluid, and (3) a source of heat. In the finer details of the *reservoir*, there is often also a reduced permeability zone or cap, or at least a partial aquiclude over the aquifer or channel network (that comprise the *reservoir*), but these elements are not essential. The hot recharge into the geothermal system is perhaps the main feature, apart from the difference in energy that distinguishes this system from its groundwater counterpart. The pressure drive that sustains the hot recharge or upflow is the buoyancy difference between the columns of descending cold and ascending hot water. This difference may be modified by topographic effects. In their natural—predevelopment—state, high-temperature liquid-dominated reservoirs must be considered as a dynamic body before any downhole measurement can be properly evaluated or interpreted.

2.4. CONVECTIVE SYSTEMS: VAPOR DOMINATED

In all the systems discussed so far, the dominant fluid has been water and the pressure distribution approximately that of a static body of water. It has also been implied that fluid can move relatively freely into, through, and out of the system.

This contrasts with vapor-dominated systems and the associated reservoir. In its natural undeveloped state, a vapor-dominated reservoir appears to contain a static column of steam, and any associated surface thermal activity consists of steam or steam-heated water with little chloride content. Four such fields are presently known to exist: The Geysers, the world's largest field by production; Lardarello; Kamojang; and Darajat. (Other hybrid fields are discussed in Chapter 12.) In liquid-dominated reservoirs, vapor-dominated zones of limited thickness are sometimes found, and in several cases, extensive vapor-dominated zones have been formed in response to exploitation—for example, Wairakei, Awibengkok, Miravalles, Ahuachapan, and Tongonan.

The natural vapor-dominated reservoir was first reported by Ramey (1970), who described a reservoir containing steam, with pressure close to steam-static and temperatures near saturation. Figure 2.7 shows pressure profiles as determined in the early period of exploration at Kamojang and Darajat, The Geysers, and Travale. Profiles have been constructed for various sections of Larderello (Celati et al., 1978) showing a mixture of vapor-dominated zones at varying pressures and liquid-dominated zones.

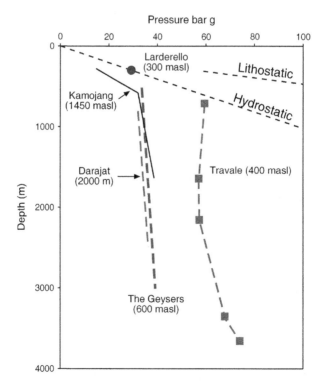

FIGURE 2.7 Reservoir pressure in vapor-dominated systems (Travale is part of greater Larderello). *Source: Allis, 2000.*

In a vapor-dominated reservoir the pressure increases only slowly with depth, implying that a low-permeability boundary exists around the sides of the reservoir, isolating the relatively low pressures inside the reservoir from regions outside the reservoir, where pressures near hydrostatic are expected (see Figure 2.7). Similarly, there must generally be poor permeability above the reservoir except in the limited areas where natural discharges occur.

2.4.1. The Conceptualized Fluid Flow System

Over the years, several models of vapor-dominated systems have been proposed. Early models proposed either a dry steam chamber or a steam zone overlying boiling water. It is now known that in the natural state, the vapor-dominated reservoir contains significant amounts of liquid water held in the rock pores. At both Larderello and The Geysers, reservoir conditions were initially those of saturated steam. Simple volumetric analysis has shown that much more steam has been produced from these fields than could be stored as steam alone. Therefore, the additional steam must have originally been stored as (immobile) liquid.

A qualitative description of the physical processes leading to the development of a vapor-dominated reservoir was first given by White and colleagues (1971) and later refined by D'Amore and Truesdell (1979). In the initial phase of development, there is a natural upflow of steam and gas from a deep boiling zone. The steam spreads laterally through the reservoir, and as it spreads, heat is lost through the caprock, and some of the steam condenses, absorbing some of the gas. This condensate migrates down through the reservoir under gravity, wetting the matrix rock and increasing the permeability by dissolving some of the rock minerals. The steam condensation results in changes in the steam chemistry with distance from the upflow zone. Liquid-mobile species present in the steam are removed with condensate, and vapor-mobile species are concentrated in the remaining steam. This model provides a good fit to the variation in steam chemistry at Larderello and The Geysers.

In a steam-dominated reservoir, the counterflow system of steam moving up and water moving down controls the fluid distribution. If the mass fluxes of steam (up) and water (down) are roughly similar, and the vertical pressure gradient is near steam-static, the relative permeability to water must be low. The flowing steam occupies most of the fracture space, and water occupies the remaining pore space. The reservoir has a saturation such that water is just mobile—that is, just above residual saturation (see Chapter 3). This implies that the mass of water in the reservoir is much greater than the mass of steam. To obtain the best match to the measured response, reservoir simulation of The Geysers requires application of dual porosity models, with liquid saturation in the matrix typically as much as 85% (see Chapter 12). Deeper drilling in Larderello and part of The Geysers has found superheated regions at over 300°C, rather than the hypothesized deep liquid (Bertani et al., 2005; Walters et al., 1991), so the water-steam counterflow may describe only the upper part of the field.

D'Amore and Truesdell also observe that the downflowing water, being acidic condensate, is chemically reactive and will in time erode the rock through which it flows. This may help to explain the high permeabilities typically found in vapor-dominated reservoirs despite the poor permeability found in cooler rock outside the steam reservoir.

Studies of the reservoir mineralogy indicates that all of the four vapor-dominated fields were at one time liquid dominated (Allis, 2000) and have somehow "boiled dry" to produce their current state. The model of Larderello-Travale (Barelli et al., 2010b) simulates this process from an initial state when the rock was filled with cold water. Thus, D'Amore and Truesdell's model represents the present state of the reservoir but not how it was created.

2.5. CONCEPTS OF CHANGES UNDER EXPLOITATION

Exploitation of a geothermal reservoir means that heat and (almost always) mass are withdrawn, and some fraction of this may be reinjected. Additional recharge fluid, hot or cold, may flow into the reservoir. Some conceptual models of reservoirs under exploitation describing the changes in distribution of heat and mass throughout the reservoir and its surroundings are described in the following sections.

2.5.1. Flow of Liquid

The simplest concept of the flow of geothermal fluid in the reservoir is the analogy of the flow of liquid water in a confined aquifer. If the reservoir is at a fairly uniform temperature, the flow is isothermal. If there is a distribution of temperature or if there is reinjection of cooled fluids, it is necessary to compute the motion of thermal changes along streamlines. Figure 2.8 illustrates such a flow.

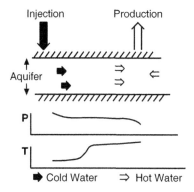

FIGURE 2.8 Flow of liquid in an exploited liquid reservoir.

Distinctive geothermal features are introduced if the reservoir is not a confined aquifer of a conductively heated system, but is part of an active geothermal field. The reservoir may have homogeneous permeability and great thickness, such as those found in the Imperial and Mexicali Valleys in North America. More usually in fractured rock, little is known about the reservoir thickness except that it is probably large. In addition, the fact that the reservoir is part of an active hydrological system makes it more likely that fluid recharge will occur from other parts of the system.

The great thickness of the reservoir causes few conceptual problems. It means that reservoir modeling must describe three-dimensional rather than two-dimensional flow, but transmission of pressure and thermal changes is conceptually the same. One possible complication is that the reservoir may be unconfined rather than confined, with a free surface above the zone of withdrawal. With the possible entry of surface groundwaters, or with reinjection of cooled fluids, the efficiency of thermal sweep through the reservoir becomes important. This is particularly true of a fractured reservoir, where there may be preferential flow along a few paths of high permeability.

In this liquid model, if there is a net mass loss of fluid in the reservoir, there will be a resultant decline in pressure. If the reservoir contains compressed liquid, this pressure decline causes expansion of water and the rock matrix; if there is a free surface, then there is a fall in this surface. Rocks cooled by the advance of colder water account for the net heat loss.

2.5.2. Liquid-Dominated Reservoirs with Boiling

In high-temperature geothermal reservoirs, a decline in pressure caused by exploitation may initiate boiling in part or all of the reservoir. In this case the changes in the reservoir caused by exploitation will include changes in the steam/water ratio, as well as pressure and temperature changes.

One concept is to regard the reservoir as uniformly mixed, containing steam and water throughout—the simplest two-phase lumped parameter model. A second is to assume that water drains from the upper parts of the reservoir, forming a "steam cap"—a vapor-dominated zone overlying a liquid-dominated zone. The first approach ignores gravity; the second assumes gravity is dominant. Both assume steam and water in thermal equilibrium. Figure 2.9 shows the drainage model of the formation of a steam cap. In its initial state there was a near-hydrostatic pressure profile. With production, pressure in the reservoir falls, and above the level of production the vertical gradient is less than hydrostatic. Water will now drain downward. As the water drains, liquid saturation decreases and steam becomes more mobile. Steam then drains upward. Where there is appropriate vertical permeability, the two phases, steam and water, segregate with time. In the upper part of the underlying liquid-dominated region, boiling occurs as the pressure declines, and steam is formed. This steam also drains upward. With time, there is usually sufficient

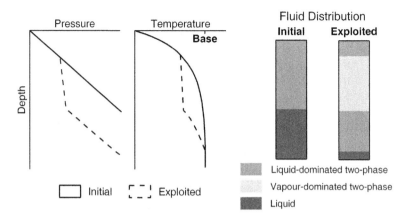

FIGURE 2.9   Fluid distribution in natural and exploited states of a liquid-dominated reservoir.

segregation that nearly all mobile water drains out of the top of the reservoir, and a distinct vapor-dominated region forms at the top—a *steam cap*. The top of the steam cap is normally defined by a capping stratum, a layer of lower permeability. Note that thermal equilibrium between steam and water is required throughout the two-phase zones, but compressed liquid remains at the bottom of the reservoir.

The assumption that the boiling fluid will segregate into vapor-dominated and liquid-dominated zones depends on the vertical permeability and the vertical extent of boiling conditions. High vertical permeability and a small temperature range of boiling conditions are expected to give more rapid segregation of the liquid and steam phases.

In addition to the boiling effects, there are likely to be a lateral or vertical inflow of hot or cold water, together with associated thermal and chemical changes. Under production conditions, the net mass loss from the boiling reservoir is a result of the replacement of liquid by steam and possibly a fall of a free surface on the groundwater above if that groundwater is connected to pressures in the deeper liquid reservoir. The net heat loss of the reservoir is the cooling of the rock as a result of water boiling and the advance of peripheral cooler waters.

2.5.3. Vapor-Dominated Reservoirs

The essential components of a vapor-dominated reservoir are the stored steam and immobile (or nearly immobile) water, the heat stored in the rock, an overlying condensate layer, and a possible deep zone of boiling brine. The boundaries of the reservoir, at the sides and top, must have poor or very poor permeability to prevent the reservoir from being flooded with water. Some boundaries are effectively impermeable; others have low permeability,

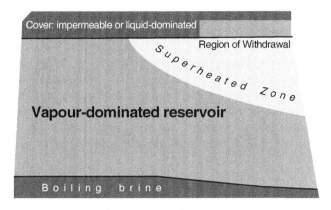

FIGURE 2.10 Fluid distribution in a vapor-dominated reservoir under production.

allowing some communication with liquid-dominated regions above or beside the reservoir. Under exploitation, the immobile water is gradually evaporated (using the heat stored in the rock), and the vapor-dominated reservoir becomes depleted of water, eventually forming a dry (superheated) zone, as shown in Figure 2.10. As production continues, the superheated zone around the area of exploitation expands into the (saturated) vapor-dominated region. There may be a recharge of steam from the deep boiling zone if one exists and possibly also from the condensate layer. The net heat and mass loss are as for the previous case.

2.6. CONCLUSIONS

In this chapter, the first analyses of reservoir information have been presented, dealing with reservoirs in their undisturbed state. The information obtained at this stage is limited but important to collect and evaluate, as it often cannot be obtained once exploitation is underway. In high-temperature convective systems, the flows stimulated by exploitation may overwhelm the original natural flows. These natural flows, continuing for tens or hundreds of thousands of years, have established the heat and fluid reserves available to be exploited.

In simple terms, it is the vertical movement of hot fluid up and through the various types of geothermal systems that established the geothermal reservoirs. To the geothermal developer, the system as a whole and the source of the heat energy are probably of little relevance. It is the flow and the way in which it relates to the reservoir's hydrogeologic structure that set the form for the reservoir. The preceding discussion concentrated on the role and form of these flows in the two main types of high-temperature geothermal reservoirs: liquid dominated (both liquid and two-phase) and vapor dominated.

For both types of reservoirs, simple conceptual models have been described that can be useful in both developing an understanding of these reservoirs and estimating some of their vital field parameters. An understanding of these basic possible models provides a starting point to appreciate more complex real systems.

The discussion in this chapter has been in general terms. In later chapters there is more focus on specific issues. Consequently analyses will be more detailed and take into account a wider range of data. Physical and structural information should not be the only items considered when conceptualizing a reservoir. All the data are relevant, and it is essential to bring together information from many disciplines to build the most robust conceptual models.

Simple Quantitative Models

3.1. INTRODUCTION

This chapter examines the various simple models that illustrate the important processes taking place in geothermal reservoirs in the natural state and under exploitation. These models, though simple, all depend only on the laws of fluid flow and for many cases provide an adequate explanation of real situations. The models discussed in this chapter are particularly useful for illustrating such basic concepts as "recharge" and "pressure stabilization by boiling." Applying these concepts and simple models to real systems is discussed in later chapters.

For the most part, the porous medium through which the fluid flows is considered homogeneous and rigid. Geothermal reservoirs, particularly those found in volcanic rocks, are frequently highly fractured, and for many purposes the fracturing is sufficiently dense and pervasive on a field scale such that the medium is considered homogeneous. For example, interference tests (see Appendix 1) often fit the line source solution very nicely. Liquid-filled reservoirs can have a storativity that corresponds to a porosity thickness value that shows the entire thickness of the reservoir participates in the interference. On the local scale, and in particular in the immediate vicinity of a wellbore, the fracture geometry may affect fluid flow significantly. If different fluids are present in the reservoir—steam and water, or waters of contrasting temperatures—fractures may act as preferential flow paths.

Geothermal Reservoir Engineering. ISBN: 978-0-12-383880-3 DOI: 10.1016/B978-0-12-383880-3.10003-4

The reservoir rock is assumed to be rigid, but in fact the rock will contract when fluid pressure or temperature declines. At the ground surface these effects are usually widespread across the whole area of the geothermal reservoir and are typically on the order of a fraction of a meter over a nominal period of around 10 years of production. Such compaction of the matrix does not appear to have any significant effect on the flow of fluid through the reservoir. Consequently, for reservoir engineering purposes, the matrix can reasonably be taken as rigid when computing fluid flow through the rock. When modeling subsidence, the changes in pressure and temperature can be used as input to a rock mechanics model of subsidence, but this does not feed back to sufficient change in permeability or porosity to affect the fluid flow.

3.2. SIMPLIFICATIONS AND CONCEPTS OF STORAGE

If the spatial distribution of permeability, porosity, and fluid throughout a reservoir was known, the evolution of any reservoir could be accurately simulated using the full equations of flow. However, such detailed knowledge is neither available nor necessary. As with other complex physical systems, most of the complexities are not particularly important, and geothermal reservoir behavior can often be represented by simple idealizations. The choice of such idealizations is, of course, a matter of informed judgment.

The complexities that occur in a geothermal reservoir are of two forms: the spatial variation and the variation in the thermodynamic state of the fluid. There are thus two alternative paths for developing these models:

1. Simplify the fluid and explore spatial variation of the reservoir—pressure transients.
2. Simplify the geometry and explore the thermodynamic properties of the fluid and rock—lumped-parameter models.

Some concepts are common to both approaches. Most prominent among these is the concept of storage: the ability of the reservoir to store and release fluid and heat in various ways, in response to pressure changes or fluid flow.

3.2.1. Closed Box of Single-Phase Fluid

Consider a box of volume V containing single-phase fluid in the pore space. The boundaries of the box are sealed except for the discharge of fluid at a known rate q m^3/s or $W = \rho q$ kg/s. The fluid is assumed to be uniformly mixed throughout and can be adequately described by a single set of thermodynamic variables, which are most conveniently pressure and temperature. Then conservation of mass and energy in the box gives:

$$V\frac{d}{dt}(\varphi\rho) = -W \qquad (3.1)$$

$$V\frac{d}{dt}[(1-\varphi)\rho_m U_m + \varphi\rho U] = -WH \tag{3.2}$$

Expanding for changes in pressure and temperature shows that the temperature change is small. The temperature changes cause small, second-order effects that can be ignored. Single-phase flow is nearly isothermal, a result long known in petroleum and groundwater systems. Then, assuming precise isothermal conditions, Eq. (3.1) gives:

$$V\varphi\left(\frac{\partial\rho}{\partial P}\right)_T\frac{dP}{dt} = -W \tag{3.3}$$

or

$$\frac{dP}{dt} = -\frac{q}{S_V} = -\frac{W}{S_M} \tag{3.4}$$

where $S_V = V\varphi c$ is the "reservoir storage coefficient," being the volume of fluid that is stored (or released) per unit pressure change, and $S_M = V\varphi\rho c$ is the mass that is stored or released per unit pressure change.

Here it is assumed that the reservoir rock is rigid. If the rock compressibility is significant, then $S_V = V(c_m + \varphi c)$. The reservoir storage coefficient divided by the reservoir volume (i.e., φc or $c_m + \varphi c$) is the compressibility of the rock and fluid taken together, per unit volume of reservoir. The compressibility per unit pore volume is c, the fluid compressibility. In other modes of storage, the fluid compressibility per unit volume is denoted by c_t, the fluid compressibility for all modes.

If the reservoir is in the form of a uniform aquifer of thickness h, the storage coefficient per unit area is the storativity S:

$$S = \varphi ch \tag{3.5}$$

This is the volume of fluid stored, per unit area of aquifer, per unit pressure increase.

Note that in this single-phase case, there is essentially no interaction with thermal effects. The concepts of storage refer to the storage of fluid mass or volume. The gain or loss of heat by the reservoir is incidental. If fluid is removed, so is heat as the sensible heat of that fluid, but the heat loss does not play a part in the pressure changes.

If the fluid is a gas and the fluid density varies strongly with pressure, it is best to use the mass change. The imperfect gas law gives the pressure in the box as:

$$(\varphi V)(P/Z) = (\varphi\rho V)\,RT/M \tag{3.6}$$

Here, $\varphi\rho V$ is just the total fluid mass in the box. Note that temperature in Eq. (3.6) is absolute, not Celsius. This implies that P/Z varies linearly with the mass in the box

$$\frac{d}{dt}\left(\frac{p}{Z}\right) = -W\frac{RT}{M\varphi V} \tag{3.7}$$

or

$$S_M = \frac{M\varphi V}{RTZ} \tag{3.8}$$

ignoring variations in Z.

3.2.2. Box with Water Level (Unconfined Aquifer)

Consider a box partly filled with water above which is a more compressible fluid such that the pressure of the overlying fluid may be considered to be constant. This is common as a groundwater aquifer, where the fluid above the free surface is air, or as a petroleum reservoir, with a gas-oil interface. In the geothermal context this can arise as the interface between a liquid-dominated and vapor-dominated zone (loosely, a steam-water interface) or from a horizontal interface between two-phase fluid and liquid water.

The problem of fluid withdrawal from the lower zone (the liquid-dominated layer) and the consequent changes in pressures within that layer are now considered. The upper compressible region is assumed to be at unchanged pressure.

Pressures in the liquid-dominated zone are assumed to be in hydrostatic equilibrium—that is, it is assumed that the time scale of pressure changes is long compared to the time taken for vertical equilibration within the layer. Then $dP/dz = -\rho_w g$. If the pressure falls by an amount ΔP, a fall in water level (or the liquid-gas interface) of an amount $\Delta P/\rho_w g$ results. If the reservoir area is A, the change in total fluid volume is:

$$\Delta q = -A\varphi \Delta S \Delta P/\rho_w g \tag{3.9}$$

or

$$\frac{dP}{dt} = -\rho_w g q/A\varphi \Delta S \tag{3.10}$$

where ΔS is the saturation change at the water level. For water-air, $\Delta S = 1$ if the rock above the water level is dry, but for a vapor-dominated liquid-dominated interface, ΔS may be less than unity. Then the reservoir storage coefficient, aquifer storativity, and compressibility are:

$$S_V = A\varphi \Delta S/\rho_w g, \quad S_M = A\varphi \Delta S/g \tag{3.11}$$

$$S = \varphi \Delta S/\rho_w g \tag{3.12}$$

$$c_t = \varphi \Delta S/\rho_w g h \tag{3.13}$$

The thickness used in Eq. (3.13) is the thickness of the liquid-dominated aquifer, not the total aquifer. The density used is that of liquid water. As in the previous case, the thermal properties of the fluid are not important.

A liquid-dominated aquifer with a free surface usually would be assumed to contain only liquid water. Note that Eqs. (3.9)–(3.13) require only that the vertical pressure gradient be hydrostatic and can thus apply to the long-term behavior of an aquifer at the boiling point.

3.2.3. Box of Two-Phase Fluid

A box containing a steam-water mixture is now considered (Grant & Sorey, 1979). The steam and water may be in any proportions provided that both are present and in contact and remain so. As before, it is assumed that the fluid in the box can be represented by a single pair of thermodynamic variables, which are taken to be pressure (temperature) and saturation. If fluid is withdrawn from the box so that the pressure falls by an amount ΔP, temperature falls by:

$$\Delta T = \Delta P / \left(\frac{dP_s}{dT} \right) \tag{3.14}$$

where $\frac{dP_s}{dT}$ is evaluated along the saturation line. This fall in temperature means that the rock is cooled, so an amount of heat Q is released from the rock:

$$Q = V \rho_t C_t \Delta T \tag{3.15}$$

where

$$\rho_t C_t = (1 - \varphi)\rho_m C_m + \varphi S_w \rho_w C_w \tag{3.16}$$

is the volumetric heat capacity of the wetted rock. This heat is released from the rock and transferred to the fluid, vaporizing the water. The fluid gains volume to replace the fluid withdrawn, with the volume increase being equal to the volume of the vaporized steam, less that of the steam as water:

$$\Delta V = \frac{V \rho_t C_t \Delta T}{H_{sw}} \left(\frac{1}{\rho_s} - \frac{1}{\rho_w} \right) \tag{3.17}$$

Dividing by ΔP gives:

$$\varphi c_t = \frac{\rho_t C_t}{H_{sw}} \frac{\rho_w - \rho_s}{\rho_w \rho_s} \frac{dT_{sat}}{dP} \tag{3.18}$$

$$S_V = V \varphi c_t, \quad S = \varphi c_t h \tag{3.19}$$

The temperature change must follow the saturation curve. If S_M is desired, the density ρ_t of the expelled fluid must be used:

$$S_M = V \varphi c \rho_t \tag{3.20}$$

This density depends on the steam-water ratio in the expelled fluid or, equivalently, its enthalpy. Note that the derivation of the two-phase compressibility did not depend on this density. The temperature drop in the box causes steam to form and a consequent gain of fluid volume. This volume gain depends only on the amount of fluid that changes phase. The fluid remaining in the pore space may have any steam-water ratio, provided that both phases are present and remain in contact.

Here, the two-phase compressibility differs from the previous two cases in its dependence on the thermal properties of the reservoir and the contained fluid and is directly proportional to the volumetric heat capacity. The reservoir storage coefficient S_V or S_M is proportional to the heat capacity of the entire volume V of the box.

3.2.4. Comparing the Different Compressibilities

There are significant differences between the values of different compressibilities. Consider an aquifer 500 m thick at 240°C with 15% porosity and volumetric heat capacity $\rho_t C_t = 2.5$ MJ/m^{3}K. For single-phase fluid:

$$c_w = 1.2 \times 10^{-9} \text{ Pa}^{-1}, \quad c_s = 3 \times 10^{-7} \text{ Pa}^{-1}$$

For an unconfined aquifer, with $\Delta S = 1$:

$$c_t = 4 \times 10^{-8} \text{ Pa}^{-1}$$

For two-phase

$$\varphi C_t = 1.4 \times 10^{-6} \text{ Pa}^{-1}, \quad c_t = 9 \times 10^{-4} \text{ Pa}^{-1}$$

There is a factor of 10^4 between the largest (two-phase) and smallest (liquid) compressibility.

3.2.5. Boxes with Different Zones

A real reservoir is seldom uniform in its fluid properties throughout its complete volume. Consequently, more complex models containing some regions of different characteristics are needed. We have already seen one such model: a box or aquifer containing liquid overlaid by vapour. This is perhaps the most common model and is more complex than a single uniform box. This model is not discussed further here except to observe that the response of such a box to fluid withdrawal from either zone is determined by conservation of mass and energy, as in the simpler preceding cases. Each zone responds as an isolated box that is coupled by the mass and energy flow between them. Such flows in this model are normally an upflow of steam or a downflow of water.

In the case of the unconfined aquifer, some of the water remains in the rock matrix as the water level falls and later drains down to the liquid zone under gravity. The storage provided by this mechanism exceeds that provided by compression within the liquid layer. In response to a pressure drop in the liquid layer, boiling may occur, and the resulting steam may flow upward, providing a source of steam recharge to the overlying vapor-dominated zone. The vapor-dominated zone also expands downward with the falling steam-water interface ("water level"). The amount of steam upflow depends on the vertical extent of boiling conditions in the lower zone or, equivalently, on the temperature distribution within that zone. It can readily be seen that many variants are possible. Pressure in the steam zone may rise or fall depending on the mass and energy balance in the steam, movement of the boundary of the steam zone, steam gain from deeper boiling, steam loss by condensation at cooler boundaries, steam loss to shallower zones or to surface discharge, and well discharge from the steam zone.

Models of vapor-dominated reservoirs frequently take the form of a vapor-dominated zone overlying a liquid-dominated zone. Fluid withdrawal in such models is restricted to the vapor zone. Contrasting zones also occur in pressure transient models. In such cases, drawdown may create a two-phase zone around the well within a liquid reservoir or a dry steam zone around the well within a two-phase reservoir.

3.3. PRESSURE TRANSIENT MODELS

The simplest model is a vertical well, circular in cross section, that fully penetrates a uniform horizontal aquifer of infinite radial extent that is sealed above and below, as shown in Figure 3.1. There is no spatial variation of rock properties (especially permeability). The only spatial variations of pressure (and temperature and saturation, if relevant) that need to be considered are those pertaining to radial distance from the well. The fluid in the aquifer is in vertical equilibrium with depth at all times, so there are no effects due to gravity.

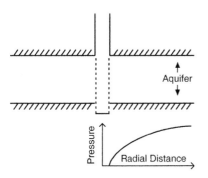

FIGURE 3.1 Basic well model for pressure transients.

3.3.1. Single-Phase Aquifer Fluid

Darcy's law in the radial (axial) form is:

$$v_r = -\frac{k}{\mu}\frac{\partial P}{\partial r} \tag{3.21}$$

In similar form, the conservation of mass equation is:

$$\varphi\frac{\partial \rho}{\partial t} + \frac{1}{r}\frac{\partial}{\partial r}(r\rho v_r) = 0 \tag{3.22}$$

or

$$\varphi c p \frac{\partial P}{\partial t} = \frac{1}{r}\frac{\partial}{\partial r}\left(\rho\frac{k}{\mu}r\frac{\partial P}{\partial r}\right) \tag{3.23}$$

It is assumed that the compressibility is constant and that changes in the viscosity may be ignored in comparison with changes in pressure. This gives:

$$\frac{\varphi\mu c}{k}\frac{\partial P}{\partial t} = \frac{1}{r}\frac{\partial}{\partial r}\left(r\frac{\partial P}{\partial r}\right) \tag{3.24}$$

This is the diffusivity equation, with the hydraulic diffusivity:

$$\kappa = \frac{k}{\varphi\mu c} \tag{3.25}$$

Many solutions of this equation are available from the literature on heat conduction, which obeys the same equation.

If a well begins withdrawal at time $t = 0$, at a constant rate q or $W = \rho q$, the pressure in the aquifer is given as a function of radial distance and time by:

$$\varDelta P = P - P_o = -\frac{q\mu}{4\pi kh}E_1\left(\frac{\varphi\mu c r^2}{4kt}\right) = -\frac{Wv}{4\pi kh}E_1\left(\frac{\varphi\mu c r^2}{4kt}\right) \tag{3.26}$$

where

$$E_1(x) = \int_x^\infty \frac{1}{y}e^{-y}dy \tag{3.27}$$

$E_1(x)$ is tabulated by Abramowitz and Stegun (1965). (The function $E_1(x)$ is denoted by $-Ei(-x)$ in the petroleum literature.) For small values of the argument x, or long time:

$$E_1(x) \sim -\ln(x) - \gamma \tag{3.28}$$

where $\gamma = 0.5772$ is Euler's constant. Using this asymptotic for Eq. (3.26) gives:

$$P - P_o = -\frac{Wv}{4\pi kh}\left\{\ln(t) + \ln\left(\frac{4k}{\varphi\mu c r^2}\right) - \gamma\right\} \tag{3.29}$$

The pressure at any point changes linearly with the logarithm of the time. In Eq. (3.26) the parameters of the aquifer and fluid enter in two groups:

$$\frac{\mu}{4\pi kh}$$

and

$$\frac{\varphi\mu c}{k} = \frac{\varphi ch}{(kh/\mu)} \tag{3.30}$$

By suitable observation of the pressure change, it may be possible to fit an observed history to theory and identify the two parameter groups kh/μ, the transmissivity or mobility-thickness, and φch, the storativity. (k/μ is called the mobility.) If the fluid viscosity μ is known, the permeability-thickness kh can be identified. Thus, in principle, two parameter groups are identifiable. One, the storativity, measures the aquifer's capacity to store fluid, and the other, the transmissivity, measures its ability to transmit fluid.

3.3.2. Flow of a Dry Gas

Assuming that compressibility, density, and viscosity are all functions of pressure, the pressure transient equation is:

$$\frac{\varphi\rho c}{k}\frac{\partial P}{\partial t} = \frac{1}{r}\frac{\partial}{\partial r}\left(\frac{r}{v}\frac{\partial P}{\partial r}\right) \tag{3.31}$$

This is the full form of the pressure transient equation for gas flow. Approximate solutions for this equation are possible for large pressure changes by using pseudopressure techniques (described in Appendix 1).

3.3.3. Unconfined Aquifer

As just described, the storativity in the unconfined aquifer relates to the fluid level change in the aquifer. The compressibility becomes $c_t = \varphi/\rho gh$. When this substitution is made, the pressure transient formulation of Eqs. (3.24)−(3.30) remains valid. As compressibility increases, so does the time scale of pressure change.

3.3.4. Two-Phase Aquifer

When water and steam coexist throughout the aquifer, the compressibility is altered to the two-phase value given by Eq. (3.18). In addition, the fluid mobility is altered if both phases are mobile. (The most common form of two-phase transients is in a vapor-dominated system, where only steam is mobile.) Darcy's law for horizontal flow is $u = -(k/v)\,\nabla P$. In two-phase flow, there is mass flux of both steam and water. The total mass flux u_t is given by the sum of the fluxes of the two phases:

$$u_t u_w + u_s = -k \left\{ \frac{k_{rw}}{v_w} + \frac{k_{rs}}{vx_s} \right\} \nabla P \qquad (3.32)$$

Defining

$$\frac{1}{v_t} = \frac{k_{rw}}{v_w} + \frac{k_{rs}}{v_s}$$

and

$$\frac{1}{\mu_t} = \frac{k_{rw}}{\mu_w} + \frac{k_{rs}}{\mu_s} \qquad (3.33)$$

restores the similarity in form to single-phase flow.

The density of the flowing fluid is given by:

$$\rho_t = \mu_t / v_t \qquad (3.34)$$

The density can also be found directly from the enthalpy H_t of the flowing fluid. In an actual well test, H_t is the enthalpy of the fluid flowing to the well or the enthalpy of the fluid discharged by the well. The steam fraction by mass in this flow is $X = (H_t - H_w)/H_{sw}$, and the water fraction is $1 - X = (H_s - H_t)/H_{sw}$. The specific volume of the flowing fluid $1/\rho_t$ is the sum of the volumes of the steam and water fractions:

$$\frac{1}{\rho_t} = \left[\frac{H_t - H_w}{\rho_s} + \frac{H_s - H_t}{\rho_w} \right] \frac{1}{H_{sw}} \qquad (3.35)$$

This completes the definition of fluid and aquifer properties for the three possible types of aquifer that can be encountered in a geothermal reservoir: liquid, dry steam, or two-phase fluid. Assuming that in each case the fluid properties (compressibility, viscosity, and density) remain constant or effectively so, it is possible to elaborate on the aquifer structure and to explore possible variations in permeability near the well through their influence on well tests. Standard analysis techniques are reviewed in Appendix 1.

3.4. SIMPLE LUMPED-PARAMETER MODELS

Simple models that reflect some of the physical processes active in geothermal reservoirs are now considered.

3.4.1. Basic Model

The simplest basic model that is applicable to a real geothermal reservoir is shown in Figure 3.2. In this model, the reservoir is described as a single box

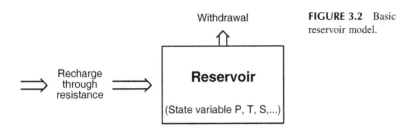

Withdrawal

FIGURE 3.2 Basic reservoir model.

Recharge through resistance

Reservoir

(State variable P, T, S,...)

containing homogeneous rock and fluid. Fluid is withdrawn from the box, and the box is recharged from external sources. The mass flow withdrawn from the box is W, and the recharge is W_r. Such recharge may be from deeper fluid, laterally adjacent aquifers, or overlying groundwater. Note that the definition of recharge depends on the definition of the box. If a new model is considered, which is larger, the volume inside the box now contains some of what was outside the original box, and at least some of what was previously recharged becomes an internal flow.

If the box undergoes a pressure change ΔP, it expels a fluid mass $S_M \Delta P$. Equating this to the net fluid loss $(W - W_r)\Delta t$ gives conservation of mass for the block:

$$S_M \frac{dp}{dt} + W - W_r = 0 \tag{3.36}$$

In order to be able to calculate the changes in pressure, there must be some rule or equation that defines the recharge flow W_r. The simplest hypothesis assumes that recharge is proportional to the pressure difference between the box and the recharge source. Assuming that the initial state was in equilibrium, this is equal to the total pressure drop in the box:

$$W_r = \alpha(P_o - P) \tag{3.37}$$

This assumption does not allow for possible transient pressure changes associated with variations in the recharge rate. More complex models allowing such effects are possible. However, if the model is significantly dependent on such transient effects, too much of the dynamics have been moved outside the box, and a different and larger model is probably indicated. Combining Eqs. (3.36) and (3.37):

$$W = \alpha(P_o - P) - S_M \frac{dP}{dt} \tag{3.38}$$

The response to a discharge at constant rate W, beginning at time $t = 0$, is an exponential approach to equilibrium:

$$P_o - P = \frac{w}{\alpha}\left[1 - e^{-\frac{t}{\tau}}\right] \tag{3.39}$$

$$\tau = S_M/\alpha \qquad (3.40)$$

where τ is the system time constant or relaxation time. For short times, pressure falls linearly with time:

$$P_o - P = Wt/S_M = qt/S_V \qquad (3.41)$$

Within this time scale, recharge has negligible effect on the pressure, and the pressure drop resembles that of a closed box with no recharge. The recharge, if any, cannot be determined until there are observations over a period of time comparable with the system time constant τ. For some parameters, observation time is needed in order to determine reservoir parameters—a requirement that cannot be circumvented by taking into account variations in flow rate or by applying a large flow rate. Over a short time the reservoir storage coefficient can be determined but not the recharge. For time values significantly in excess of the relaxation time, the pressure stabilizes at a value determined by a balance with the recharge:

$$P_0 - P \approx W/\alpha \quad t >> \tau \qquad (3.42)$$

This balance is independent of the storage coefficient; fluid withdrawn is replaced by the recharge. Sarak and colleagues (2005) give a more detailed discussion of the errors involved in fitting lumped-parameter models. A simple model can be more constrained by adding more physical properties to model, such as temperature or fluid chemistry (Lopez et al., 2008; Türeyen et al., 2009).

3.4.2. Reservoir with Change of Fluid Type

A box may undergo a change of storage mechanism as its fluid is depleted. If the box is initially full of compressed water, continued withdrawal of water may lower the pressure at the top of the box to saturation or to atmospheric pressure, causing a free surface to form. In this case the storage coefficient changes discontinuously from a confined to an unconfined reservoir. This transition is common in the analysis of cold groundwater reservoirs.

Another possibility is a transition from liquid to two-phase conditions. The vertical pressure distribution in water may be approximately hydrostatic. Suppose the temperature increases with depth such that the pressure everywhere exceeds saturation by the same amount. There may be a specific pressure, or a small pressure range, over which the entire fluid content of the box passes from compressed liquid to boiling. This is more likely if the vertical extent of the reservoir is small. Under these circumstances the compressibility of the fluid increases greatly from compressed liquid to two-phase.

In both of these cases there is a transition time at which the reservoir storage coefficient and fluid compressibility increase by a large factor. As a consequence, at that time the rate of pressure decline decreases by the same large factor. Because this transition is caused by the commencement of boiling within the reservoir, it is referred to as pressure stabilization by boiling. In practice, a reservoir often contains various regions that are hydraulically connected but have different temperature distributions. As the overall pressure declines, rather than a sudden transition, the boiling region progressively expands, and the reservoir storage coefficient that is averaged over all regions changes progressively rather than discontinuously.

A transition to lower compressibility is also possible. Suppose a vapor-dominated reservoir is initially two-phase, with steam in the fractures but water in the matrix pore space. Under drawdown the water boils, providing recharge to the steam in the fractures. If depletion causes the liquid content to completely evaporate, there is a transition from the two-phase state to superheated dry steam. Compressibility decreases, and the rate of pressure fall with fluid withdrawal is significantly increased (Brigham & Morrow, 1977). As with the previous case, in a real reservoir, variation in liquid saturation across the reservoir means that dry regions of the reservoir expand with drawdown, rather than a sudden drying of the entire reservoir, and the transition is again progressive rather than sudden.

3.4.3. Cold Water Recharge

In the previous basic model, it was assumed that the pressure in the box was controlled by the mass balance alone. That is, it was implicitly assumed that either the recharge fluid was identical in composition to the discharged fluid or the reservoir was entirely liquid. In the latter case, cold water can replace hot water.

Under many circumstances a significant part of the recharge is likely to be water from the groundwater system surrounding a geothermal reservoir or from groundwater directly above the high-temperature reservoir. In this case, the recharge water will be colder than the reservoir fluid, and as it enters the reservoir the volume of hot fluid must decrease.

If the reservoir is wholly liquid, cold water entering from the edge pushes hot water ahead of it toward the production area, which is the pressure sink. A unit volume of cold water displaces the same volume of hot water (ignoring the effects of water being reheated by passage through the rock) and so is equivalent to the same volume of hot water for the purposes of maintaining pressure. Pressure in the reservoir is controlled simply by the volume of liquid in the reservoir. This will not be the case if the reservoir contains two-phase fluid because the compressibility and two-phase storage mechanisms depend on an energy balance as well as a mass balance.

Consider a box containing two-phase fluid. Recharge is introduced at a rate W_r kg/s, with enthalpy H_r. We now ignore discharge in order to concentrate on the effects of the cold recharge alone. Conservation of mass and energy give:

$$V\frac{d}{dt}[\varphi S_w \rho_w + \varphi S_s \rho_s] = W_r \tag{3.43}$$

$$V\frac{d}{dt}[\varphi S_w \rho_w U_w + \varphi S_s \rho_s H_s + (1 - \varphi)\rho_m U_m] = H_r W_r \tag{3.44}$$

From these two equations dS_w/dt can be eliminated to give the change in pressure. Defining:

$$H_n = H_w - \frac{\rho_s H_{sw}}{\rho_w - \rho_s}$$

the pressure decreases if the recharge enthalpy is less than H_n (Pruess et al., 1979). The enthalpy H_n corresponds to water a few degrees cooler than reservoir, temperature, so any recharge significantly cooler than reservoir temperature will result in a pressure fall in a two-phase reservoir. This occurs because when cold water is added to a steam-water-rock mixture, the added water is heated to reservoir temperature, and the heat required to do so is gained by condensing steam with a loss of fluid volume.

This example shows that the concept of recharge in a geothermal system embraces two ideas: resupply of fluid and pressure support. In a liquid aquifer, recharge of liquid means both that fluid is added and the pressure is supported. Recharge of water to a geothermal reservoir implies addition of fluid to the reservoir and thus an increase in the reservoir's fluid content. If the reservoir is single-phase, additional fluid implies an increase in pressure. In a two-phase reservoir, recharge may increase or decrease pressure, since the pressure depends on an energy balance as well as mass balance. If cold water mixes with two-phase fluid, pressure will fall, not rise. What happens in a particular case depends on the details of how the fluids mix. In a real reservoir containing two-phase conditions, injection may produce a liquid zone around the point of injection, so cool injected water is screened from any two-phase fluid. Water injection can provide pressure support if the injection is into superheated regions but not if it is into two-phase conditions (see following).

3.5. STEAM RESERVOIR WITH IMMOBILE WATER

A particularly simple case of a two-phase system is a porous medium containing water below residual saturation. Figure 3.3 shows a typical relative permeability for homogeneous porous media. When the pore space is fully occupied by water, the relative permeability to water is 1, and the relative permeability to steam is zero. As the liquid saturation decreases, relative permeability to water decreases and relative permeability to steam increases. There is a saturation value, known as the residual saturation, greater than zero, where the relative permeability to water is zero. In this case, water is coating the pores, but the connected passages are all filled with steam and only steam is able to flow. The water is then

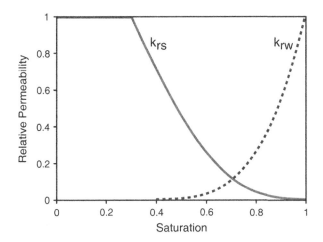

FIGURE 3.3 Relative permeability to water as a function of liquid saturation.

immobile; it cannot move because its relative permeability is zero. The fluid flowing in the reservoir is steam, which is at saturation temperature due to its contact with the water. Observations of the produced fluid would indicate that the reservoir is "dry" because only steam enters the wellbore. The water reveals its presence in the reservoir through the saturated state of the steam present over a range of flowing downhole pressures.

3.5.1. Equation for Steam Flow

Assuming that the immobile water does not impede the flow of steam ($k_{rs} = 1$), Darcy's law for the flow of saturated steam is identical to that for dry steam:

$$u = -\frac{k}{v}(\nabla P - \rho_s g) \tag{3.45}$$

Conservation of mass and energy give:

$$\varphi \frac{\partial}{\partial t}(S_w \rho_w + S_s \rho_s) + \nabla \cdot u = 0 \tag{3.46}$$

$$\varphi \frac{\partial}{\partial t}(S_w U_w \rho_w + S_s U_s \rho_s + (1 - \varphi)\rho_m U_m) + \nabla \cdot (H_s u) = 0 \tag{3.47}$$

The two-phase pressure transient equation is valid and particularly simple:

$$\frac{\varphi \rho c_t}{k} \frac{\partial P}{\partial t} = \frac{1}{r} \frac{\partial}{\partial r}\left(\frac{r}{v} \frac{\partial P}{\partial r}\right) \tag{3.48}$$

The two-phase compressibility c_t depends on the pressure and (weakly) on the saturation. It can be taken as a function of pressure only. Likewise the other fluid parameters—density, viscosity, and temperature—are functions of pressure alone. With the modified compressibility, Eq. (3.48) is identical in form to the equation for flow of a dry gas. Consequently all standard gas reservoir techniques can be used for analysis of steam pressure transients and steam well performance, with suitably modified parameters.

Equation (3.48) remains valid provided that water is present in the reservoir. Saturation changes can be calculated from pressure changes using Eqs. (3.46) and (3.47). Since the enthalpy H_s of saturated steam is nearly constant over the temperature ranges of interest (180–250°C), the divergence terms $\nabla \cdot \boldsymbol{u}$ and $\nabla \cdot (H_s \boldsymbol{u})$ can be eliminated, giving:

$$\frac{\partial}{\partial t}[\varphi H_s(S_w \rho_w + S_s \rho_s) - (1 - \varphi)\rho_m U_m - \varphi S_w \rho_w U_w - \varphi S_s \rho_s U_s] = 0 \quad (3.49)$$

Similarly, if there is a box containing steam and immobile water, and steam is withdrawn, conservation of mass and energy are:

$$\varphi V \frac{d}{dt}[S\rho_w + (1 - S)\rho_s] = -W \quad (3.50)$$

$$V \frac{d}{dt}[\varphi S\rho_w U_w + \varphi(1 - S)\rho_s U_s + (1 - \varphi)\rho_m U_m] = -H_s W \quad (3.51)$$

which also gives Eq. (3.49). If the initial temperature and liquid saturation are denoted by T_o and S_{wo}, respectively, there is a simple relation between temperature and saturation:

$$(1 - \varphi)\rho_m C_m(T_o - T) = \varphi S_{wo}[\rho_w H_{sw}]_o - \varphi S_w \rho_w H_w \quad (3.52)$$

3.5.2. Dry-Out

The temperature T_d of dry-out is of considerable significance in the immobile water model of a steam-dominated reservoir. This is the temperature at which the liquid saturation in the medium reaches zero. There is no longer water present in contact with the steam, and the steam is truly "dry"; as pressure continues to decline, the steam becomes superheated. From the preceding equation, this is given by:

$$(1 - \varphi)\rho_m C_m(T_o - T_d) = \varphi S_{wo}[\rho_w H_{sw}]_o \quad (3.53)$$

This derivation from the full equations of mass and energy shows that the results are valid for any flow configuration, provided that the flowing fluid is steam and further provided that heat conduction, such as from matrix to fracture, is not important. In principle, dry-out can be observed by progressively opening a well and determining at what downhole pressure the steam entry becomes superheated. This superheating is local to the well and depends on the drawdown. Note that Eqs. (3.49)–(3.53) are not valid if water is also mobile in the reservoir, as is usually the case in the initial state of a vapor-dominated reservoir.

The attainment of dry-out has a double significance. On the reservoir scale, dry-out marks depletion of the reservoir's fluid reserves. Because water is so much denser than steam, nearly all of the fluid mass in the unexploited "steam" reservoir is stored as water. When this fluid reserve has been used, little producible fluid remains in that part of the reservoir, which is now extensively superheated, not just locally around the well. Steam may still flow from other parts of the reservoir less affected by dry-out. On the local scale, drawdown near a well can create a dry zone so superheated steam enters the well.

Typical values for φS_{wo} in vapor-dominated reservoirs are less than 10% (Truesdell & White, 1973; D'Amore & Truesdell, 1979). This means that dry-out is normally attained during exploitation. This simple reservoir model is not valid in a fractured medium, with different fluid saturations in the matrix and fractures.

3.5.3. Cold Water Injection

If the reservoir fluid has been depleted, but the rock is still hot, water injection may help to reduce the decline in steam supply. When the reservoir is dry and the steam is superheated with temperatures above saturation for the pressure, all of the available heat in the rock above the saturation temperature cannot be recovered. However, this heat could potentially be extracted by adding water to the system through reinjection so the heat stored in the rock can be recovered by boiling. A major injection project at The Geysers (after much of the liquid reserves had been depleted) resulted in an increase of steam production relative to previous trends (Stark et al., 2005). In this case, the water injection produced a net increase in steam flow from the production wells.

Considering a box of rock and dry steam, with temperature T and pressure P, and inflow W of water and enthalpy H, conservation of fluid mass and total energy gives:

$$\frac{d}{dt}(\varphi \rho V) = W \tag{3.54}$$

$$\frac{d}{dt}[(1 - \varphi)\rho_m U_m]V = -W(H_s - H) \tag{3.55}$$

and using the gas law, Eq. (3.6):

$$V\frac{dP}{dt} = W\left[\frac{P}{\varphi\rho} - \frac{P}{T}\frac{(H_s - H_r)}{(1 - \varphi)\rho_m C_m}\right] \tag{3.56}$$

The second term, due to the cooling, is negligible. The addition of cold water raises the pressure in a dry steam reservoir. This conclusion is no longer valid once saturation conditions are reached. After saturation conditions have been attained, further addition of cold water causes pressure to decline because the cold water lowers temperature and, thus, pressure.

3.6. RESERVES

This section discusses a simple means of making a rough assessment of what—heat or fluid—is stored in a geothermal reservoir, the processes of heat and mass transfer during exploitation, and how much of the resource may potentially be available for use. Several cases are illustrated with the example of a 1 km^3 box of porous rock at 240°C.

3.6.1. Available Energy

There is an important defect in the simple estimation of the heat content of rock and fluid above a given base temperature. Heat cannot be converted into electricity with 100% efficiency, not even by a thermodynamically perfect engine. The efficiency of conversion is limited by the second law of thermodynamics. Given a fixed amount of heat, a larger fraction can be converted to electricity if the heat is at a high temperature than if it is at a low temperature. Geothermal heat at 150–350°C in the reservoir and 100–250°C at the wellhead means that the fuel available to the power station is a low-grade heat resource compared to the fuel for a fossil or nuclear-fueled thermal plant in which the actual or nominal temperature of the heat supply can be above 1000°C.

As a practical example, consider a geothermal plant exploiting a reservoir of liquid water from which steam is separated at 150°C (5 bar a). The heat above 150°C—the heat reserve—can be converted to electricity at some efficiency—say, 20%. Water at 240°C will sustain discharge and generation. If the reservoir water is mixed with an equal amount of water at 150°C, the result is water at 195°C. This water may not sustain discharge at sufficient wellhead pressure to maintain flow through the well and turbine. Although the mixing process did not change the amount of heat above 150°C, the capacity to turn this into electricity has been degraded. When waters of different temperatures mix, energy is conserved, but entropy increases, and the ability to turn heat into electricity is reduced.

The concept of the usable energy can be quantified by the concept of available energy or "exergy" (DiPippo, 1997). This is the part of the energy contained in the fluid that could be turned into electricity or work by an ideal heat engine. It is defined by:

$$e = H - H_o - T_o(s - s_o) \tag{3.57}$$

where s is the specific entropy of the fluid, and the subscript o refers to conditions at the reject temperature. This temperature would most usefully be the temperature of the steam condenser (if there is one) or of the plant exhaust. The total available energy stored in any system is the total amount of useful work that might be obtained from it by a thermodynamically perfect engine.

The most useful measure of the efficiency of a geothermal power station is the useful work produced (electricity) as a fraction of the total available energy in the reservoir. Geothermal generation commonly attains such a second law efficiency similar to other means of electricity generation. The low thermal efficiency of geothermal generation cannot be fairly compared to that of coal or nuclear generation, since such stations consume a higher-grade fuel. Comparisons between different types of generation and different fuels should only be made fairly by using second law efficiencies. See, for example, Quijano (2000) or Aqui and colleagues (2005) for an energy analysis of producing fields.

3.6.2. Stored Heat Estimate

The simplest assessment of a reservoir is the "stored heat" method. Similar approaches are also known as the "volumetric" and "mass in place" methods. In principle, these methods are variations of calculating the amount of "useful" heat stored in the reservoir volume. The stored heat method is described by Muffler (1977, 1978), Muffler and Cataldi (1978), and more recently in AGEG (2008).

$$Q_{tot} = \int \rho_t C_t (T - T_o) dV = V \rho_t C_t (T_r - T_o) = A h \rho_t C_t (T_r - T_o) \quad (3.58)$$

where T_r is the average reservoir temperature and T_o is a reference temperature that is the endpoint of the thermodynamic process utilizing the fluid. Normally, this would be the reject temperature of the power plant. Sometimes average local air temperature is used, which is similar to the reject temperature. V is the reservoir volume, and A and h are its area and thickness, respectively. Another relevant temperature is the cut-off temperature T_C. This is the temperature below which there is no economic value in the fluid—the temperature at which wells cease to flow or it becomes uneconomic to pump them. This defines the outer limits of the resource. Sometimes the cut-off temperature is used as the reference temperature—for example, in Pastor and colleagues (2010). This significantly reduces the total amount of heat, and a higher recovery factor (see following) is then appropriate.

The factors required for the calculation are:

1. $\rho_t C_t$, the specific heat of the wetted rock
2. The temperature difference between the reservoir temperature and the reject temperature of the plant using the fluid
3. Porosity
4. The reservoir volume

The specific heat of the rock does not vary greatly among different rock types.

Assuming that all the heat in the reservoir can be produced by circulating fluid at reservoir temperature, the heat that is available for use by the plant is the heat between this reservoir temperature and the temperature at which waste fluid is rejected by the plant.

For the purposes of illustration, consider a reservoir that consists of a box of volume 1 km^3, at 240°C, with porosity 15%, and with specific heat of the rock $\rho_t C_t = 2.5$ MJ/m^{3}K. This case is used in later sections to illustrate the amount of heat in the reservoir and the amount that is produced by different physical processes. The heat in the rock, above a reject temperature of 15°C, is 5.6 × 10^{17} J. If the pores are water filled, this adds 1.2 × 10^{17} J. If they are 100% steam filled, with no liquid, at saturation pressure this adds 7 × 10^{15} J. In either case, the rock is the principal store of heat.

The reservoir volume can be a more difficult parameter to specify. In principle, it is the entire volume from which heat can be withdrawn by circulating fluid during the lifetime of the project. In practice, of course, this is not known at the exploration stage. There may be barriers within the reservoir or unknown extensions into recharge areas that determine the effective reservoir volume. Most reservoirs contain zones of hot but impermeable rock; with limited permeability, no fluid can flow from these zones, and they cannot contribute to the reservoir volume available to produce energy. Defining the base of the reservoir can also be difficult, even when many wells have been drilled. There have been two approaches to defining the reservoir volume:

1. Draw isotherms using available well data, and assume that the entire volume within the minimum temperature at which production is possible is the reservoir.
2. Define areas around successful wells to determine a "proven" reservoir volume that is assumed to be productive.

The first approach gives a maximum reservoir volume, and the second, a minimum. The second approach is more consistent logically, since many reservoirs have a peripheral region that is hot but impermeable, and in a few only a small part of the reservoir is permeable. If the permeability is so low that pressure drawdown does not propagate into it, the region cannot contribute to production and is not part of any reserve.

In addition, a thickness for the reservoir is needed to complete the volume definition. Following the second approach for defining area, a consistent method is to define the top and bottom of the reservoir as the top and bottom of the interval in which significant permeability has been found—permeability good enough that fluid can be withdrawn by production wells.

Because of the uncertainty about reservoir size and the certainty that there will be some problems within the reservoir that prevent total recovery of the heat in it, the "recovery factor" r is introduced. This is the fraction of the heat in the reservoir that may be recoverable:

$$Q = r \int \rho_f C_f (T - T_o) dV \tag{3.59}$$

The total amount of power that may be generated can then be calculated by assigning a conversion efficiency η. This can be computed in a variety of ways, but the best is by consideration of the proposed power plant.

The total amount of electrical energy that may be recovered is then:

$$E = \eta Q = \eta r \int \rho_t C_t (T - T_o) dV \qquad (3.60)$$

An alternative approach is to use the second law of efficiency by writing:

$$\eta = \eta_A \times \left(1 - \frac{T_o + 273}{T_{WH} + 273} \right) \qquad (3.61)$$

The expression inside the brackets is the efficiency of a thermodynamically perfect heat engine with a supply of fluid at wellhead temperature T_{WH} and reject temperature T_o. η_A is the second law efficiency and is commonly around 40−50% with present technology. Second law efficiencies vary much less than first law (thermal) efficiencies do, making the result less sensitive to assumptions about plant design. Equation (3.61) is equivalent to calculating the stored available energy and then applying the second law efficiency.

It is usual to express the reserves in MW_e-yr or as MW_e for some set period, often 30 years. This then gives the power capacity in MW_e as:

$$MW = \eta r \int \rho_t C_t (T - T_o) dV / (L_F \times 30 \times 10^6 \times 3.05 \times 10^7)$$

$$= \eta r \int \rho_t C_t (T - T_o) dV / (L_F \times 9 \times 10^{14}) \qquad (3.62)$$

L_F is the plant load factor. This shifts the uncertainty to the *recovery factor*, for which there is usually minimal field data. Models of flow in fractured media (Nathenson, 1975) were used to support an average value for the recovery factor of 25%. This value was adopted by the U.S. Geological Survey (USGS), which applied the method to a countrywide assessment of geothermal reservoirs in the United States, using assumed reservoir areas and depths that roughly correspond to the preceding first assumption. It was also suggested that the recovery factor could increase with porosity.

3.6.3. Validation of the Stored Heat Method

There was initially no practical validation of this assumption, and in the 1990s, it became apparent from "post-audits" that some stored heat assessments had resulted in significant overestimates by multiples of up to five (Grant, 2000; Stefansson, 2005; Sanyal et al., 2002, 2004).

Sanyal and colleagues (2002) made the first validation of recovery factors by comparing stored heat assessments against numerical simulation and actual reservoir performance, and they found that factors of 5—10% were "a more reasonable range of values." Sanyal and colleagues (2004) reevaluated the USGS assessments (Muffler, 1978), and in that case they found that the total resource was only one-third of the original estimate and that recovery factors should lie in the range of 3—17%, with a mean of 11%. Similarly, Williams (2004) found recovery factors closer to 10% than 25% for three fields in the United States and also observed that the recovery factor varied strongly between fields and the application of a constant value could lead to significant errors. In contrast, Sarmiento and Björnsson (2007), reviewing past reserve estimates, found that stored heat estimates in the Philippines gave results that were roughly in line with subsequent performance. This may be because the use, in these original estimates, of the cut-off temperature as the reference temperature makes the calculation significantly more conservative.

In summary, the practical evidence suggests that recovery factors should be around 10% but could be larger or smaller by a factor of at least two. Past application of stored heat estimates has been validated by later experience in some cases, but in other cases it has been shown to produce overestimates. There are few reported cases of underestimates.

Stored heat calculations ignore possible recharge to the reservoir from the natural flow or possible increased recharge due to drawdown, although sometimes the power that could be generated from this flow is added on to the final power estimate. In most fields such recharge is relatively small, so this is not an important issue.

One approach sometimes taken to address the uncertainties in stored heat calculations is a Monte Carlo simulation. For each parameter, such as reservoir area, depth, porosity, average temperature, and recovery factor, a distribution is assigned with a range of values. Then the Monte Carlo simulation makes a large number of random picks within each distribution to calculate the final value, thus producing a probability distribution for the final stored heat estimate. When a probabilistic approach is taken, it is normal to regard the 10% value (i.e., 90% chance that this value will be exceeded) as proven and the 50% value as probable. It is important to ensure that the distributions chosen are realistic. For example, if a range of 5%—20% is chosen for the recovery factor, the distribution should be skewed with a most likely value of 10%. It is also important that the mean or modal values are realistic; the use of the probabilistic approach does not compensate for systematic overestimation. For example, some fields have been explored and abandoned. Therefore, in a field in which a successful well has not been drilled, there is a finite possibility that there is no (economic) resource. Any probabilistic assessment must then include a finite possibility that the resource capacity is zero.

Garg and Combs (2010) show a worked example of how greatly a Monte Carlo assessment can be altered by parameter variations. Bjornsson (2008) reviews resource estimates for Momotombo, Nicaragua, that exceeded actual generation by a factor of three or more. Sarak and colleagues (2009) showed that cognitive bias can cause significant underestimation of the uncertainty in reserves estimates, and Onur and colleagues (2010) discussed some of the error sources.

Further adjustments must be made when only part of the resource is accessible: the part of the potential resource that can be accessed by wells and for the effects of injection. The flow pattern imposed by the production and injection defines the volume of rock through which fluid will pass to the production wells, and it is only from this volume that heat can be extracted. This implies in general that any reserves lying beyond the planned injectors are stranded, lost to production, and reserves outside the available land area are probably also inaccessible, depending on the geometry of the resource and wellfield. In either case, only the accessible and producible area, and the volume beneath it, can be counted into any reserve estimate.

In summary, while estimating the total heat reserves is common practice, frequently not all the relevant factors are included in this assessment, and the use of Monte Carlo methods does not compensate for significant errors in assumptions about resource size or recovery factor.

3.6.4. Power Density

A method commonly used in the early stages of exploration and development to estimate resource capacity is simply based on area. Each successful well is taken as proving the area within a 1 km circle or square around the well, and the total area so defined is taken as proven. This area is then multiplied by a power density—typically $10-15$ MW/km^2. Grant (2000) used actual performance of some fields to define the power density as a function of temperature, increasing with higher temperatures. The method is equivalent to stored heat with a fixed reservoir thickness, but it has the advantage that it is referred to actual performance to provide calibration. Sanyal and Sarmiento (2005) argued that the power density varied greatly between fields, while Sarmiento and Björnsson (2007) quoted some Philippines values similar to Grant's. As with stored heat estimates, some of the potentially productive resource areas may need to be discounted to allow for reinjection.

3.6.5. Fluid Reserve

In the simplest depletion situation the pore fluid of a geothermal reservoir is removed. Assuming no reinjection, the total fluid producible from the reservoir is limited to the total initially present. In the case of the preceding 1 km^3 box, if it is liquid-filled, the total amount of pore fluid is 0.15 km$^3 = 1.2 \times 10^{11}$ kg. This is a maximum estimate of the amount of fluid that can be produced from the box.

Reserve estimates for vapor-dominated reservoirs have been attempted by estimating the amount of water in place. The reliability of such estimates is limited by the uncertainty in the liquid saturation. In practice, the best estimates of the initial liquid content come from numerical simulations by history matching production performance and are not available until many years of production and pressure response data are available.

As reinjection of separated brine and steam condensate is now nearly universal, there is no risk of depleting the liquid content of a liquid-dominated reservoir, and the initial fluid content of the reservoir is not a constraint. The total heat content, and the efficiency with which it can be transferred to circulating liquid, is the effective limit on production.

3.6.6. In Situ Boiling (Intergranular Vaporization)

In many geothermal reservoirs, withdrawal of fluid stimulates boiling in the pore spaces. This means that the producing wells are exploiting a two-phase reservoir. The removal of any fluid, water, or steam means that additional boiling of water occurs to make up the volume of lost fluid, and this boiling cools the reservoir. An extreme case is when only steam is produced and boiling is the only mechanism acting to replace it. This is, in essence, the reservoir containing steam and immobile water, although there is also the possibility of producing steam from above a liquid-dominated zone.

Considering again the 1 km^3 box at 240°C, being partly saturated with water, assume that a flowing wellhead pressure of 10 bar has been set. Steam in the reservoir can be withdrawn down to a minimum pressure of 10 bar, where the corresponding saturation temperature is 180°C. From Eq. (3.51) dry-out at 180°C requires that $\varphi S_o = 10\%$, or 70% saturation for 15% porosity. The initial water saturation must be at least 70% to keep the reservoir wet down to 10 bar. Assuming 70% initial saturation, the initial heat content of the box is 6.4×10^{17} J. When abandonment occurs at 10 bar, the now dry rock contains 4.1×10^{17} J, so 2.3×10^{17} J of heat has been removed in 8.2×10^7 kg of steam.

If there were additional water present such that abandonment could be set at 150°C or 5 bar (that is, a minimum production WHP of 5 bar), additional steam could be produced by lowering the reservoir pressure to 5 bar. A total of 3.4×10^{17} J could be extracted, carried in 1.2×10^8 kg of steam.

3.6.7. Cold Sweep

Production by cold sweep is at the opposite extreme from the depletion of a liquid reservoir. In this case the reservoir is subject to recharge by water, usually colder, equal in volume to the amount of withdrawal. This is approximately the actual situation for a reservoir with a binary plant: all the produced water is returned by injection. (Because the cooled water has lower specific volume, there is still a net volume loss to the reservoir and a net pressure drop.) The result is that there is withdrawal of hot water at one place and replenishment by the entry elsewhere of cold water that "sweeps" in from the injection well locations. It is assumed that the porous medium is homogeneous and the permeable paths are sufficiently finely distributed so that rock and fluid are in thermal equilibrium. This is a major assumption that, in practice, is seldom valid except in reservoirs hosted in sandstone. Proceeding with this assumption, if V is the Darcy velocity, defined by Eq. (A3.10), a particle of water moves with velocity V/φ, since the water flows only through the fraction φ of the rock that is occupied by water. Then the temperature changes are given by the equation for conservation of energy (A3.5). For steady flow $\nabla . \boldsymbol{u} = 0$, and:

$$\rho_f C_f \frac{\partial T}{\partial t} + \rho_w C_w V . \nabla T + K \nabla^2 T = 0 \tag{3.63}$$

If the heat conduction term can be ignored, this equation is simply soluble:

$$T = f(r - V_{th}t) \tag{3.64}$$

where

$$V_{th} = \frac{\rho_w C_w}{\rho_f C_f} V \tag{3.65}$$

is the thermal velocity—that is, the velocity at which a change in temperature moves. The particle velocity is also called the chemical velocity, as it is the rate at which water of different chemistry advances.

Now consider a block of porous rock that is initially at a uniform temperature and filled with water. If hot water is removed at one side and cold water replaced at the other, there will be a steady flow across the block. As water flows through the rock, the initial pore fluid flows out. Particles of the entering fluid advance behind the initial fluid at the chemical velocity. The interface between the two is called the chemical front. The thermal front advances more slowly at the thermal velocity. Thus, behind the chemical front is recharge water that has been heated to the original temperature of the rock. Following that is the thermal front, behind which is water at the recharge temperature and rock cooled to this temperature.

The produced water simply corresponds to the same profile: the first to be produced is the original pore water, then recharge water reheated to the temperature of the original reservoir, and finally cold recharge water. In this model, all of the heat in the rock is produced in the form of water at the rock's original temperature. The heat is "swept" out of the rock. Considering again the 1 km^3 block, all the 6.8×10^{17} J of heat stored in the initially water-saturated rock is produced as $240°C$ water, a total of 6.6×10^8 kg. Flashed at $150°C$, this provides 1.2×10^8 kg of steam—the same amount as can be produced by intergranular vaporization. If a second flash stage were added, the extraction by cold sweep would deliver more steam to the turbines than by in situ boiling. Without the second flash stage, the two means of heat extraction are comparable.

The ratio λ of the thermal and chemical velocities is:

$$\lambda = (V/\varphi)V_{th} = \frac{\varphi \rho_w C_w}{\rho_f C_f} = \frac{\varphi \rho_w C_w}{(\varphi \rho_w C_w + (1 - \varphi)\rho_m C_m)} = \qquad (3.66)$$

Note that λ does not depend on the reservoir or recharge temperatures. In the case of our 1 km^3 block, this ratio is 0.19. The chemical front moves at five times the speed of the thermal front. This ignores the effect of heat conduction, which thickens the front by a few meters a year. More importantly, it also ignore is the effect of fracturing in the medium.

3.7. FRACTURED MEDIA

Most geothermal reservoirs consist of fractured rock. The bulk of the pore space is in blocks of porous rock, but the bulk of the permeability is in the fractures between these blocks. It is common because it is simpler and easier to assume that the medium behaves as a homogeneous one. There are nevertheless situations in which the fractured nature of the rock cannot be ignored. Notably, engineered geothermal systems (EGS) consist of nothing else but fractures with fluid heated by the surrounding rock.

Instead of a thermal front advancing smoothly to the producing well, sweeping heat from the rock, there is preferential return of cold water to the producer. Some heat is transferred by conduction from the adjacent rock to the water flowing in the crack, so the advancing cold water is heated to some extent. However, cold water will reach the production well ahead of the thermal arrival time expected for flow through a homogeneous medium. For example, Figure 11.7 shows calculations by Nakanishi and colleagues (1995) of the temperature at a production well in a fractured medium compared to the uniform sweep result obtained in a homogeneous porous medium. It can be seen that the crucial variation is with the crack spacing. The closer the crack spacing, the more heat is drawn from the matrix and the slower the cooling of the produced water.

3.7.1. Thermal Effects

To calculate heat transfer from the matrix to fractures to produce the results in Figure 11.7, the actual fracture network is idealized as a regular array of parallel fractures. Each crack has thickness $2h_f$, and each rock has a layer thickness of $2h_b$. The heat transfer mechanism is fluid flow in the fissure and conduction in the blocks. Let ∇_2 denote the gradient in the plane of the fissure, z the coordinate normal to the fissure, and u the mass flux density in the fracture. It is assumed that the fracture width is so small that temperature and fluid flow are distributed uniformly across it. Then a heat balance on the fluid in the fissure gives:

$$h_f \left[\rho_f C_f \frac{dT_f}{dt} + C_w u \cdot \nabla_2 T_f \right] = Q \tag{3.67}$$

where Q is the heat transfer from block to fissure per unit surface area of fissure. In the block conservation of heat gives:

$$\rho_b C_b \frac{\partial T_b}{\partial t} - K \frac{\partial^2 T_b}{\partial z^2} = 0 \tag{3.68}$$

The temperature in the block is considered only to vary normal to the distance from the fracture; transverse variation is ignored. This is because the fissure is a much better transport medium along the plane of its orientation. Boundary conditions are needed for Eq. (3.66). The middle of the block is a plane of symmetry:

$$\frac{\partial T_b}{\partial z} = 0 \quad \text{at } z = h_b \tag{3.69}$$

and at the fissure

$$T_b = T_f \tag{3.70}$$

and

$$K \frac{\partial T_b}{\partial z} = Q \tag{3.71}$$

Heat transfer within the block is by conduction, with thermal diffusivity $\kappa_b = K/\rho_b C_b$. A parameter of importance is the block relaxation time $\tau_b = (2h_b)^2/\kappa_b$. For $t \gg \tau_b$ there is approximate thermal equilibrium across the block, and the medium behaves like a homogeneous one. For $t \ll \tau_b$ the adjacent fissures have no effect on each other, and each fissure behaves in isolation like a single fissure in an infinite medium. The time (roughly) for heat to diffuse across the block is τ_b.

The results illustrated in Figure 11.8 show how the fractured medium spreads the heat stored in the reservoir rock to the discharge fluid. Instead of the ideal case of a homogeneous cold sweep, the same total amount of heat is produced at a lower average temperature. Available energy extraction by fracture flow is always less than that by cold sweep, since it involves heat conduction in the blocks.

3.7.2. Dual Porosity Theory: Models by Barenblatt, and Warren and Root

The first fractured medium theory was developed for pressure transients in groundwater (Barenblatt et al., 1960) and in petroleum (Warren & Root, 1963). This is a double-medium or double-porosity theory. Figure 3.4 shows this idealized reservoir.

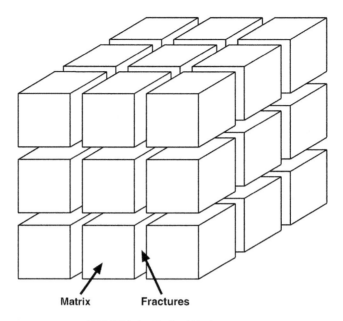

Matrix   **Fractures**

FIGURE 3.4   Idealized block structure.

The porous medium is conceived as two interpenetrating media with fluid transfer between them. It is convenient to define the fracture and matrix properties with respect to the total (fracture plus matrix) volume. k_f and k_m are the contributions to the total medium permeability due to the fractures and matrix $k = k_f + k_m \approx k_f$. Likewise, φ_f and φ_m are the void spaces per unit reservoir volume in fracture and matrix, respectively. If $\varphi' \ll 1$ is the fraction of the reservoir occupied by fractures, the permeability of the fracture itself is k_f / φ' and its porosity is φ_f / φ'. The pressure equation for flow in the fractures is then:

$$\frac{k_f}{u} \nabla^2 P_f = \varphi_f C_f \frac{\partial P_f}{\partial t} + Q \tag{3.72}$$

where Q is the volume transfer per unit reservoir volume from matrix to fracture.

There remains the specification of the pressure distribution in the block and thus the transfer term Q. One approach introduces a pressure to represent the average pore pressure in the matrix and assumes that the fluid transfer term is proportional to the difference between fracture and matrix pressure, and matrix fluid mobility—that is, it assumes quasi-steady flow:

$$Q = \frac{k_b}{u} \frac{P_f - P_b}{L^2} \tag{3.73}$$

where L is a representative block dimension. Conservation of mass in the block gives:

$$Q = \varphi_b c_b \frac{\partial P_b}{\partial t} \tag{3.74}$$

Equations (3.71) and (3.72) allow us to solve for the difference $P_f - P_b$ and thus:

$$Q = \alpha \varphi_b c_b \int_0^t \frac{\partial Pf(t')}{\partial t} G(t - t')dt' \tag{3.75}$$

$$G(t) = e^{-t/\tau_b} \tag{3.76}$$

where $\tau_b = \mu \varphi_b c_b L^2/k_b$ is the block relaxation time and α is a geometric constant. If allowance is made for the unsteady pressure distribution within each block, the function G is replaced by a more complex expression (Najurieta. 1980). The simple form of G used here is inaccurate at short times.

3.7.3. Dual Porosity Theory: Elaborations

The preceding model has three major limitations:

1. The matrix fluid is single-phase.
2. The fracturing divides the reservoir into blocks of uniform size.
3. The simple transfer function does not correctly represent the transient flow between matrix and block.

These limitations have been addressed in a wide variety of methods, all using reservoir simulators of various types.

The limitation to single-phase fluid is easily removed by using a simulator. The assumption of uniform block size is clearly a gross simplification, since fractures tend to be distributed with a large degree of randomness within the reservoir. Extensive work at The Geysers has shown that the fracture pattern is self-similar—that is, there is the same pattern of larger and smaller fractures at different scales from fault patterns to jointing (e.g., Sammis et al., 1992; Williams, 2007). A better representation of the transfer between matrix and fracture has been addressed in a number of ways, including explicit computation of the fluid distribution within the matrix, which is more accurate but significantly more time-consuming. An example of the latter is the MINC (Multiple Interacting Continua) formulation of Pruess and Narasimhan (1985) and Pruess and colleagues (1999). Each block is represented by a series of nested layers, providing a coarse discretization within the block, as shown in Figure 3.5.

EGS experiments are to some extent experimental validation of modeling methods for fractured media, but as yet no EGS has run for a long enough time to provide a full test. Most numerical reservoir simulators now provide the ability to include dual-porosity as a standard option.

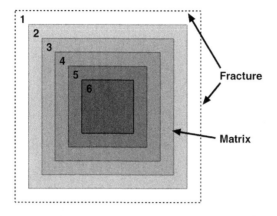

FIGURE 3.5 Multiple Interacting Continua (MINC) element. *Source: Pruess & Narasimhan, 1985.*

3.7.4. Dispersion

Dispersion is a problem that arises in heat and tracer transport but not in pressure transients. The different fractures are very unlikely to be of equal size, so fluid will move faster along some than others. Consequently, particles of

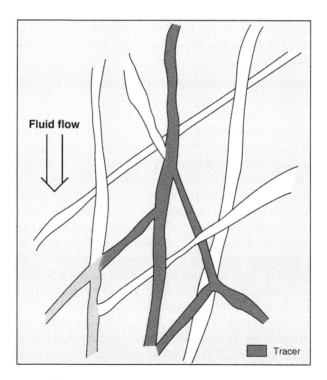

FIGURE 3.6 Dispersion of tracer in a fracture network.

fluid move relative to one another rather than advancing as a smooth front, as illustrated in Figure 3.6.

The problem is a familiar one in groundwater (Bear, 1972). The random medium creates a dispersivity κ_D, which is given approximately by:

$$\kappa_D = VL \tag{3.77}$$

where V is the fluid velocity and L is a characteristic length of the medium, such as the size of a block. This dispersion greatly exceeds that due to molecular diffusion in geothermal systems as is seen in tracer tests.

3.8. CHEMICAL FLOW MODELS

The variations in chemical constituents provide an indication of the changes experienced by sampled fluid in its past flow. Chemical geothermometers, relying on temperature-dependent equilibria with differing time constants, provide an indication of temperatures for distances correspondingly back along the flow path.

The variations in solute content also provide an indication of the extent to which produced fluid has undergone mixing or boiling. Variations in solute content are the principal means by which a hypothesized deep-source fluid may be identified. If there are two different fluids, with concentrations f and F of two different chemical species, or heat or gas, and the two fluids mix with fraction x of the first, then for the mixed fluid:

$$f = xf_1 + (1 - x)f_2$$
$$F = xF_1 + (1 - x)F_2$$

This defines a straight line; plotting different f against F for different mixing proportions gives a straight line between two endpoints. The endpoints are $x = 0$ and $x = 1$, the concentration of each original fluid contributing to the mix. A mixing diagram—a crossplot of the concentration of one species against another—of two fluids produces a straight line between the two end members. The samples may not include any pure end member sample, so the end member may lie further along the trend line. The traditional chloride-enthalpy plot is an example: a crossplot of enthalpy against chloride.

If the rising water boils in a steady-state flow, a steam-water mixture at lower temperatures is formed. All the chloride remains in the liquid phase so that:

$$f = f_b \frac{H_{sw}}{H_s - H_b} \tag{3.78}$$

which is (nearly) a linear relation between chloride and enthalpy such that the chloride content increases with decreasing liquid enthalpy. Given an assembly of water samples from different temperatures, a single-source water temperature can be identified as the minimum temperature that permits the observed range of chloride-enthalpy data.

The use of the mixing-dilution model is valid for the chemical variations in the initial steady-state reservoir conditions. Under exploitation there will be heat transfer between rock and fluid, altering the chloride-enthalpy relation. The heat transfer from rock to fluid increases (or sustains) fluid enthalpy without the addition of chloride, giving the appearance that hot water or steam of zero chloride content has been added to the system. For example, if waters created by a dilution process are produced by cold sweep, the chloride-enthalpy becomes:

$$\left(\frac{df}{dH}\right)_{sweep} = \frac{1}{\lambda}\left(\frac{df}{dH}\right)_{mixing} \tag{3.79}$$

λ is defined by Eq. (3.66). Mixing production fluids with injection water introduces a different mixing trend, since the injectate is a high-chloride fluid of relatively low enthalpy.

3.9. APPLICABILITY OF THE MODELS

In this chapter, the physical processes identified as relevant to geothermal systems have been catalogued, and simple idealizations of such processes have been described. It is not suggested that any of these simple models are directly applicable to any particular reservoir. Rather, these models develop concepts that can be applied to some aspects of a reservoir's behavior. For example, regarding the discussion of the effects of a change of reservoir fluid type, no actual reservoir has been observed to undergo such a change in its pressure history, but the contrast in the behavior of the different types shows the importance of correctly identifying the actual physical processes present.

It is a somewhat surprising observation that many of the techniques used in geothermal reservoir engineering, and particularly in resource assessment, are theoretical and poorly validated by experiment or published experience, although considerable unpublished material is available. Measurements on wells and well models are wells tested by practice, with many of the fitted models validated against later behavior. But reservoir assessment and performance are not as well supported. There are relatively few published papers where methods are well described and the results can be checked or were checked against later performance.

Interpretation of Downhole Measurements

4.1. INTRODUCTION

This chapter introduces well testing and measuring and the unique problems encountered when interpreting the measurements made in a geothermal well. With petroleum and groundwater, running a pressure-temperature survey in a well usually results in straightforward data, but this is not the case with geothermal measurements. This section introduces downhole temperature and pressure measurements and their interpretation, production testing, test design, and reporting.

In most cases it is not possible to directly measure the downhole characteristics that are needed to assess a geothermal resource. Instead, an interpretation or inference is made from the information that is available from the well test program. Interpretation of well test data involves bringing together a range of information from the drilling operations, geology, downhole measurements, and production tests. Knowledge of other nearby well characteristics is also useful. Often the set of test data that are available will be incomplete. There may be errors in the measurements, or the well conditions may not be stable. The end result of these factors is that interpretation of well test data is an imperfect science. Different interpretations can be derived using the same data, and within the limitations of the available data, both may be "correct." This is due to a combination of physical factors: the permeability

Geothermal Reservoir Engineering. ISBN: 978-0-12-383880-3 DOI: 10.1016/B978-0-12-383880-3.10004-6

distribution, the state of the reservoir, and the well design. In most geothermal reservoirs the permeability is not constrained within a uniform aquifer but is distributed as an array of fractures, varyingly pervasive through the rock, that intersect the well at a few (almost random) points; the wellbore usually has an open hole section of up to 2000 m, and the reservoir fluid is not static.

The consequences can be seen by considering a simple example: an exploration well drilled into the central upflow of a geothermal field. This well has permeability at a few depths, scattered along the open hole section. Outside the well there is a distribution of pressure $P(z)$ and temperature $T(z)$ with depth. The geothermal field has a natural flow rising from depth eventually reaching surface to discharge as springs and thermal features. For example, Figure 12.28 shows the distribution of pressure and temperature with depth in the center of Mak-Ban. There is a vertical pressure gradient of 8.2 bar/100 m compared to hydrostatic (for the temperature) of about 7.4 bar/100 m. Darcy's Law gives for the vertical flow density v_z:

$$v_z = \frac{k}{\mu} \left(\frac{dP}{dz} - \rho g \right)$$

or

$$\frac{dP}{dz} = \rho g + \frac{\mu}{k} v_z$$

All the water properties are those for the pressure and temperature at that depth. The vertical pressure gradient exceeds hydrostatic by an amount needed to drive the natural flow upward to the surface. If the well contains a column that is in thermal equilibrium with the reservoir, the pressure gradient in the wellbore is hydrostatic for the reservoir temperature, and it cannot be in equilibrium with the reservoir pressure over the entire open interval. If the wellbore pressure balances reservoir pressure near the center of the open interval, it will be less than reservoir pressure in the lower part of the well and greater than reservoir pressure in the upper part. If there are two or more permeable zones, fluid will flow into the well at the lower zone(s), up the well, and exit at the upper zone(s). The fluid in the wellbore is no longer static and will not be in thermal equilibrium with the surrounding reservoir. Instead there will be a thermal equilibrium between heat conducted into or out of the well and heat carried by the flow in the well. The details of this example are particular to a well drilled into the upflow part of the reservoir. In the particular case of Mak-Ban, the wells typically contain two-phase upflows, as shown in Figure 4.5.

In other parts of the reservoir, wells may typically exhibit downflows. In an exploited field, the fluid distribution is highly disturbed and this can affect temperature-pressure profiles measured in the wellbore in various ways. Rather than being a passive monitor of the reservoir pressure and temperature, the well must be considered as a pipe penetrating the reservoir, with connections to the reservoir at various depths. Fluid flows along the wellbore between these points (later referred to as "feed zones") depending on the permeability of the various feed zones and the pressure imbalance between the wellbore and the reservoir at the different zones. For these reasons it is important when designing a well for monitoring reservoir conditions that the open hole length be small and, ideally, has a single feed zone.

The well test examples used here are obtained from geothermal wells in different fields in several different countries. The wells and data have been selected because they make good examples of the various interpretation methods that can be used. In practice, the data interpretation is usually more difficult than in the examples due to missing information, unstable well conditions, transient changes in pressure and temperature, noisy data, or an incomplete well measurement program.

Each well is an individual and deserves special attention to get the best information about the underground resource. Sometimes there is only one opportunity to obtain certain information during a well's lifetime, and if the appropriate measurements are not made in that period, then the opportunity has passed and can never be recovered.

4.2. OBJECTIVES OF THE WELL TESTING PROGRAM

The physical measurements made in geothermal wells provide the primary information from which ideas about the underground resource can be developed. The essential measurements are pressure, temperature, and flows made both in the wellbore and at the surface. Downhole logging techniques other than pressure-temperature-flow are available that can provide additional information about rock properties, such as porosity and density, and help to determine detailed information about fracture zones, such as fracture density and orientations that assist in understanding the resource.

As a new geothermal field is explored, developed, and put into production, the objectives of the well testing program change. In the exploration stage, the measurement program is building a picture of the resource in its natural (undisturbed) state, the extent of the resource, and the distribution of temperature, pressure, and rock characteristics (permeability, porosity, fracture patterns, etc.) using the following parameters:

- Temperature distribution
- Pressure distribution
- Permeability distribution
- Reservoir state (liquid/vapor dominated, single-phase/two-phase)
- Fluid chemistry
- Noncondensable gas content

Some of this information is obtained from a combination of the downhole and surface measurements program. For example, to obtain gas content and fluid chemistry samples that are representative of the reservoir, the well must be discharged for some time to remove contamination resulting from the drilling process.

In individual wells some of the important characteristics are the following:

- Location of feed zones (depth)
- Feed zone temperatures
- Undisturbed formation temperature profile
- Feed zone pressures
- Fluid type (downhole)
- Production capacity: mass flow, steam flow, fluid enthalpy, gas content, and reservoir fluid chemistry
- Injection capacity: flow/pressure characteristics

When large-scale production starts, the measurements program becomes focused on changes from the "baseline" values with time and forecasting future field performance:

- Production temperature change (at individual feed zones)
- Reservoir pressure change
- Production flows and enthalpies (at surface)
- Changes in chemistry and gas content of the produced fluids
- Changes in well performance (permeability)
- Impact of production on resource
- Impact of injection on resource and production
- Individual well problems (cool inflows, blockages, rundown, casing damage, etc.)

4.2.1. Multidiscipline Approach

The overall measurements, evaluation, and development program is interactive and has inputs and contributions from a group of related scientific and engineering disciplines. To have the most effective well evaluation program, there must be a team approach combining the skills of the various science and engineering disciplines involved in geothermal development:

- Geology
- Geophysics
- Geochemistry
- Subsurface measurements
- Production testing
- Drilling
- Surface plant design and operations
- Environmental

The actual field measurements that are made in exploration wells are generally as follows:

Measurements taken while drilling

- Temperature buildup tests
- Formation pressure at permeable zones
- Permeability tests at loss of circulation zones

Measurements taken on well completion

The completion test involves conducting temperature and pressure surveys in the well while injecting cold water:

- Feed zone locations, permeability, formation pressure, injectivity/productivity
- Overall well permeability

Heating

After well completion, the well is shut and allowed to heat up to full formation temperature, and pressure-temperature profiles are measured in the well:

- Formation temperatures/vertical temperature profile in reservoir
- Reservoir pressure: feed zone pressure
- Location of feed zones

Production Test

In the case of injection wells, this would be an injection test:

- Measure production potential (surface measurements to obtain mass flow, enthalpy, gas content, and chemistry)
- Measure downhole response to production (temperature and pressure/ permeability)
- Confirm feed zones locations and feed temperatures using downhole surveys with well flowing
- Monitor other wells for pressure response (interference)

Usually not all the preceding methods will be used in one program. The actual methods will be controlled by local factors such as well drilling technique (water-air-mud), availability of test equipment, and well availability (e.g., during drilling).

In addition to making measurements to determine the characteristics of the geothermal resource, sometimes additional measurements are made to assist in the evaluation of conditions inside the wellbore and casing (to help in understanding changes in well performance and assist with design of procedures to fix the problem). An example is finding the location of mineral deposition or scaling formed as a result of production or injection by using callipers or measuring the drift clearance in the wellbore using go-devil or gauge rings. Casing problems such as breaks, casing deformation, and corrosion can be measured using specialized tools such as internal casing callipers or electronic casing evaluation instruments that monitor the casing wall thickness and deformation/corrosion of the internal and external surfaces.

4.3. WELL MODELS

The object of measurement interpretation is to deduce from measurements made within the dynamic fluid inside the wellbore the properties of the reservoir around the well. Ideally, a complete interpretation provides a complete model of the well, including the following:

- Location and thickness of permeable zones
- Permeability of these zones, expressed as injectivity, productivity, and/or transmissivity
- Reservoir pressure at each zone
- Reservoir temperature over the entire well depth
- Reservoir temperature or enthalpy of fluid at each permeable zone
- Any mechanical variation of the well: wellbore diameter, suspect casing, blockage, deposition

Usually only part of this information can be determined. Note that reservoir temperature may be found over the whole interval, since conductive heat flow into the well may control well temperatures, but reservoir pressure can only be determined at permeable zones, since it is only at these depths that the well communicates with the reservoir fluid pressure.

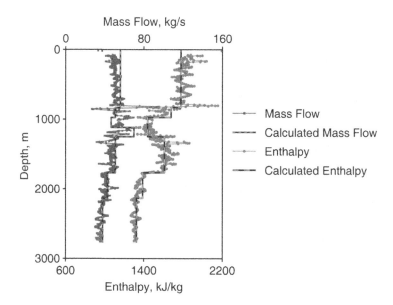

FIGURE 4.1 Mass flow and enthalpy in modeled well. *Source: Acuña & Acerdera, 2005.*

TABLE 4.1 Well Feed Zones

Depth, m	Flow, kg/s	Enthalpy, kJ/kg
831	5	2791
975	9	2791
1116	−28	N/A
1250	19	977
1768	8	2791
2134	5	1861
2775	38	1326

Source: Acuña & Acerda, 2005.

Figure 4.1 and Table 4.1 show an example of a well model being used to match discharging spinner measurements. Seven feed zones, including one thief zone and one cold zone, are needed. Downhole enthalpy, shown in Figure 4.1, is the enthalpy of the flowing mixture as computed from the pressure gradient and flow rate, using the drift-flux model for two-phase flow (Hasan & Kabir, 2002) and allowing for frictional pressure drop.

Acuña (2003) reports that using such a well model provides better forecasts of future steam flow than decline analyses. A detailed well model helps to diagnose problems such as a cold intrusion at one feedzone and then plan repairs. For an example, see Torres and Lim (2010).

4.4. SOME BASIC WELL PROFILES

4.4.1. Conductive versus Convective

The simplest distinction made in temperature profiles is between conductive and convective profiles. When rock is impermeable, heat is transported by conduction. This produces a characteristic profile where temperature increases linearly with depth; the gradient will change if there is a change in thermal conductivity of the rock.

Convection by contrast is a far more efficient means of heat transport than conduction. Once there is some permeability in the rock—and the required permeability is much less than what is needed for economic well performance—the fluid motion controls the temperature distribution. Convective profiles can take a considerable variety of forms, with isothermal sections, inversions, boiling sections, and mixtures of all of these. Figure 4.2 shows temperature profiles from two wells at the engineered geothermal systems (EGS) project at Soultz, France (Genter et al., 2009). There are three sections on the profile. The

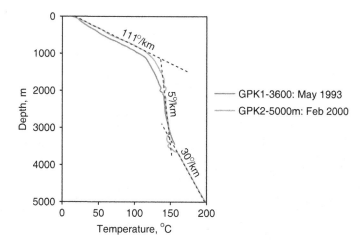

FIGURE 4.2 Temperature profiles in GPK-1 and GPK-2. *Source: Genter et al., 2009.* © *Geothermal Resources Council.*

first kilometer has a high gradient and linear profile, indicating conductive transport. Then from 1 km to 3.3 km there is a much lower gradient, which is attributed to a convective system along faults and fissure zones. Finally, below 3.3 km there is again a high linear gradient, indicating conductive heat transport and consequently lower permeability in the surrounding formations.

4.4.2. Isothermal

An isothermal profile is a section of the well where the temperature is constant or nearly constant with depth. This can reflect circulation of fluid in a section of the wellbore or interzonal flow (without boiling), or it may be that the reservoir itself has isothermal temperatures due to convection. Figure 4.3 shows two profiles in well NG13 at Ngawha. During injection there are inflows to the well at 960 m (Figure 4.3 PT1). At the inflow depth, the temperature increases rapidly over a short interval as the cold water being injected into the well mixes with the hot inflow. The mixture flows down the well and exits at a lower feed zone at 1600 m. In the shut-in condition, the temperature profile looks similar to Figure 4.2, but in this case the isothermal section between 960 and 1600 m is due to interzonal flow in the wellbore between feedzone at these depths (confirmed by spinner data), while above 960 m the isothermal profile is attributed to convection in the reservoir. Below 1600 m the wellbore temperatures are again conductive. The pressure profiles show that during injection, the pressure at 960 m is less than the stable shut pressure and more than stable shut pressure at 1600 m, confirming that during injection there is inflow at the upper zone and outflow at the lower zone (without the need for spinner information). The stable shut pressure is in fact not reservoir pressure, since even when shut there is still a downflow between the two zones.

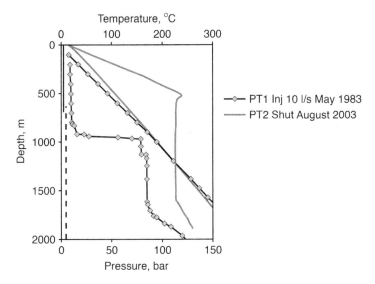

FIGURE 4.3 Temperature profiles in NG13 during injection and stable shut. Note that between 1983 and 2003 there has been no significant change in the reservoir pressure ($\ll$ 1 bar) *Source: Top Energy, Personal Communication.*

The water flowing down the well gains or loses some heat by conduction to the surrounding formation, but this amount is generally negligible unless the flow is small—in the order of one liter per second or less. There is also a slight heating due to the adiabatic (isentropic) compression of the fluid with greater depth and pressure.

4.4.3. Boiling Curve

A boiling point temperature profile is a column of water that is at the boiling point for the pressure, allowing for the effect of dissolved gas. Figure 4.4 shows two such profiles from well WK24 at Wairakei. These boiling point profiles are produced by an upflow in the well. In the 1955 profile, boiling water enters the well near 580 m and flows up, continuing to boil as it ascends. Boiling water exits the well at a shallow feed point between 350 and 400 m, and the steam, together with noncondensable gas, rises into the casing. In the 1958 WK24 profile, liquid water enters the well near the well bottom and flows upward, boiling at 450 m. The boiling profile in both 1956 and 1958 obscures the detail of the shallow reservoir temperatures, which were inferred from heating measurements in WK24 and measurements in adjacent wells.

4.4.4. Two-Phase Column

A more extreme upflow is shown in Figure 4.5. Here, two-phase fluid enters the well at a deeper feedzone and flows up the well, producing a column of

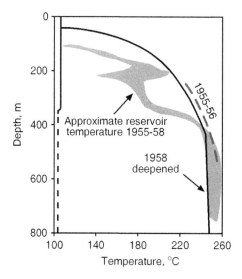

FIGURE 4.4 Stable downhole temperatures and reservoir temperatures in WK24. *Source: Contact Energy, Personal Communication.*

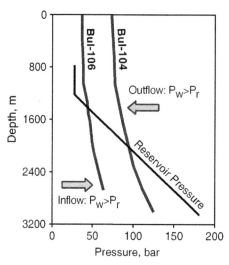

FIGURE 4.5 Downhole profiles in well with two-phase upflow. *Source: Menzies et al., 2007.* © *Geothermal Resources Council.*

two-phase fluid throughout the complete wellbore. Water, and possibly steam, exits at a shallow feed, and some steam continues up the wellbore to condense in the casing if shut or to bleed. In this case both reservoir temperatures and pressures are completely obscured.

4.5. GAS PRESSURE AT WELLHEAD

When there is internal upflow in a well, and water boils, there is a flow of steam (and noncondensable gas) into the upper part of the well and into the casing, depressing the steam-water interface, with the result that a pressure is developed at the wellhead. When the shallow, cased-off formations are cold, heat is lost to the cold formations, and with time, the steam condenses and the remaining gas accumulates in the upper part of the casing. This process is the source of gas pressure buildup, which is frequently observed in high-temperature boiling reservoirs and can also occur in a reservoir with quite low gas content.

4.6. UNUSUAL OR MISLEADING WELL PROFILES

4.6.1. Temperatures in Well WK10, Wairakei

This example shows an extreme case of the difference between measurements in the well and the state of the reservoir. WK10 was a peripheral shallow well at Wairakei that was drilled in 1951. Figure 4.6 shows temperature profiles measured over a period of 20 years.

The successive measurements show an apparently clear history: temperature has declined steadily over time. Since the well is on the edge of the field, this is not surprising, and it seems that it is showing the steady inflow of cooler waters from the field edge. However, this is quite wrong. The bottomhole temperatures were measured during drilling (marked as "BHT" in Figure 4.6). Drilling was done only during daylight hours, and the bottomhole temperature was measured each day after approximately 12 hours of heating overnight. These measurements should be close to the true formation temperature, since the bottom of the hole has been subjected to little circulation to cool the formation. The drilling reports also make it clear that the formation conditions were relatively cool before encountering hot and gassy conditions at 250 m. The downhole profile in 1951 shows much hotter fluid in the well, except at

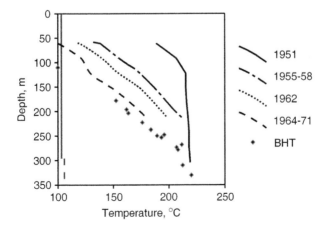

FIGURE 4.6 Temperatures measured in WK10. *Source: Contact Energy, Personal Communication.*

well bottom. What happened in 1951 was that hot fluid was flowing up the well from near well bottom and exiting between 100 and 150 m (presumably at a casing break). With time, reservoir pressure falls due to exploitation, and the upflow weakens. Conductive heat losses become more important as the upflow decreases. Eventually, by 1971, the measured temperatures are close to the original drilling bottomhole temperatures. Thus, the steady fall in measured temperature with time does not reflect a change in formation temperature outside the wellbore; the formation temperature did not change. Rather, the change in temperature reflects a fall in pressure driving the upflow in the well.

4.6.2. Reservoir Pressures at Matsukawa

Matsukawa is a relatively small low-permeability field in Japan and actually a small part of the much larger Hachimantai geothermal system. Wells in Matsukawa produce dry or superheated steam, and when shut, they contain a column of steam with varying pressures in the different wells.

Hanano and Matsuo (1990a, b) described the initial data in the field. Starting in 1952, wells were drilled and found to have a local two-phase zone at around 300 m. The reservoir has been described as vapor-dominated (see Fukuda et al., 2005). There were no downhole pressure measurements in the 1950s, and temperature profiles were only measured during heating when the wells contained water with a water level below the wellhead. The original feedzone pressures were calculated using water levels together with the temperature profiles (Hanano & Matsuo, 1990a, b). Celati and colleagues (1977) demonstrated that this method was an effective way to reconstruct pressure in a vapor-dominated reservoir. Using these pressures, the reservoir pressure profile was developed, which showed the undisturbed reservoir has a liquid pressure gradient. Figure 4.7 shows the results of this analysis. Although the wells discharge dry or superheated steam and (once heated) stand

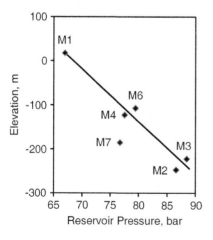

FIGURE 4.7 Original reservoir pressures in Matsukawa. *Source: Hanano & Matsuo, 1990a.*

with a column of steam when shut, the reservoir pressure gradient shows the reservoir is initially liquid-dominated. The pressure profiles in the heated wells are a column of steam, but the reservoir pressure is quite different. The pressure gradient is close to hydrostatic, indicating a liquid-dominated reservoir—a striking contrast between the pressure gradient in the reservoir and that measured in each well.

4.6.3. Liquid-Gas-Liquid Profile

Figure 4.8 shows a pressure-temperature profile that appears to have a vapor section between two liquid sections. Profiles of this form are encountered occasionally, usually in vapor-dominated reservoirs or steam zones, while injecting cold water. In this case the well is under injection at 40 kg/s, and four passes were made using a pressure-temperature-spinner (PTS) instrument: two up and two down. For all of these passes the profile remains essentially unchanged. The data show that inside the perforated liner, there is a gas zone sandwiched between liquid zones both above and below. In this case the reservoir has a segregated steam zone above a liquid zone, and in this well there is a feed at 600 m in the steam zone, which is indicated by the temperature increase during injection at this depth. During injection, steam and gas are flowing into the wellbore. Steam is condensed, leaving the gas, which forms a bubble inside the perforated liner, forcing the injected water to flow down around the liner annulus. The casing diagram shows that the top of the bubble is just above the top of the perforated liner. The spinner data in the gas interval were very noisy, indicating that some of the water was also showering down the

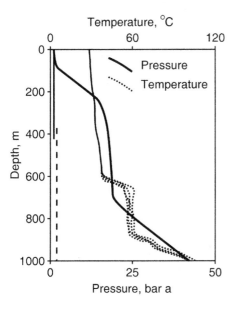

FIGURE 4.8 Well profile at Lihir, injecting 40 kg/s. *Source: Lihir Gold, Personal Communication.*

inside of the perforated liner. Below 700 m, water is flowing down the well to a loss zone just above 900 m. Within the casing, the pressure gradient is near hydrostatic or less than hydrostatic. This reflects the counterflow in the casing—water descending while gas bubbles rise to the surface. In this case an unusual pressure profile can be explained by considering the fluids in the reservoir, the fluids in the wellbore, the feedzone depths, and the configuration of the casing and perforated liner.

Downhole Measurement

5.1. INSTRUMENTS

The starting point for the reservoir engineering program is collection of data from wells. Scientific and engineering staff involved in data interpretation should have some familiarity with the reliability of the field data and, where relevant, its interpretation, since this will influence judgments that are made when using the data. The basic measurements are pressure and temperature. Together with knowledge of the steam tables and the effects of noncondensable gas, the downhole pressure-temperature data, perhaps with downhole spinner/ wellbore flow information, can be used to assess fluid state and flow within the wellbore and in the formation. Over the last few years, advances in technology have resulted in greatly improved reliability, accuracy, and range of equipment available for downhole measurements. Following is a brief outline of the methods commonly used in geothermal well testing.

The most common downhole measurements made in geothermal wells are temperature and pressure logs or profiles. These are normally measured over the full depth of the wellbore, with the well remaining in the same condition throughout the run. The well may be shut, flowing, or under injection. Until the 1990s, most of the instruments for making these measurements were mechanical recording temperature and pressure gauges that were capable of measuring 15−20 data points in one survey. Over the last 10 years, new electronic pressure-temperature-spinner (PTS) instruments capable of operating at elevated

Geothermal Reservoir Engineering. ISBN: 978-0-12-383880-3 DOI: 10.1016/B978-0-12-383880-3.10005-8

temperatures have become the standard logging equipment. These new instruments can operate as either e-line with surface readout or run using wireline in memory mode, and they are capable of simultaneously measuring several thousand pressure-temperature-flow data sets in a single survey.

Logging techniques developed for oilfield application are also finding use in geothermal wells. The most directly applicable to geothermal work are calipers and formation imaging instruments, although other techniques with high-temperature capability are being continually added to the suite of available logging methods. The fluid in the wellbore may limit the type and value of logging. For example, where the wellbore quickly fills with steam or gas, logs relying on sound transmission (e.g., cement bond and other sonic logs) cannot be used.

Formation imaging tools designed to look for fractures and geological structure found in the wellbore are now available. These imaging tools, using either ultrasonic or microresistivity methods, can map the full circumference of the wellbore and after processing provide a three-dimensional picture of the wellbore, mapping formation bedding and the orientation and characteristics of fractures or fault zones. These interpretations can then be combined with "traditional" geothermal interpretation to identify which of the structures mapped by the imaging log actually match zones where fluid is produced or lost from the wellbore and map fracture zones that do not produce fluid.

Other downhole "measurements" that are made include "go-devils" (sinker-bar or gauge-ring surveys), casing evaluation logs, and various types of downhole sampling for fluid and solid samples. Go-devils are plug gauges that are used to determine the drift diameter of well casings and liners. Although at the most basic level of technology, these measurements are simple and inexpensive and provide vital information to assist with the interpretation of casing damage or the buildup of mineral deposits within the wellbore. Various kinds of samplers are available to take samples of wellbore fluids or of solid materials, such as calcite, that have been deposited in the wellbore.

The logging instruments and logging program for a particular test will depend on a series of factors:

- Expected downhole conditions (mainly temperature)
- Test objectives
- Test program
- Well design
- Type of fluid in well
- Temperature rating of the logging equipment, as well as the heating characteristics of the well
- Availability of instruments/equipment

For tests made on completion of a well, it is useful to have a real-time readout of test data at the surface so the test program can be immediately adjusted to improve the final data quality where unforeseen changes occur. Once the feedzones have been located, most downhole logging objectives can be achieved using memory PTS tools. For wells where sustained high temperatures are expected at the wellhead and downhole (more than ~280°C), memory tools run on wireline are usually preferred (and more reliable) because of temperature effects on the e-line conductor and the cable head.

5.2. GEOTHERMAL WELL DESIGN

Geothermal wells are different from those completed for oil and gas exploration in several significant ways:

- The complete open hole section is exposed to the potentially productive formation.
- The well is completed with a slotted liner.
- The geothermal reservoir pressure is not usually exactly hydrostatic for the reservoir temperature.
- In the volcanic geothermal environment found in most high-temperature geothermal resources, the permeability found by wells occurs as discrete zones.
- There are often several permeable zones in a well, and these may be spaced several hundreds of meters apart.

One of the most significant differences between geothermal and oil/gas wells is that in geothermal wells the potentially productive formations are exposed to the wellbore throughout the complete open-hole section, while in oil and gas wells the producing zone is usually confined to a small well-defined vertical interval. The open-hole section in geothermal wells may be as great as 2000 m. This long open-hole section allows communication between the well and the formation wherever the well intersects a fractured zone, and where several permeable zones are encountered, there are usually interzonal flows. To further confuse interpretation of downhole pressure-temperature data, often there are circulation cells within the wellbore, usually driven by a combination of well completion design (variation of casing size and perforation details) and the diameter of the drilled hole (Allis & James, 1980; Boyd, 2009)

Most geothermal wells are completed with a perforated liner over an openhole section 1000–2000 m below the production casing. The liner is regular well casing with either slots or drilled holes. As a rule of thumb, the slot configuration is 15–20 mm diameter holes with a density sufficient to provide a perforated area over one meter at least equivalent to the cross-sectional area of the liner. Liners are necessary in many geothermal wells to prevent the fractured and altered formation from collapsing into the wellbore during discharge. In competent formations and where there is little pressure drawdown during production, a perforated liner may not be needed. This decision, however, must be made at the time of well completion. The presence of the liner in the wellbore precludes most electrical and sonic logging or other tests where the logging tools must be in direct contact with the formation. If such logs are needed, they must be done before running the perforated liner.

5.3. TEMPERATURE-PRESSURE INSTRUMENTS

Downhole instruments range from relatively simple mechanical tools which measure only temperature and pressure to high temperature electronic instruments protected in a sophisticated Dewar flask with an ability to measure a wide range of physical parameters of the wellbore fluid and the surrounding rock.

5.3.1. Mechanical Instruments

The first downhole pressure instrument for use in oil and gas wells was the "Amerada Bomb" that was created in 1929 by Geophysical Research Corporation. Although hundreds of design improvements have been made to these gauges over the years, the basic principles of mechanical downhole gauges have undergone little change from the original instruments. Amerada pressure gauges began to be applied in New Zealand geothermal wells in about 1960. Until the advent of high-temperature electronic downhole instruments in the 1990s, the Kuster KPG series gauge (essentially the same design as the Amerada GRC RPG) was the most widely used instrument for making temperature and pressure measurements in geothermal wells.

The Kuster KPG or Amerada GRC RPG gauges (http://www.kusterco.com) are very robust and provide good results at minimum cost for routine measurements. These instruments are small in diameter (32 mm diameter and about 1.9 m long) and can be run in nearly all wells where there are no serious casing problems or other blockages. The gauges are very reliable and require minimal maintenance for temperatures up to 300°C. In some fields they are used regularly in wells up to 350°C, although at these temperatures more maintenance is needed. With careful calibration and field procedures, the accuracy can be in the order of ±1°C for temperature and ±0.2% for pressure gauges, although in practice the accuracy of field data is often somewhat less than this.

One of the first instruments designed for measuring temperatures in geothermal wells was the "geothermograph," which was developed in 1950 by the Department of Scientific and Industrial Research for use in the exploration of the Wairakei field. This instrument used a bimetal probe attached to a stylus that was deflected by the differential expansion of the probe. The stylus scribed a trace on a smoked plate that could later be interpreted to provide temperatures at each depth station in the well. In New Zealand, the geothermograph was superseded by the Kuster KT mechanical temperature gauge in the mid-1970s, and 20 years later the Kuster mechanical instruments were themselves superseded by electronic instruments.

5.3.2. Electronic Downhole Instruments

Since the mid-1980s, various electronic simultaneous PTS (pressure-temperature-spinner) instruments capable of operating at high temperatures have become available. These logging tools can transmit the downhole data to the surface via a single-conductor armored cable for real-time surface readout ("e-line"), or they can store the information in memory to be downloaded when the tool is brought to the surface. The downhole electronics are contained inside a Dewar flask to maintain the internal instrument temperatures below about 100°C, which in geothermal conditions limits the downhole logging time to 3 to 6 hours, depending on the tool heating rate (a function of the external temperature, design of the Dewar flask, internal electronics package, internal heating rate and time). Electronic instruments have the potential for improved accuracy when compared with the older mechanical instruments. However, it is difficult to exactly quantify the accuracy, since it depends on several factors: These include the primary sensing element, design, temperature response of the electronic circuits, and quality procedures associated with operating the instruments. Absolute accuracy should be better than $\pm 1°C$ for temperature and better than ± 0.1 bar for pressure. As discussed below in 5.5.1, determining the exact instrument depth or level in the reservoir also impacts on the overall accuracy of measuring the actual wellbore pressure at a particular depth and relating this data between wells.

5.4. DOWNHOLE FLOW MEASUREMENTS

Downhole flow measurement has been standard practice for many years in the groundwater and petroleum industries. Many techniques have been developed, but the most common is spinner profiling, although tracer injection and temperature profiling can also be used.

Measurement of flow in the wellbore using spinner profiles during injection or production is now routinely used in the geothermal industry. These logs are also useful for assessing interzonal flows while the well is shut, and can be done at minimal additional cost above that for a "regular" pressure-temperature profile. For a test programme where spinner/flow evaluation is being attempted, it is most important to maintain stable and constant well status during any set of spinner logs.

For normal geothermal well completions, interpretation methods appropriate for individual wells fall into two general categories: liquid-filled wells with fluid velocity of less than about 5 m/s and steam or two-phase wells with fluid velocity of more than 5 m/s. For the first category, it is essential to use the crossplot method with at least two sets of down-up logs at different speeds. For steam wells, a single successful spinner profile may be sufficient to evaluate the flow profile.

In practice there are some other factors critical to the success of spinner logging:

- The spinner bearing friction must not change during the logging program. (This can occur through poor maintenance or by incorrect adjustment so that friction varies with temperature.)

- The spinner impeller itself may be damaged during the test program. (This includes abrasion, physical damage, or "contamination" by materials in the injected water—usually plant material—or by rock fragments, which may totally stop the impeller.)

The methodology for interpreting spinner logs is given later in this chapter.

5.5. SOURCES OF ERROR IN DOWNHOLE MEASUREMENTS

In addition to the accuracy of the actual measuring instrument, there are several other potential sources of error in downhole temperature-pressure data. When logs are being compared—say, a PTS log against a formation image or fracture log—errors of a few meters can become important in correctly identifying an image feature as productive or unproductive.

5.5.1. Depth Measurement

The depth of the instrument in the well is measured using a counter to measure the wireline at the surface. The calibration of the counter system needs to be checked from time to time to verify that depth measurement errors are minimal; the counter accuracy should be within 1 part in 1000. A surveyed calibration range can easily be set up where the depth counter accuracy can be checked over several hundred meters on the ground surface. In some cases the depth counter can be checked against accurately known features in a well casing design or by using a casing collar locator.

The same reference point for measuring well depths must always be used, and if a non-standard reference point is used for some reason, it must be noted in the log report. When a well has been completed, depths are measured relative to either the casing head flange (CHF) or from cellar top or some other fixed reference point. During drilling, the depths are usually measured relative to the rig kelly bushing (RKB - or rig floor), since the CHF may not be installed until late in the drilling program. All measurements while drilling must later be corrected to the final CHF datum. Sometimes the difference between the rig floor and CHF or cellar top datum is as much as 10 meters, equivalent to 1 bar.

5.5.2. Thermal Expansion of the Wireline

The length of the wireline used to convey downhole logging instruments will vary according to temperature. Often the wireline will be several meters "longer" when it is being recovered from a well due to thermal expansion. In deep hot wells, the wireline expansion must be allowed for to obtain the best accuracy in pressure measurements. The expansion of the wireline is given by:

$$\Delta L = \alpha \int (T_w - T_o)dz \tag{5.1}$$

$$= \alpha L(T_{av} - T_o) \tag{5.2}$$

where T_{av} is the average downhole temperature over the length of the cable and T_o is the surface temperature (at which the cable length is measured). α is the coefficient of expansion, which is 1.7×10^{-5} per degree C for 316L stainless steel (and is similar for other materials used for logging high-temperature geothermal wells). The thermal expansion of the wireline is much greater than elongation due to stress. For example, for a 1000-meter length of 0.072-inch 316L stainless wireline with a 20 kg load, the elongation is about 0.4 meters, while the thermal expansion for the same length at 250°C would be more than 4 meters.

By comparing the fine detail on spinner logs at different injection rates or injection spinner logs against discharging logs, it is often possible to observe that the logs are displaced by a few meters relative to each other due to thermal expansion of the wireline. Grant and colleagues (2006) give an example of such displacement. Logs in discharging wells can be displaced by over 10 meters compared to injection logs.

5.5.3. Well Deviation

Many wells today are purposefully deviated from vertical, and even wells that are nominally vertical usually deviate to some extent. When plotting downhole data, it must be clear whether the information is measured depth (MD) along the wellbore or true vertical depth (TVD). For assessment of pressure data and comparing information among wells, it is best to use the elevation of the measurement point (together with temperature correction for wireline expansion).

5.5.4. Well Stability

When a temperature or pressure profile is measured in a well, the complete set of data may take up to 4 or 5 hours to obtain. Unless the well is being intentionally manipulated to obtain some transient information, the well state must be the same throughout the test program; otherwise, conditions in the wellbore may change, making interpretation of the data difficult or, more likely, of no value. For example, where a well is shut in with a high gas pressure, any significant leaks at the wireline gland (or elsewhere in the surface equipment) will result in declining gas pressure during the downhole survey, which in turn will cause the water column deep in the wellbore to start moving upward. Where temperatures are already close to the boiling point, boiling can start and the wellbore fluid can change to two-phase during the survey. It is important to have a record of the wellhead pressure during any survey as part of the quality control to ensure that conditions are truly stable. Where electronic gauges are used for well logging, it is important to check the complete recorded log from the time the instrument enters the well until the time it is recovered. The downlog from wellhead to bottom should be the same as the uplog if the conditions in the well are stable. (Usually only the downlog needs to be retained for normal use once the original data has been verified.)

Where a well is flowing or under injection, conditions must be stable throughout the downhole run. This is evident by stable wellhead pressure and from surface flow rate measurements made during the survey. Having a continuous flow record is preferable. While drilling, this can usually be obtained from the mud logger or for a discharge run by datalogging the separated steam and water flows or lip pressure, depending on the flow measurement equipment being used. For the same reasons, and where temperature and pressure profiles are made by separate runs, the well must remain in exactly the same state for each run so the thermodynamic conditions for different runs can be compared.

5.5.5. Instrument Lag

Where mechanical downhole instruments are used, the pressure and tempera-ture response is scribed on a metal chart while the instrument is held stationary at fixed stations in the wellbore. This chart is later interpreted to obtain the pressure or temperature values at each station. As long as the instrument is held sufficiently long at each station, the measured values can reliably be assigned to the station depth (after allowing for uncertainty in depth as just discussed).

Modern electronic instruments provide greater accuracy and frequency of measurement. However, the reported data can still be subject to errors asso-ciated with data recording and processing. For most electronic pressure-temperature logging, the pressure-temperature profiles are recorded while the tool is moving (at rates of 0.5−1.5 m/s), and there is the opportunity for data to become offset or shifted with respect to depth. Such errors can be checked by comparing different up and down passes recorded in the same log or by comparing details of logs at different injection rates. Some PTS logging tools include a casing collar locator (CCL) so that reducing different up and down logging runs to a common datum is straightforward, since the log can be referenced to casing or liner connections. However, there may still be some remaining error between runs at different temperatures due to the thermal expansion of the liner.

Figure 5.1a shows part of a temperature log in a well during cold water injection. Temperatures in the down runs are displaced about 10−15 meters below the up runs. This is due in part to the time lag in the temperature sensor, since this takes a few seconds to equilibrate. There is also a difference depending on which direction the flow is passing the tool, since the down runs are more consistent than the up runs. In this case the difference is not due to

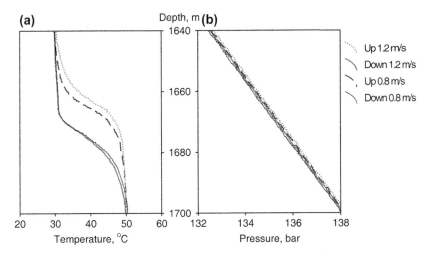

FIGURE 5.1 Temperatures, (a), and pressures, (b) measured during injection. *Source: Kawerau Joint Venture, Personal Communication.*

cable stretch, since comparing pressures on the up and down passes, also shown in Figure 5.1b, shows only small differences at the same depth, with a maximum displacement of 3 m. (The irregularities in the pressure-depth plot reflect the resolution of the pressure sensor.) There may also be some lag in the time between measurement and recording (where the downhole data and depth measurements are merged), which can vary depending on whether logging is real time or memory mode. The data displacement will generally be proportional to the logging speed and for most purposes can be ignored. If a time lag correction is required, the log is displaced by a fixed time, where the time is found as that which produces the best agreement between up and down logs run at the same speed. Because these logs are used to identify permeable zones (in Figure 5.1 there is an inflow where the temperature rises), this correction can make a difference of several meters.

5.6. DESIGNING A DOWNHOLE MEASUREMENT PROGRAM

Almost all downhole pressure-temperature logs today are made using electronic instruments. The main factors to consider when using these tools are the internal heating rate and possibly the memory capacity limiting the time the instruments can remain downhole. Where well temperatures are greater than 300°C, the logging program may need to take into account the performance of cable insulation when using e-line systems because the insulation can break down in sustained high temperatures.

Different information can be obtained from the downhole measurement program by making measurements with the well in different states. The well may be *open* (with no pressure at the wellhead), *shut* (with a steam or gas pressure), on *bleed* (very small discharge), *flowing* (large discharge rate: 100–200 t/h), or under *injection* (usually cold water). After completion of drilling, the state may be *heating*, which is the same as shut, except that the well is recovering in temperature after being cooled by the drilling process.

To enable the best interpretation of the downhole data to be made, it is important that the *state* of the well during the downhole run, the *wellhead pressure* during the run, and the *flow rate* and enthalpy for flowing or injection runs are recorded with the downhole data. It may also be useful to know the type of fluid at the wellhead—for example, bleeding water or bleeding steam.

TABLE 5.1 Completion Test Program (Assumed well depth 2500 m, casing depth 1000 m, main feedzone 2000 m)

Activity	Depth, m		Tool Speed, m/s	Pump Rate
	From	To		
Set up, check pump				Rate 1
Run in	0	2500	1.5	
	2500	1000	−1.5	
	1000	2500	0.7	
	2500	1000	−0.7	
To feed	1000	2000	1.0	
Transient				Rate 2
Wait nominal 1 hour				
To bottom	2000	2500	1.5	
	2500	1000	−1.5	
	1000	2500	1.5	
	2500	1000	−0.7	
	1000	2500	0.7	
To feed	2500	2000	−1	
Transient				Shut
Wait nominal 1 hour				
Run out	2000	2500	1	
	2500	0	−1	

A completion test program needs to cover all the measurements to provide a definition of profiles in the well and permeability, so it must include multiple passes up and down the well at different flow rates and pressure transients between flow rates. Table 5.1 shows a basic completion test program. The program includes spinner profiles at two flow rates, a heating profile after stopping injection, and pressure transients between flow rates. Spinner passes at a minimum of two speeds are needed for each flow rate, and there must be sufficient waiting time after each pump change to record the transient and preferably for pressure and temperature conditions in the complete wellbore to stabilize. Total time downhole must be within the tool capability, and temperatures below the main feed must be assumed to be possibly reservoir temperature. Logging speed for the down runs should be adjusted so they are not within 0.2 m/s of the theoretical fluid velocity, and the logs should start at about 30 meters above the top of the perforated liner (i.e., within the cemented casing). This allows the spinner performance and injection rates to be checked.

A program for carrying out downhole logs in a flowing well follows the same pattern, but it is usually simpler. An ideal set of measurements includes profiles at two flow rates, again with passes at two different velocities. The stabilization time between flow rates may be much greater than on injection, so it may not be possible to measure multiple flow rates. Also, the entire well will be hot, and this may limit the time that the logging instruments can remain downhole. An output or deliverability test may be conducted over several days, with the well being stabilized at different rates for a day or two. Then a flowing profile can be measured at each rate.

5.7. SPINNER MEASUREMENTS

A spinner is an impeller that is used to measure fluid velocity. The fluid flow causes the impeller to turn, with the frequency being proportional to the relative velocity between the tool and fluid:

$$f = (V_t - V_f)/C \tag{5.3}$$

or

$$V_f = V_t - C \times f \tag{5.4}$$

where V_f is the fluid velocity, V_t is the tool velocity, and C is the pitch of the impeller in meters per cycle. It is assumed that the spinner measures a representative fluid velocity. If the wellbore is deviated, the tool must be centralized, since otherwise it will lie on the bottom side and may not measure a representative value. Figure 5.2a shows a typical set of good-quality spinner measurements. Here, there are two sets of down and up logs at 0.8 and 1.2 m/s, respectively. In this case, it is immediately apparent that there is a major loss

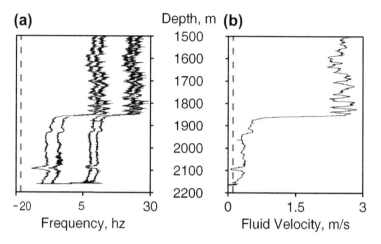

FIGURE 5.2 Spinner data in well under injection. (a) Frequency data. (b) Interpreted velocity. *Source: Rotokawa Joint Venture, Personal Communication.*

zone at 1840−1865 m, where the frequency changes significantly, while there is no significant consistent change above 1840 m.

5.7.1. Crossplotting and Tool Calibration

For analysis, spinner data are grouped into *stations*, or short intervals. Figure 5.3 shows *crossplots* of data in two plots of frequency against tool velocity for data within a short interval or station, which in this case is a 1-meter interval. Figure 5.3a shows a linear relation between frequency and tool velocity. The line extrapolates to zero spin at a velocity of 2.6 m/s. This is the interpreted fluid velocity (because in ideal conditions the spinner would not rotate if the logging tool is moving at the same velocity as the fluid in the

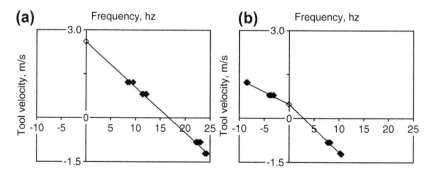

FIGURE 5.3 (a) Crossplot at 1573.4−1574.4 m. (b) Crossplot at 2000−2001 m. *Source: Rotokawa Joint Venture, Personal Communication.*

wellbore). For many geothermal wells a simple spreadsheet method can be used to obtain the fluid velocity—at least as a preliminary estimate using the following procedure:

- Check spinner frequency data for null values (and delete these; the impeller may be stuck due to wellbore debris).
- Sort data by depth (over the interval where there is at least one up and one down profile).
- Check some crossplots of frequency versus log speed at several depths to see how many data points are required to obtain a reliable crossplot.
- Using one of the built-in spreadsheet functions, extrapolate a regression over a fixed number of data points to obtain the fluid velocity at the zero frequency intercept using log speed (y values) and spinner frequency (x values).
- Plot depth versus interpreted fluid velocity.

This approach may provide a reasonable flow profile for many geothermal spinner logs, but in detail the spinner data can be somewhat more complicated.

Figure 5.3b shows data collected at another station where there is both positive and negative frequency data. There is a different linear relation for positive and negative frequency. That is, the effective pitch of the spinner is different depending on whether the flow is incident on the tool from above or below. (Since the impeller is usually located at the bottom of the PTS instrument, flow from above and below the impeller is not symmetrical.) This "bilinear" behavior is due to the configuration of the impeller itself and flow around the impeller housing and tool body, and it is quite common. Equation (5.4) is then modified:

$$V_f = V_t - Cf \quad f > 0 \tag{5.5}$$
$$V_f = V_t - Df \quad f < 0 \tag{5.6}$$

Note that fitting a single line to the data in Figure 5.3b would result in an estimated fluid velocity that is erroneously low.

A best estimate of the fluid velocity can be found by regression (Grant & Bixley, 1995). The data are grouped into stations. Let i denote the ith station, and j the jth data point within it. Equations (5.5) and 5.6) are written as

$$V_{fi} = V_{tij} - Cf_{ij} + \varepsilon_{ij} \quad f > 0 \tag{5.7}$$
$$V_{fi} = V_{tij} - Df_{ij} + \varepsilon_{ij} \quad f < 0 \tag{5.8}$$

where ε_{ij} is the error. A regression is then performed, minimizing the total error, to find the velocity values V_{fi} and the calibration constants C, D. Spinner data are usually quite noisy. The fitted velocity is smoothed by the grouping of data into stations, and station size can be increased to reduce the noise at the expense of lowered resolution. To make a bilinear spinner analysis requires data at two or more different positive and negative tool velocities. If there is only one pass up and down, or if all the frequency data have the same sign, a linear model must be used. Accuracy of the calibration can be checked by comparing computed velocity within the casing against the expected velocity from the measured well flow rate.

If a regression is not done, the next best alternative is to calibrate the tool by doing passes up and down in the casing (when setting the spinner logging program, continue each profile into the cemented casing about 50 meters above the top of the perforated liner), crossplot the data to determine the effective pitch of the spinner, and then use Eq. 5.2 to convert the frequency to fluid velocity. This procedure also provides a cross-check that flow remains the same between the different spinner passes. As a last alternative, the theoretical impeller pitch can be used, but this is not desirable because the actual pitch often differs from this value, especially at low-frequency values, due to friction.

In a well with high fluid velocity—say, a discharging steam well—fluid velocity is much larger than the tool velocity, and it is not possible to calibrate the tool by fitting the data. A known tool calibration must be used or simply the frequency data plotted, since the effect of tool velocity will be small.

The spinner data collected at different tool velocity should produce profiles with frequency displaced by a constant amount, this amount being the pitch times the difference in velocity. The profiles in Figure 5.2a do show this. On careful inspection it can be seen that the profiles below 1860 m have a different displacement between the two positive frequency and the two negative frequency profiles compared to those above 1860 m. This reflects the different positive and negative calibration. If there is a feature on one profile that is not present on profiles at other velocities and there was no change of tool velocity within the profile, this indicates a problem with the tool. Often, for example, the impeller can get blocked or slowed by debris.

5.7.2. Wellbore Radius Effects

Figure 5.2b shows the fitted fluid velocity profile from the data in Figure 5.2a. There is a clear loss zone from 1840–1870 m, where fluid velocity decreases from 2.5 to 0.5 m/s, so 80% of the flow is lost at this interval. The other loss zone is near the bottom of the hole, at 2100–2150 m. There is a lot of variation on the velocity profile. This is real, and it is caused by changes in wellbore diameter. The fact that it is real can be seen by comparing profiles at different flow rates: The variations due to wellbore diameter appear at all flow rates, whereas inflows and outflows change with the well flow rate. The dip in velocity just above 2100 m must indicate a significant enlargement near here. Above 1840 m there are irregular variations. There is often significant enlargement of the wellbore above the expected drilled diameter due to erosion of the formation, particularly when drilling with aerated fluids. Often the wellbore is enlarged close to permeable zones, most likely due to increased fracturing of the rock in these areas. By comparing velocity profiles at different injection rates, it can be possible to separate the effects of wellbore enlargement and inflow/outflow (Grant et al., 2006)

Figure 5.4a shows an interpreted fluid velocity and temperature log for well MK11 while injecting 45 kg/s of cold water. The spinner frequency data from 4 up/down profiles was processed with 1 m stations. The well was drilled using an air-water mixture for the circulation fluid to balance formation pressure. The temperature increase from 23 to 60°C indicates inflows broadly spread over the interval 800−1300 m, with no sharply defined zone. Using the enthalpy balance method (section 7.2.3) and assuming an inflow temperature of 230°C, the inflow is relatively small at 10 l/s. In contrast to this relatively small inflow, there are large variations in fluid velocity. There is a velocity peak at the casing shoe. There is usually turbulence at the liner top/casing shoe, which can make it quite difficult to detect a feed zone near the shoe. A similar maximum velocity at 1010 m, 1530 m, and 1810 m suggests that the well is tight on the liner at these points and there is basically the same fluid flow at these points. The flow of 55 kg/s − 45 kg/s pumped into the well plus 10 kg/s inflow − corresponds to a velocity of 2.8 m/s, shown as a dotted line in Figure 5.4a. The maximum measured velocity is in good agreement with this. The large decreases in velocity 800−950 m and 1050−1430 m correspond to major enlargement of the

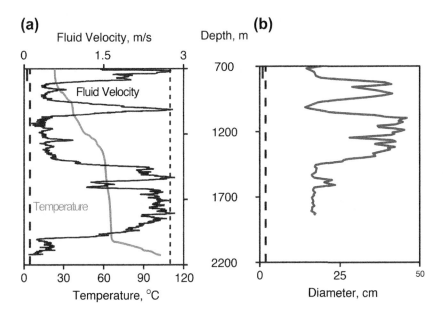

FIGURE 5.4 Spinner log in MK11. (a) Fluid velocity and temperature under injection at 45 kg/s. Dotted line corresponds to a flow of 55 kg/s inside 7-inch liner, corresponding to injected flow of 45 kg/s plus additional inflow of 10 kg/s. (b) Interpreted wellbore diameter. *Source: Tuaropaki Power Company, Personal Communication.*

wellbore. Below 1840 m, velocity decreases steadily down to 2000 m. This corresponds to the major loss zone, and the temperature confirms that 2000 m is the bottom of the loss zone. The loss appears to be widely spread across the interval 1840–2000 m, with about half the loss concentrated within 1960–2000 m, but details could be confused by further wellbore diameter variations. Deviation in the well can cause some bias in spinner results because the liner and the tool lie on the lower side of the wellbore and may not intercept a representative flow mixture. The magnitude of this bias is unknown but appears to be present in some cases.

Using the data down to 1840 m, it is possible to compute the wellbore radius, since down to this depth there appear to be no outflows. The total cumulative inflow can be calculated from a heat balance (Equation 7.2). Assuming an inflow temperature $T_r = 230°C$, based on measured temperature in the first heating run, and an injected flow of 45 kg/s that is at 23°C, the total flow at a depth above the first loss is given by a heat balance

$$W(z) = 45 \times \frac{(H(T_r) - H(23))}{(H(T_r) - H(T(z)))} \tag{5.9}$$

and the wellbore diameter D is given by

$$w(z) = \pi[D(z)/2]^2 \rho V \tag{5.10}$$

Figure 5.4b shows the computed diameter in cm. This calculation cannot be continued below 1840 m because velocity decreases due to fluid loss.

5.7.3. High Fluid Velocity

The preceding discussion relates primarily to measurements made with liquid in the wellbore. When the fluid velocity is large compared to the tool velocity, the change in frequency between up and down passes is relatively small, and crossplotting is not effective as a means of calibrating the tool. Figure 5.5 shows a spinner run in a discharging steam well. Because the contrast between the two passes is small, the calibration must be taken from some other test and applied to the frequency data to give fluid velocity. The pressure profile shows the steam-water interface below 1400 m, and the spinner indicates no significant flow below the first inflow at 1250 m, with a second inflow at 1100 m.

5.7.4. Data Problems

Figure 5.6 illustrates runs with data problems. Figure 5.6a shows a stage test with no liner. Only one pass up and down was done in most of the open hole. A second up and down pass was done in the upper part of the well. The frequency data are inconsistent in part. The circles enclose data that show the

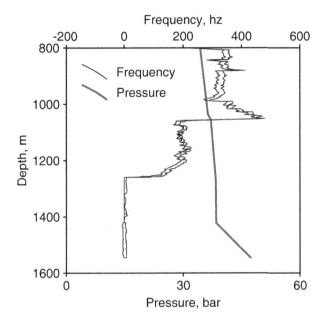

FIGURE 5.5 Discharging profile in well MBD-2, Wayang Windu. *Source: Star Energy, Personal Communication.*

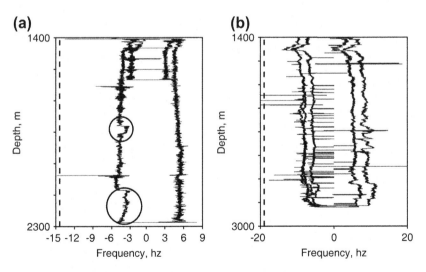

FIGURE 5.6 (a) spinner run with problem segments (b) spinner run with outlier data. *Source: Kawerau Joint Venture and Rotokawa Joint Venture, Personal Communication.*

features that are not repeated in the other pass. This implies that there are questions about the instrument or flow rate. Occasional data outliers should be ignored and deleted from the data file before analysis. In this case the noise level on the data is typical and not unusually large, but the circled excursions represent some instrument problem other than random noise. Figure 5.6b shows another profile with questionable data. There are many outliers. In particular there are many near-zero values that are spurious—most likely the impeller was stalled due to rock particles, or the frequency counting circuit was not performing correctly. These "suspect" data values need to be filtered out before processing a flow velocity interpretation, except in the interval 1400–1600 m, where there are genuine values near zero. There are also fewer high outliers, which should be removed. In this case the data cleaning is done by filtering the data and removing values outside a particular range.

Measurements During Drilling

6.1. GENERAL

Geothermal fields are dynamic systems extending from depths of several kilometers to the surface. Although the formations overlying the deep hot resource may not be of immediate interest for production purposes, their hydrological properties are of crucial importance in the long-term performance of the reservoir under any major exploitation scheme. Often during the drilling of a well, the opportunity to obtain invaluable information from these shallow formations is overlooked in the rush to get to the productive high-temperature resource as quickly as possible. Pressure measurements made at permeable zones in the shallow formations as they are encountered while drilling or during unforeseen delays in the drilling are vital to obtaining this data. There is often a complicated hydrology in the shallow formations overlying a geothermal reservoir, with perched aquifers, steam-heated zones and lateral flows of groundwater. For one example see Dench (1980), also shown in Grant et al. (1982) If a well is completed with an extensive open interval through such a region, there is often a vigorous downflow in the well. Such downflows make the pressure measurements inaccurate since there will be some drawdown in the upper inflow zones and buildup in the lower feedzones accepting water. Temperature profiles are also completely inaccurate when such downflows are present. Pressures in the shallow aquifers can be accurately measured only in wells open to that depth and that are not open to any other aquifer. With good planning, pressure measurements in the shallow aquifers can be done while drilling a deep well; otherwise dedicated shallow wells are needed to acquire these data.

Temperature buildup tests are sometimes used when exploring new fields to assist in the determination of production casing depths and to give indications of bottomhole conditions when the well is approaching the programmed total depth.

Geothermal Reservoir Engineering. ISBN: 978-0-12-383880-3 DOI: 10.1016/B978-0-12-383880-3.10006-X

6.2. PRESSURE

Whenever there is a circulation loss (or gain) while drilling, there is a potential opportunity to measure the true formation pressure. A circulation loss means that a connection between the reservoir and the wellbore has been established. If the depth to this permeable zone is known, the pressure at the permeable zone can be calculated using the density of the drilling fluid together with the fluid level below the surface when no fluid is being pumped. Preferably, the pressure is measured using a downhole pressure gauge.

However, in practice it is not usually possible to fit downhole measurements into the drilling operation when a loss of circulating fluid has occurred, since the primary concern at this time is to ensure that the well is kept under control and there is no collapse of the wellbore or a blowout. In practice it is better to program tests into the drilling program, targeting specific formations or known aquifers for testing.

6.3. SIGNIFICANCE OF DRILLING LOSSES

A sudden loss or gain of circulating fluid during drilling indicates that a connection between the wellbore and the formation permeability/fracture network has been established. The degree of permeability that is indicated depends on the balance between formation pressure and pressure in the wellbore at the permeable zone. Consider a well being drilled with water in a reservoir with hydrostatic pressure from ground surface. Measurements in other wells have established a reservoir pressure gradient:

$$P = 0.083 \times z + 1$$

where P is the pressure in bars, and z is the depth in meters. Assume that while drilling at 1500 m depth using cold water (15°C) there is a loss of 19 l/s. Reservoir pressure at this depth is $0.083 \times 1500 + 1 = 125.5$ bar. The pressure due to a static column of water in the wellbore is $0.098 \times 1500 + 1 = 148$ bar, so there is a pressure difference of 22.5 bar between the wellbore and the reservoir at the 1500 m loss zone. The pressure difference is likely to be greater than 22.5 bar because of the increased density of the circulating fluid due to the drill cuttings being returned to the surface. The loss is 19 l/s, giving an injectivity of 0.9 l/s.b. Although the quantity of the loss is significant because of the large pressure difference between the wellbore and the formation, it does not correspond with high permeability, and in this case the injectivity index value of 0.9 l/s per bar is unlikely to be sufficient to sustain an economic production flow (see Figure 7.11).

Estimates of injectivity based on changes in circulation fluid returns during drilling only provide indicative injectivity values. The permeable zones are likely to be wholly or partly blocked by drilling fluids and cuttings, and the injectivity may change with time as these materials are removed by washing of the hole, by thermal stimulation with continued cooling, or by discharge.

The interpretation is different when drilling with fluids other than water. If mud is used as the circulating fluid, the natural fracture permeability will almost certainly be impaired.

If downhole pressure measurements are available while drilling, the combination of the downhole pressure record and drilling loss against depth can be used to give the injectivity of the well as a function of depth. Ideally, a profile of the well can be determined with injectivity of each loss zone measured, although in practice, after the first major feed zone has been encountered, the interpretation of subsequent changes in circulation returns is difficult to assign to particular depths.

6.4. TEMPERATURE

An indication of formation temperature during drilling can be helpful to guide decisions about casing depths and final completion of a well.

6.4.1. Temperature Buildup

Estimating stable formation temperatures from transient data is a useful tool during the early stages of a geothermal exploration program. The method has applications in a range of circumstances:

- To determine the appropriate depth at which to set the production casing—in particular to ensure that formation temperatures are sufficiently high. For most high-temperature fields, a minimum stable formation temperature at the production casing would be at least 230°C.
- To check temperatures as drilling proceeds, to check for temperature reversals, and so forth.
- As an immediate indication of final temperatures in a new exploration area.
- Where steam is encountered in the production zone, the shallower temperatures cannot be measured after completion because the wellbore fills with steam. Temperature buildup tests can provide information about the shallower formations in this situation (where the buildup measurements can be made after cementing the production casing but before drilling ahead into the reservoir).
- For operational reasons, it may not be possible to allow sufficient time for a well to fully heat up before discharge; temperature buildup data can be used to estimate the fully heated formation temperatures.

Various techniques have been developed to obtain true formation temperatures from transient data (Dowdle & Cobb, 1975; Barelli & Palama, 1981). Most of the older methods required long shut-in periods to obtain accurate predictions. Roux and colleagues (1979) published a method where reliable estimates of final temperatures could be obtained using data measured over a few hours during a pause in drilling. The procedures recommended by Menzies (1981) using this method has been tried in many fields in the Philippines and New Zealand and found to be simple and effective. This method is described in more detail in Appendix 1.

The theory behind the method makes the following assumptions:

- Cylindrical symmetry exists, with the wellbore as the axis—that is, 2D heat flow.
- Heat flow is by conduction only.
- Thermal properties of the formation do not vary with temperature.
- The formation can be treated as though it is radially infinite and homogeneous with regard to heat flow.
- No vertical heat flow exists in the formation.
- The presence of mud cake is disregarded.
- The temperature at the formation face is instantaneously dropped to some value and is maintained at this value throughout the circulation period.
- After mud circulation ceases, the cumulative radial heat flow at the wellbore is negligible.

In practical terms this means that the method should be used in the following circumstances:

- Tests should be made some distance above the bottom of the wellbore—say, 50 m minimum.
- Tests should be on an impermeable section of hole—that is, conductive heat transfer only.
- There should be no fluid movement in the wellbore by interzonal flow, internal circulation, or crossflow. In some circumstances fluid may be draining from the wellbore if there is a large difference between the wellbore and formation pressures.
- The circulation (drilling ahead) and heating times should be roughly equal; usually 15 to 20 hours is sufficient.
- Tests should be in an open hole.
- The tests should not be made at depths where there have been multiple cycles of cooling.

The well pressure should also be measured during the buildup surveys to check that the pressure is constant. If the formation is overpressured, slow leakage of well fluid will result in fluid movement within the wellbore, invalidating the temperature buildup data where the fluid is moving.

6.4.2. Temperature Buildup Example

Well MK4 has been drilled and cased $9^5/_8$ inches to 651 m depth. Drilling has continued ($8^1/_2$ inches diameter) to 1054 m at 8 meters/hour. A temperature

TABLE 6.1 Measured Temperature Data for Well MK4 at 934 m

Time t_{obs} (Hours:Minutes)	Wellbore Temperature (°C)	Shut-in Time Δt (Decimal Hours)	Horner Time $\dfrac{t_c + \Delta t}{\Delta t}$
17:46	43	2.77	6.44
18:35	49	3.58	5.21
19:44	50	4.73	4.19
20:34	59	5.57	3.71
21:46	64	6.77	3.23
22:34	72	7.58	2.99
00:15	75	9.25	2.63
01:00	81	10.03	2.5
02:15	82	11.15	2.35
02:54	86.5	11.9	2.27

buildup test is scheduled to check formation temperatures. The drill bit reached 1054 m at 15:00 hours on April 5, circulation was stopped, and the drill pipe was run out of the well. A set of temperature measurements are made at 934 m, as shown in the first two columns of Table 6.1.

Procedure

1. Calculate the circulation time (t_c) at the buildup depth. This is the time that the zone has been cooled by drilling fluid. This time can be best obtained from the drilling logs at the rig. In this case the bit was at 934 m at 5 minutes before midnight, 23:55 hrs on April 4. Therefore, the circulation time was 15 hours and 5 minutes, $t_c = 15.08$ hours (decimal hours).
2. Calculate shut-in time: Δt, the observation time less the time of stopping circulation.
3. Calculate dimensionless Horner time $\dfrac{t_c + \Delta t}{\Delta t}$.
   These values are tabulated in columns 3 and 4 in Table 6.1.
4. Plot results on semi-log, find the straight-line section of the buildup plot, and calculate slope (m°C per log cycle) and intercept of the straight line, T^*_{ws} at Horner time $= 1$ (Figure 6.1).
   In this example, the intercept, T^*_{ws} is 138°C, and the slope, m, is 140°C per log cycle.

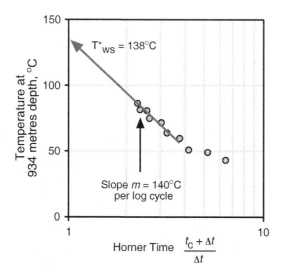

FIGURE 6.1 Semilog plot of temperature buildup for well MK4 using data in Table 6.1. *Source: Tuaropaki Power Company, Personal Communication.*

5. Calculate the correction factor, since the intercept T^*_{ws} underestimates the final stable formation temperature. The final stable formation temperature, T is calculated by:

$$T = T^*_{ws} + mT_{DB}(t_{PD})$$

where m is the slope of the Horner straight line and T_{DB} is the dimensionless correction term, which depends on t_{PD}, the dimensionless circulation time. The t_{PD} value is calculated from circulation time, formation thermal characteristics, and wellbore size. To calculate t_{PD}

$$t_{PD} = \frac{Kt_c}{C\rho r_w^2}$$

where K is the thermal conductivity of the formation, C is the specific heat, and ρ the density.

Some properties for typical volcanic rocks in the geothermal environment taken from the Wairakei field are given in Table 6.2. The Waiora formation is a low-density, often highly porous pumice breccia and may not be typical of the more dense formations found at depth in many other geothermal fields.

Assuming properties for the Waiora formation at Wairakei and using wellbore diameter 8.5 inches ($r_W = 0.1$ m):

$$\frac{K}{C\rho r_w^2} = \frac{1.56 \times 3600}{0.18 \times 4186 \times 1600 \times 0.11^2} \ hr^{-1} = 0.466 \ hr^{-1}$$

For the Wairakei Ignimbrite the value is $0.434 \ hr^{-1}$

TABLE 6.2 Thermal Properties of the Wairakei Ignimbrite and Waiora Formation Rocks at the Wairakei Geothermal Field for Saturated Rock

Formation	Thermal Conductivity, K W/m.K	Specific Heat, C J/kg.K	Density kg/m^3
Waiora Formation	1.56	750	1600
Wairakei Ignimbrite	2.11	800	2200

Source: Hendrickson, 1975.

Thus, for geothermal wells drilled at 0.22 m diameter, a value of 0.4 hr^{-1} would be reasonable for $\left(\frac{K}{C\rho r_w^2}\right)$. To calculate the dimensionless circulation time factor t_{PD}

$$t_{PD} = \left(\frac{K}{C\rho r_w^2}\right)t_c = 0.4 \times 15.08 = 6.03$$

Using Eq. A1.48 from Appendix 1, obtain T_{DB} using the t_{PD} value for the appropriate range of Horner times, the range of values $\Theta = \frac{t_c + \Delta t}{\Delta t}$ defining the straight line is 2–4 (see Figure 6.1), taking the midvalue of 3 and $t_{PD} = 6.03$:

$$T_{DB} = 0.03\Theta^{1.678}t_{PD}^{-0.373} = 0.097$$

Therefore, the predicted stable formation temperature at 934 m in well MK4 is:

$$T = T_{WS}^* + mT_{DB}(t_{PD}) = 138 + 140 \times 0.097 = 152°C$$

6.5. STAGE TESTING

When developing a new resource and there is already an understanding of the relationship between downhole injectivity and production flow rates, stage testing can be used to assist with deciding at what point drilling individual wells should be terminated. To decide when to stop drilling, stage tests are carried out from time to time as drilling proceeds to determine if the target permeability value has been achieved. A target permeability can be defined in a field where a number of wells have been tested, both for downhole performance and by production or medium-term injection testing so statistics of well performance can be established. Then economic criteria can be used to optimize the drilling program (Grant, 2008).

As with most other well testing operations, it is very important to record the background information for all the factors likely to impact on the interpretation of the test data. When making measurements in a partly completed well, it is important to record the actual state of drilling at the time: drilled depth, casing depths, and activities prior to measurement. Similarly in a well where the hole has been modified by deepening, sidetracking, or forking (more than one open leg in one well), it is important to ensure that it is recorded which well track the measurements refer to and what the state of the well is at the time of measurement. To avoid confusion, it is essential to have a clear naming convention for wells. For example, in well EX1, redrills should be named differently, such as EX1RD1, EX1RD2, and so on, and similarly sidetracks. In a multilateral well there should be a name for each leg—for example, EX1L1, EX1L2, and so on. Without a clear system to identify which penetration measurements are under consideration, it can be very difficult when reviewing data to correctly identify where the measurements were actually made.

6.6. THE DRILLING OF RK22

RK22 illustrates the use of stage testing in deciding at what depth to terminate drilling or if a sidetrack is required to achieve the target injectivity value. The well was designed to be completed with $13^{3}/_{8}$-inch casing and a $10^{3}/_{4}$-inch perforated liner, with a target injectivity of 8 kg/s.b. The $13^{3}/_{8}$-inch casing was set at 1392 m and drilling continued with a $12^{1}/_{4}$- inch bit. Figure 6.2 shows the change of injectivity with depth based on the stage tests and final completion testing.

6.6.1. Drilled Depth 2428 m

The first stage test was carried out at drilled depth 2428 m. Using pressures monitored at the wellhead, an incremental injectivity of 0.3 kg/s.b was found and drilling continued.

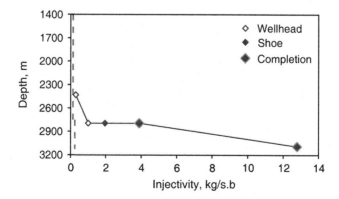

FIGURE 6.2 Tested injectivity of RK22 against drilled depth. *Source: Rotokawa Joint Venture, Personal Communication.*

6.6.2. Drilled Depth 2791 m

A second injectivity test, again using wellhead pressure values, showed a slight increase of injectivity to 1 kg/s.b but still far short of the target value. The injectivity test was repeated using a downhole gauge located at the casing depth (1390 m), and the transient record from this test is shown in Figure 6.3. There is a prior trend of declining pressure (as normally found in the case of a partially cooled well soon after stopping drilling where the pressure measurement point is some distance above the actual permeable zone). The pressure change was recalculated as deviation from this trend giving an injectivity of 1.9 kg/s. This value is still significantly less than the target value, and drilling should continue. However, it was not possible to continue drilling with the full $12^1/_4$-inch diameter, and the $10^3/_4$-inch perforated liner was run and a normal completion test carried out before drilling ahead.

Figure 6.4 shows the fluid velocity profile during the completion test while injecting 30 l/s. The velocity shows considerable variation, mostly due to changes in wellbore diameter. There is a single well-defined loss zone near the bottom of the well at 2695–2740 m, shown by the abrupt falloff in the velocity profile below 2695 m. The injectivity measured 200 m above the loss zone was 3.9 kg/s.b. The associated pressure transient again shows some rebound in the as the flow rate is reduced, normally an indicator of interzonal flow in the wellbore and good permeability. However, this possibility is discounted here because the pressures at the permeable zones are all significantly above the stable formation pressures. There is also the same problem that the pressure was not stable before the transient.

The apparent increase in the injectivity measured in the three tests that were made while the drilled depth was 2791 m, from 1.0, to 1.9, to 3.9 kg/s.b may be due in part to cleaning of the wellbore in the time between the

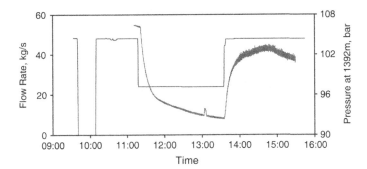

FIGURE 6.3 Pressure transient during stage test measured at 1392 m and drilled depth 2791 m. *Source: Rotokawa Joint Venture, Personal Communication.*

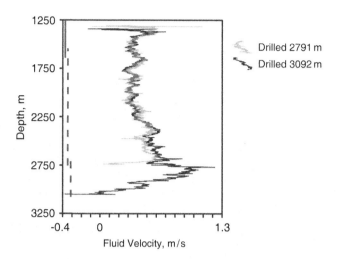

FIGURE 6.4   Fluid velocity profile during injection, drilled 2791 m and at final depth of 3092 m. Injection rate was the same for both profiles at 30 l/s. *Source: Rotokawa Joint Venture, Personal Communication.*

different tests but is most likely a result of stimulation of the permeability where cold water is being injected into high-temperature formation.

6.6.3. Drilled Depth 3092 m

Because the injectivity had not reached the target value, drilling was continued at a smaller diameter of $8\frac{1}{2}$ inches using water for circulation fluid. During this stage of drilling, there were no returns to the surface, and therefore it was not possible to assess any indication of permeability change without doing downhole pressure measurements. The well was deepened to 3092 m, and a 7-inch perforated liner was run from 2754 to 3092 m.

After running the perforated liner, a second completion test was performed. Spinner data from this test confirmed the permeable zone at 2695–2740 and showed a new zone at 2820–3045 m had been intercepted. Pressures showed the injectivity had increased to 13 kg/s.b. Fluid velocity profiles for the two completion tests are shown in Figure 6.4. Above 2750 m the two profiles are almost identical. Comparison of the two velocity profiles shows that at 2791 m the well had just penetrated the upper part of the permeable feature. The target injectivity had been achieved, and the well was completed at 3092 m.

In this case it was fortunate that acceptable permeability was encountered when the drilling conditions were close to the maximum rig capacity. Normally the stage testing methodology is applied to terminate drilling at some point before the programmed depth is reached when acceptable permeability has been encountered, either for production or injection. In the RK22 case, if the target injectivity had not been achieved when the drilled depth was 3092 m, a decision to look for better well performance would have been required. The following options could be considered in this process:

- If the injectivity is very poor, plug back and drill a new sidetrack.
- If the injectivity is moderate, drill a second leg, retaining the first—that is, a multilateral completion.
- Stimulate by cold water.
- Use an acid treatment.
- Accept that the well failed and move to a more prospective location.

Well Completion and Heating

7.1. INTRODUCTION

This chapter discusses the measurements that may be made during and immediately after completion of drilling and during the period of heating as the conditions in the wellbore gradually approach steady state. It is implicit through much of the discussion that the reservoir conditions are liquid dominated, but this is not a severe restraint because much of the discussion transfers with little alteration to wells in vapor-dominated reservoirs. It is also implicit that reservoir permeability is largely associated with fractures rather than the primary permeability of the bulk of the reservoir rocks. Note that in this chapter the injectivity index units of l/s.b and kg/s.b are treated as equivalent, since during well completion, injection flows are usually volumetric measurements for cold water.

7.1.1. Objectives of Completion Testing

The design of the test program during and on completion of drilling varies according to the original well objectives: For exploration wells, the program tends to be more "scientific," using a wider range of test methods to explore reservoir characteristics, whereas for production wells the test program is very focused on determining characteristics relevant to likely production or injection capacity.

     For most wells the ultimate test is the capacity of the well to deliver fluid to the surface or its capacity for injection. However, there is considerable other

Geothermal Reservoir Engineering. ISBN: 978-0-12-383880-3 DOI: 10.1016/B978-0-12-383880-3.10007-1

information that is vital to understanding the reservoir and can only be obtained in the short period between completion of drilling and discharging the well, when the well is completely shut in and gradually recovering from the drilling process. Valuable information, such as real formation temperatures, that is vital to understanding the relationship between the deep hot reservoir and overlying aquifers, or when constructing numerical simulation models, is lost or, at best, obscured by premature action to flow the well and prove production capacity.

The following local reservoir and fluid properties are of greatest interest:

- Fluid state (pressure/temperature/enthalpy)
- Reservoir state (liquid or vapor dominated)
- Reservoir permeability
- Reservoir chemistry (salinity, cation, and isotopic ratios)
- Noncondensable gas content

The associated well characteristics include the following:

- Location of feed zones in the well
- Feed zone permeability/productivity characteristics
- Near-well changes in permeability (skin)
- Physical condition of the well (casing, liner, or open hole)

7.1.2. Test Sequence

These tests are normally made after the final completion of the well, although staged tests prior to reaching total depth are sometimes made to determine if sufficient permeability has been found to terminate drilling or to test a zone that may be cased off later. The type of testing will be determined by the drilling method used, the availability of water for injection testing, and the grade of permeability encountered.

The completion test is normally performed by injecting cold water at varying flow rates and measuring pressure-temperature-spinner (PTS) profiles in the well to define the location and characteristics of feed or loss zones. In cases where a well has only one major feed zone, interpretation is straighforward. This is the exception, however, and almost all wells have multiple feed zones, requiring more sophisticated interpretation. With good data it is usually possible to assign permeability characteristics to individual feed zones and identify true formation pressure for at least one major feed zone.

The basic test program design is primarily determined from drilling results. Where highly permeable fractures are expected, a program to look for "stabilized" temperature and pressure profiles is most suitable, and for wells with poor permeability, a program designed to follow transient temperature fronts in the wellbore is usually most successful in locating the loss zones.

In the high-permeability case, the difference between the "cold" pressure gradient inside the well and the "hot" pressure gradient in the formation results in part of the upper section of the open hole being underpressured with respect to the formation. Thus, if there are any permeable zones in this underpressured zone, there will be flow of hotter fluid from the formation into the wellbore. Where a well encounters poor overall permeability during injection, the wellbore will be overpressured everywhere with respect to the formation, so all potential flows will be from the well to the formation.

The primary objectives of the completion test are as follows:

- Identify potential feed zones
- Determine the overall permeability and thus an estimate of the well's likely production (or injection) capacity
- In some cases, allocate permeability characteristics to individual feed zones

The basic measurements that are made during a completion test are temperature profiles, pressure profiles, and flow profiles (when a spinner tool is used).

All of the preceding data are obtained while injecting cold water at the wellhead using a downhole logging tool to make simultaneous pressure-temperature-spinner observations. This is preferably done using a surface readout PTS tool (so the test program can be immediately adjusted if the real-time data show unexpected changes—for example, if the spinner impeller becomes jammed and a repeat log of part of the wellbore is required). Accurate transformation of the spinner frequency data into a fluid velocity profile requires that at least two up and down profiles be made at different logging velocities over the complete openhole section for at least one of the injection rates. Logging should also extend into the production casing by 30–40 meters for data and equipment quality checks.

7.1.3. High Permeability Wells

As just mentioned, for these wells, the upper part of the openhole section is usually underpressured during cold water injection, so there are inflows of hotter fluid from the formation into the wellbore. This usually results in a characteristic stepped temperature profile, with isothermal sections between the inflow zones. Figure 7.1 shows such a profile, with two large steps (plus a third small step), representing inflows of hot fluid and with the final loss zone at 930 m marked by a change in temperature gradient.

The preferred test procedure is to measure two or three sets of "stabilized" temperature-pressure-flow profiles while injecting water at the wellhead at different rates. By changing the injection rate and, consequently, the pressures downhole, it is sometimes possible to "switch off" or "switch on" inflows from different depths and then determine the formation pressures at these zones. Using heat balance calculations (discussed in section 7.2.3 below), it may be possible to calculate individual zone characteristics such as injectivity, productivity, and formation pressure.

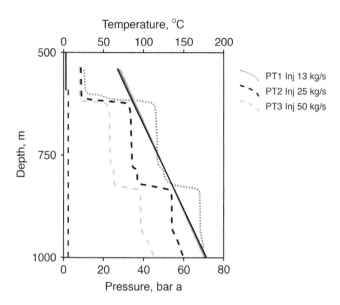

FIGURE 7.1 Stepped profile during injection in TH6. *Source: Contact Energy, Personal Communication.*

7.1.4. Low Permeability Wells

For wells with poor permeability, the wellbore pressure during cold water injection is usually greater than the formation pressure throughout the complete openhole section. Often there will be several small fluid loss zones spread throughout the well. The most successful method of locating these loss zones during cold water injection tests is to first allow a short period of heating after drilling is completed. Usually 12 hours is sufficient, although it may be convenient to allow longer times while the drilling equipment is removed off the site. After the well has partially heated, a series of temperature profiles are made while injecting water at a low flow rate. Loss zones are marked by a change of temperature gradient. Figure 7.2 illustrates this process where four temperature profiles have been measured over several hours. The well has poor injectivity. During injection there is a break in gradient near 1500 m where the main loss zone is located. Some water penetrates past this point, as shown by the slow progressive cooling of the wellbore below 1500 m. This flow is small, since the wellbore is cooling quite slowly.

In some low permeability wells, the bottom part of the wellbore will continue to heat up as cold water is injected. This shows that this part of the well is not accepting any of the cold fluid being injected and is essentially impermeable and is a good diagnostic test to identify the deepest level of permeability.

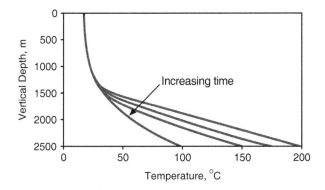

FIGURE 7.2 Temperature profiles during injection into well PK5. *Source: Kawerau Geothermal Ltd., Personal Communication.*

7.2. QUANTIFYING RESERVOIR PARAMETERS

The completion test is carried out by pumping cold water at varying rates into the wellbore. While injecting cold water and even when injection is stopped, there cannot be equilibrium between the wellbore fluid pressures and formation pressure at all the feed zones due to the difference in pressure gradient between the cool fluid in the wellbore and the hotter fluid in the formation surrounding the wellbore. Sometimes it is not possible to properly interpret the field data due to an incomplete test program, measuring pressure falloff data at the wrong depth, strong interzonal flow, or failure of downhole instrumentation during a test. Where consistent spinner data are available to construct a fluid flow profile in the wellbore during injection, the feed zone locations and the injectivity of different zones can be reliably assessed.

The choice of depth for pressure analysis is critical to obtain reliable results. For example, in a typical well with 1000 m of open hole, the difference between well and reservoir pressure can change from zero at the top to 25 bar at the bottom of the interval (about 15% of the total pressure at 2000 m depth using a 250°C temperature difference). Where the pressure measurement is made too shallow, pressure changes will be smaller than the true values, and well performance will be overestimated.

Because of all the uncertainties arising from the change of fluid properties with temperature and assessing what is really the "proper" depth at which to do the analysis, the initial approach is to calculate an overall injectivity index. This can be later refined as new data come in from detailed analysis, discharge testing, and so on. The key element in analysis is the Productivity or Injectivity Index (*PI, II*) (Craft & Hawkins, 1959), where:

$$II, PI = W/(P_f - P_w) \tag{7.1}$$

This index is assumed to stay constant with flow rate and has been found to work effectively in most liquid-fed wells. At two-phase feed zones, experience shows that this linear relationship is not usually applicable due to the variation with enthalpy of the flowing viscosity of the two-phase mixture (see Appendix 1).

7.2.1. Injectivity Test

For wells that have poor overall permeability and can be overpressured throughout their openhole depth during injection, the injectivity index can be easily calculated once the main loss zone has been identified. Usually the pressure gradient in the well while injecting cold water is significantly greater than in the hot formation, so the evaluation of the injectivity index should be made as close as possible to the zone of highest injectivity, which may not be at the same depth as the "major" loss zone, since the pressure difference between the wellbore and the formation increases with depth, favoring the deeper zones. This is particularly important where there is a large openhole length (>1000 m). In wells with poor permeability, it may take several hours or even days for the well pressures and formation pressures to equilibrate following an injection test, and care must be taken to check that the zero pressure used to evaluate the pressure tests matches stabilized pressures measured in the weeks of heating following the injection test. For very low permeability wells, it is easy to overpressure the formation, even at relatively small injection rates (as low as 10 l/s) and obtain misleading indications of permeability because the wellbore pressure is sufficient to stimulate the natural rock permeability by opening fractures.

For higher permeability wells that have inflows at the upper levels during injection (because wellbore pressure is lower than formation pressure in the upper part of the open hole), a value for gross permeability may not be very meaningful, since the value will depend on the depth at which the pressure is measured and may not take into account the additional inflowing fluid. In some cases, the inflow rate can be greater than the injection rate at the surface, so comparing the downhole pressure change with the surface injection rate may not be realistic.

Although it is the stabilized pressure at each rate that is used to define the injectivity, the pressure transient measured after changing the injection flow rate should always be checked for signs of irregularity: rebounds, cycling, severe pump noise, and, most important, unlogged changes in pump rate (for example, caused by failure of a valve in the rig pumps, which can be difficult to identify because the pump rate is usually determined by counting the stroke rate and assuming a fixed volume per stroke). If such irregularity is observed during a pump test, then the injectivity data are at best of lower accuracy or totally invalidated. The injection flow rates can be cross-checked using the fluid velocity determined from spinner data measured in the cased hole at the top of each spinner profile.

7.2.2. Low Permeability Wells

Figure 7.3 shows the completion test program, and Figure 7.4 shows the temperature and interpreted fluid velocity profiles while injecting at two rates, 25 and 35 l/s in RK18L2, the second leg of well RK18. The first leg is blocked off during this test.

The fluid velocity profiles (see Figure 7.4), which are determined from spinner logs, show marked irregularities. The two profiles have a very similar shape, with the same maxima and minima. The variations in fluid velocity above 1700 m are due to variations in wellbore diameter. If the velocity increases were due to inflows, there would be temperature rises at the inflows, which are not present. Note that the maximum velocity at 1300 and 1670 m are the same; this confirms that the velocity decrease between these depths is due to wellbore enlargement and the velocity maxima correspond to the wellbore being tight on the liner with the same fluid velocity and thus the same flow rate. The theoretical velocity for a flow of 35 l/s inside a 7-inch liner is shown. The actual maximum is 15% larger. Experience shows that an error of 10% in fluid velocities determined from spinner measurements is not unusual, so this is reasonable agreement. (Such apparent error in the spinner interpretation is often due to error in the flow measurement rather than error in the spinner itself and highlights the need to have spinner flow data within the cased part of the wellbore.) The first clear loss zone is seen at 1885−1915 m, shown by the sharp drop in velocity at 1885 m, whereas the temperature gradient does not change significantly until the second zone at 2000 m with both a velocity decrease and temperature gradient increase. It is possible that the velocity decrease at 1690 m is due to a loss zone. However, it is more likely that the length of the hole with the annulus packed is relatively short, and this velocity decrease reflects some of the flow going down the annulus. The decrease at 1885−1915 m must be

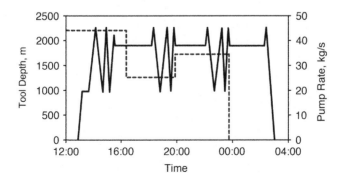

FIGURE 7.3 Completion test in RK18L2. *Source: Rotokawa Joint Venture, Personal Communication.*

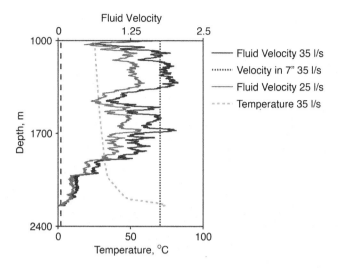

FIGURE 7.4 Injection profile at 35 and 25 l/s in RK18L2. *Source: Rotokawa Joint Venture, Personal Communication.*

a loss, since the velocity falls to a value too low to be explained by flow in the annulus unless it were greatly enlarged. A minor third zone is near 2200 m.

Figure 7.5 shows the injectivity using the pressures measured at 1900 m. There is a good linear plot of flow rate against pressure, with an injectivity of 2.5 kg/s.bar. The pressure transients between flow rates were reasonably regular, containing slight irregularities due to pump rate variations but no reversals. The final falloff is shown in Figure 7.6. A slight break in gradient (arrowed) indicates that some flow continued and was finally shut at this time. The deviation in trend is more obvious on a semilog plot. Depending on the final temperature at the feed zones, an injectivity of 2.5 kg/s.bar indicates that the well should be a moderate producer. The injectivity plot shows that pressure is 150.8 bar at zero flow and so identifies this as reservoir pressure at 1900 m.

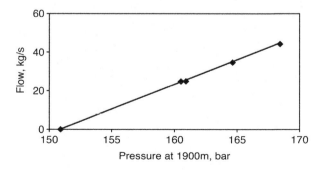

FIGURE 7.5 RK18 injectivity measured at 1800 m. *Source: Rotokawa Joint Venture, Personal Communication.*

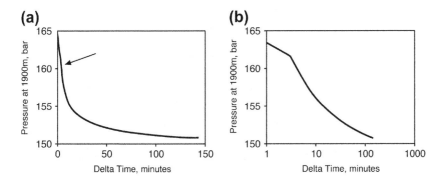

FIGURE 7.6 Pressure falloff. (a) Cartesian plot, arrow indicates break in trend. (b) Semilog plot. *Source: Rotokawa Joint Venture, Personal Communication.*

In a low permeability well, it is possible that the well was overpressured during the entire test by the water injected during drilling, so this pressure needs to be confirmed by pressures during heating.

7.2.3. High Permeability Well: Inflow Evaluation

For highly permeable multiple feed zone wells where inflows are present during cold water injection, properties of the various feed and loss zones can be evaluated by a combination of enthalpy balance methods, wellbore fluid velocity profiles, and the productivity/injectivity index concept. Where an inflow of hotter fluid—water or steam—is present during cold water injection, this is usually evident as a stepped temperature profile (see Figure 7.1). This step may not be as steep as that shown in Figure 7.1, but the isothermal section below the step usually confirms an inflow. The diagnostic test for an inflow is an *increase* in temperature in the isothermal section below the inferred inflow point if the injection rate is *decreased*. If the inflow temperature (enthalpy) is known, the inflow rate can be easily calculated by heat and mass balance because the injection rate at the surface and the downhole temperatures above and below the inflow point are also known.

Using the notation shown in figure 7.7, at the upper inflow zone the enthalpy balance is:

$$W_i H_i + W_{f1} H_{f1} = W_2 H_2 \tag{7.2}$$

and the mass balance is:

$$W_i + W_{f1} = W_2$$

Substitute and rearrange to solve for W_{f1}:

$$W_{f1} = \frac{W_i(H_2 - H_i)}{(H_{f1} - H_2)} \tag{7.3}$$

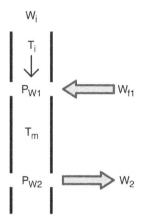

FIGURE 7.7   Schematic flow profile and parameters required to calculate heat and mass balance, injectivity and productivity for a well with hot inflow during injection. Note that the fluid enthalpies used in equations 7.2 and 7.3 correspond to sensible heat in liquid at the temperatures T_i, T_{f1} and T_m (Unless the inflow is two-phase temperatures can usually be substituted for enthalpy within the accuracy of the data).

Once the flow rates from the formation are known, it should then be possible using the injectivity/productivity concept, together with the downhole pressure measurements, to calculate either formation pressure by assuming constant productivity, or productivity of the individual feedzone. If the formation pressure gradient is already known.

At the upper zone:

$$PI_1 = \pm \frac{dW}{dP} = \frac{W_{f1}}{P_{f1} - P_{w1}} \qquad (7.4)$$

At the lower outflow zone:

$$II_1 = \pm \frac{dW}{dP} = \frac{W_2}{P_{w2} - P_{f2}} \qquad (7.5)$$

The formation pressure and pressure gradient are usually known from other wells, so a good average value for productivity-injectivity can be obtained from multiple rate tests. A three-rate test should provide sufficient data to solve the equations even if both the injectivity/productivity and formation pressure are unknowns.

In practice there are limitations to the application of this technique. These arise largely from the accuracy and reliability of the data. Where mechanical downhole instruments are used, accuracy for temperatures is typically not better than $\pm 2°C$ and ± 0.5 bar. Modern electronic instruments should have an accuracy almost an order of magnitude better than the older mechanical instruments. The field data may not be truly stable, and tests have shown that, especially during the completion test period, the inflow temperature is not constant (Bixley & Grant, 1981). The potential for error can be reduced by

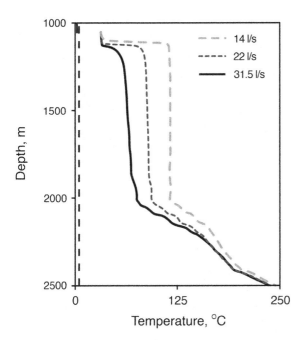

FIGURE 7.8 Temperature profiles in well RK27 during injection. *Source: Rotokawa Joint Venture, Personal Communication.*

running the tests at several injection flows to provide check data. However, the overall results must be considered together with all the other available data and the technique regarded as another tool that the engineer can use to provide a more reliable interpretation of the field information, rather than a definitive method in itself.

Consider the following example. Figure 7.8 shows temperature profiles measured during completion testing at well RK27. There is an inflow near 1100 m, and detailed examination of the temperature profiles locates the inflow zone at 1114 m, and fluid is lost from the wellbore over the interval between 1830−2000 m. The lower loss zone is marked by the change in temperature gradient, which starts at 1830 m, and the clear gradient change at 2000 m marks the bottom of the loss zone. There is a little permeability below 2000 m because the temperature decreases slightly with time, but there is no permeability at all below 2400 m because the temperature here does not change.

Pressure measurements were made at 1114 m, so this depth was used for the inflow calculation. The inflow rate at 1114 m can be calculated assuming a temperature for the inflow. In this case the inflow temperature was assumed to be 225°C, which was the temperature measured at this depth 1.5 hours after stopping injection. The inflow at 1114 m is calculated in Table 7.1.

Figure 7.9 shows the calculated inflow rate plotted against the pressure. This indicates that the incremental productivity of the upper zone is 1.3 kg/s.bar. Using the data measured during injection, the extrapolated pressure at zero flow, which should be the reservoir pressure at this depth, is 102.2 bar. Because

TABLE 7.1 Inflow Analysis of Upper Zone

Pump Rate l/s	T_i °C	T_2 °C	H_i kJ/kg	H_2 kJ/kg	W_{f1} l/s	Pressure at 1114 m (bg)
14	30.4	116	127	487	10.55	94.04
22	30.4	86.6	127	362	8.61	95.31
31.5	30.4	69.4	127	290	7.63	94.04

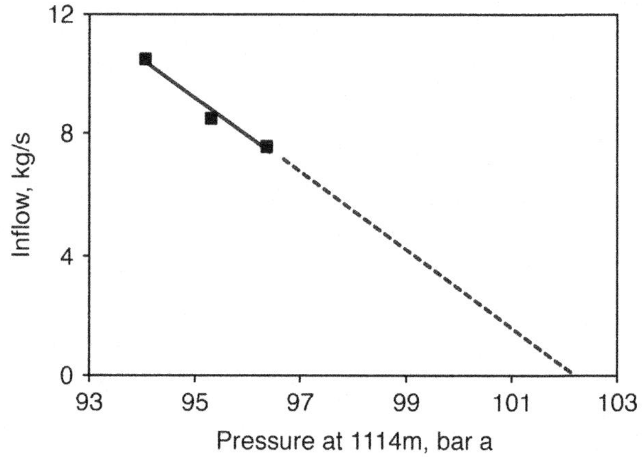

FIGURE 7.9 Productivity of upper zone. *Source: Rotokawa Joint Venture, Personal Communication.*

the injectivity plot is extrapolated a significant distance, there will be some uncertainty in this value.

Both the productivity and the computed reservoir pressure are sensitive to the assumed inflow temperature. If the inflow temperature is lower, the inflows must be correspondingly larger to produce the measured heating, thus increasing the calculated productivity of this zone. It is also possible that the inflow has excess enthalpy, since there is two-phase fluid elsewhere in the reservoir at this depth.

Note that in this case the inflow rate changes as the pump rate changes. The total flow lost at the lower zone is the sum of the pump flow and the computed inflow, and it is these rates that should be used to correctly evaluate the injectivity of the lower zone. Figure 7.10 shows the injectivity of the lower zone plotted with the pump flow and the total flow into the lower zone. Pressure was measured at 1981 m. Because the inflow decreases with rising pump rate, the actual increase in flow at the lower zone is less than the increase in pump rate.

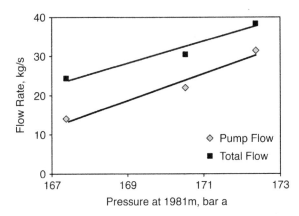

FIGURE 7.10 Injectivity of lower zone. *Source: Rotokawa Joint Venture, Personal Communication.*

Injectivity of the lower zone is 3.3 kg/s.b using the pump rate, but using the total flow, it is 2.8 kg/s.b The injectivity plot is not a tight straight line, so it is not reliable to extrapolate to zero flow to give the reservoir pressure.

Because of the uncertainties associated with actual field data, these calculations are suited to analysis using a spreadsheet, where the unknowns can be quickly found, as well as the sensitivity of the results to variation in the field data tested using a trial and error method based on experience in that particular geothermal field. It can be helpful to check the measured or calculated flow in the well against spinner measurements. It is often the case that one zone may be so permeable that there is little pressure change at it with changing flow, and careful and accurate measurements are needed to determine permeability.

It can also happen that the inflow does not change much with changing pump rate due to the rise in pressure at the lower zone at higher flow rate being counteracted by the increasing density of the cooler water column between the two zones. In this case the productivity of the upper zone can be found by using a reservoir pressure determined from measurements in adjacent wells or a pressure measured in the test well during heating at a time when temperature or spinner results show that the flow in the well has stopped.

Wells with strong interzonal flow effects can be difficult to quench. If the upper zone is inflowing high-enthalpy fluid during injection, the well can come under pressure quickly after stopping pumping because the wellbore rapidly fills with steam and gas. If the lower zone is relatively underpressured and permeable, it can be impossible to quench the upper zone when pumping water alone; at all flow rates the upper zone continues to flow, and the lower zone accepts all fluid.

7.2.4. Estimated Production

The measured injectivity of the well can be used to estimate the expected production. Figure 7.11 shows the relation between injectivity and the expected

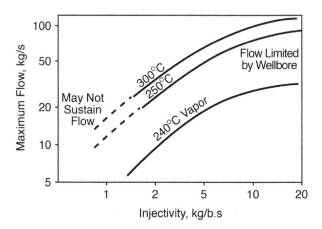

FIGURE 7.11 Relation between injectivity and expected maximum discharge of an 8-inch well.

maximum discharge for a standard well with an 8-inch casing, depending on the temperature of the main feed and assuming that the reservoir is normally pressured. This relation is approximate, with an accuracy of 50%. It is convenient for sizing equipment for the production test and for giving an early indication of the well potential.

7.3. WELLBORE HEAT TRANSFER

The dominant means of heat transfer in a geothermal reservoir is by convection. Some heat is also transferred by conduction. As fluid flows up or down a well through casing or impermeable rock, there is heat transfer through the rock to the wellbore. This causes some heating of water when it is injected into a well. This transfer can be calculated by assuming that the heat flow is entirely normal to the axis of the well (Ramey, 1962). The equation for heat conduction is then:

$$K \frac{1}{r} \frac{\partial}{\partial r}\left(r \frac{\partial T}{\partial r} \right) = \rho_f C_f \frac{\partial T}{\partial t} \qquad (7.6)$$

At distance, the temperature is the undisturbed reservoir temperature T_r, and at the wellface it is the temperature T_w of the fluid in the well. The conductive heat transfer per unit length of wellbore is given by:

$$Q = 2\pi r_w K \left(\frac{\partial T}{\partial r} \right)_{r_w} \qquad (7.7)$$

For times greater than a week this becomes:

$$Q = 2\pi K (T_r - T_w)/f(t) \qquad (7.8)$$

$$f(t) = \ln(2\sqrt{\kappa t / r_w}) - 0.29 \qquad (7.9)$$

where κ is the thermal diffusivity of the rock. It has been assumed that there is no tubing or other source of thermal resistance in the well. The thermal properties of geothermal rocks do not vary greatly, and typical values are:

$$K = 0.2 \, \text{W/m.K}, \quad \kappa = 0.1 \, \text{m}^2/\text{day}$$

Taking some typical values, suppose there is a temperature difference of 150 degrees between the well and the reservoir, the well radius is 0.1 m, and the time is 1 week.

$$f(t) = \ln(2\sqrt{0.1 \times 7/0.1}) - 0.29 = 2.35$$
$$Q = 2 \times 3.14 \times 0.2 \times 150/2.35 = 80 \, \text{W/m}$$

Over 1000 m length, the heat gain would be 80 kW, which is very small in relation to the heat produced by a typical geothermal well. One case where this heat transfer can be noticed is during injection, particularly at low flow rates. If water is injected in a section of the well with no inflow or outflow, there is a small heat gain by conduction. Equating the heat gain by the water to the conductive flux gives:

$$\frac{dT_w}{dz} = \frac{2\pi K(T_r - T_w)}{WC_w f(t)} \tag{7.10}$$

The temperature gradient is inversely proportional to the flow rate, so an increase in gradient marks a decrease in flow—that is, a loss zone.

Figure 7.12 shows an example. Water is being injected, and there is a major loss at 1000–1050 m, as shown by the fluid velocity and the very low temperature gradient above 1000 m. The gradient is computed taking data points 3 meters apart. About 20% of the flow continues below 1050 m, and below this depth there is an inverse correlation between the velocity and the temperature gradient. There is further loss starting at 1350 m, where the velocity decreases and the gradient increases.

7.4. HEATING

7.4.1. Measurements

The downhole measurement program after completion of drilling will be determined largely by the drilling techniques that have been used and by the reservoir characteristics. Wells drilled into a liquid or two-phase reservoir with generally high permeability that develop strong interzonal flow will heat very quickly over the first few days and then show only slow changes with time (in the parts of the well unaffected by interzonal flow). Low permeability wells usually heat slowly but evenly over two or three months. In some cases it can take more than a year for the well to reach stable temperatures. Most

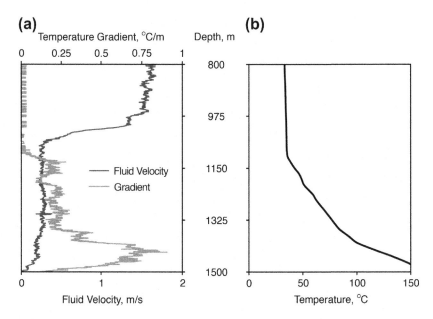

FIGURE 7.12 (a) Fluid velocity and temperature gradient, (b) Temperature. *Source: Lihir Gold Ltd., Personal Communication.*

geothermal fields develop characteristic well heatup patterns and the downhole survey program should be optimized to collect data at the best time intervals.

During the heatup period after drilling, data can be obtained that are available at no other time in the well's history. This information can only be recorded if appropriate measurements are made. This is particularly true for shallow temperature data in the cased-off section of well where the wells are prone to become steam-filled. The heating measurements consist of temperature-pressure profiles made at expanding time intervals after stopping injection. Where PTS logging capability is available, the spinner should be incorporated into the heatup program, since this can be done at minimal additional cost and provide confirmation of fluid movement in the wellbore (using a basic 0.5 m/s down and up PTS log; additional passes at different logging speeds would be needed to accurately determine any flow rate). An effective sequence for heatup surveys in exploration wells is usually 1 hour, and then 2, 7, 28, and more than 60 days, if possible (all with the well completely shut in). Generally speaking, where strong interzonal flows are present (usually high permeability wells), more data are required immediately after drilling completion, whereas for low permeability wells, longer-term data are more appropriate.

The period allowed for heating may, in practice, vary from only a few hours to several months, depending on the well's purpose. Ideally, a minimum of at least one month heating is necessary to obtain the bulk of the information. For the first exploration wells in a new field, it is most important to ensure that the well has fully heated to the stable formation conditions before discharge. For new production wells, there is often a demand to flow the well as soon as possible and prove the production potential. Following this discharge, the well will normally be kept warm by maintaining a small "bleed" flow (to minimize casing stress through multiple heating and cooling cycles). As a result of this process, the shallow temperatures inside the wellbore (perhaps the first 1000 meters) will be heated to the deep reservoir temperatures, and it may never be possible to determine true formation temperatures outside the wellbore again. Ideally, there needs to be a balance between obtaining the maximum reservoir information during the initial heatup period and proving productive capacity.

7.4.2. Pressure Control Point

The pressure pivot or pressure control point concept is crucial to understanding well behavior during the heatup period. Where a well is filled with liquid, the pressure gradient within the wellbore is controlled by the fluid temperature. In the ideal case of a well with a single feed zone, the pressure in the well at the depth of this feed zone will be controlled by, and equal to, the formation pressure. Thus, as the wellbore fluid changes temperature during the heating period, the density of fluid also changes, but pressure at the feed zone is fixed by the formation pressure, so the pressure profile observed in the well pivots about the feed depth. For this type of well, the pivot uniquely identifies this depth. In a well with two or more zones, the pivot forms between the zones—at a depth that is an average of the feed depths weighted by their injectivity/productivity indexes. For these wells there will usually be associated internal wellbore flow between zones resulting from the pressure differences.

Figure 7.13 shows heating profiles in RK17. The pressures form a pivot at 1400–1500 m between the two feed zones identified by the injection profiles. There are permeable zones at 1100 m and 1750 m. The upper zones inflow during the heating period, with a downflow to the lower zone. This internal flow rapidly heats the well, and in 17 hours the inflow at 1100 m has heated the well to 260°C, and after two weeks the deep production zones are fully heated, with a small upflow from the bottom zone at 2 weeks. There remains part of the wellbore below the deepest feed that is heating more slowly.

In a well with several feed zones, the pivot will form at a depth between the different feed zones of the well. Because the lower zones are accepting and the upper are producing and the injectivity of any one zone is often several times larger than its productivity, the lower zones that are accepting fluid will have greater weighting in forming the pivot than in their contribution to the well discharge.

There have been a number of papers reviewing the relation between injectivity and productivity. Some papers have found that injectivity and productivity are approximately equal with a wide spread (Grant, 1982;

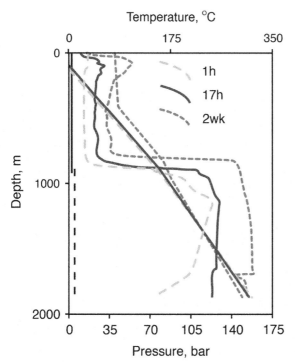

FIGURE 7.13 Heating profiles in RK17. *Source: Rotokawa Joint Venture, Personal Communication.*

Combs & Garg, 2000; Elmi & Axelsson, 2009). Others report that injectivity is an order of magnitude greater than productivity in two-phase fields (Garg et al., 1998), and still others report that a clear relationship does not exist in high-temperature fields (Axelsson & Thórhallsson, 2009). Recent wells drilled in New Zealand have shown injectivity three to five times greater than productivity, again with wide scatter.

Where other information is unavailable, the depth and pressure at the pivot point are the best indicators of formation pressure for multiple-feed wells. The temperatures and pressures measured in the wellbore during the heating period reflect internal flow or static conditions that are the result of pressure differences between the fluid in the wellbore and in the formation fracture network.

7.4.3. Temperatures During Heating

Immediately following the completion of drilling and injection testing, the formations around the well are relatively cool. Impermeable zones (and the cased-off section) have been cooled conductively, and the permeable zones have been cooled by their acceptance of water and drilling fluid. The fluid within the well can be heated by conduction through the surrounding formation and the casing the flow of fluid into and within the well, convection cells within the well, and flow across the wellbore at a single permeable zone. As always, convective processes greatly outweigh conduction as a means of heat transfer. Thus, wells with poor permeability are usually slow in heating, sometimes taking many months to fully stabilize, while permeable wells with strong internal flows may become fully heated (in some sections) within hours.

There are several characteristic features that provide information about the location of feed zones and true formation temperatures:

- Zones of persistent temperature inversion, associated with fluid losses while drilling (usually toward the bottom of the openhole interval)
- Zones of rapid heating, associated with fluid inflow
- Persistent isothermal or near-isothermal sections, indicating interzonal flow that enters and leaves the wellbore at either end of the linear profile
- Boiling-point temperature curves
- Sections of slow, conductive heating

In the cased section of the well, there are also particular features that give us useful information about the reservoir:

- Linear (conductive) gradients
- Temperature peaks and inversions
- Boiling point temperature curves (for reservoir pressures outside the casing)

To obtain the best interpretation of the data, these temperature features must be correlated with geology, drilling, and pressure information.

Figure 7.14 shows the warmup of PK5. Injection profiles were shown in Figure 7.2. The heating pressures pivot nicely just below 1500 m, which is consistent with a main loss at 1500 m and a little permeability deeper. Figure 7.15 shows the warmup of PK1, a well of low permeability with an injectivity of 0.7 kg/s.b. As with most low permeability wells, little circulating fluid was lost to the formation during or completion testing. The warmup shows the typical smooth and featureless warming of an impermeable well, with the exception of the sharp change in gradient just above 1200 m, which in this case corresponds to the top of the perforated liner at 1175m. Usually the perforated liner is installed with one or two joints of unperforated liner at the top, and frequently a convection cell will develop here, producing an anomaly in the otherwise smooth temperature profile. This effect is also seen in Figure 7.13.

Gas Pressure Buildup

In some wells, noncondensable gases rapidly accumulate in the cased-off section when the well is shut. The pressure of this column is sometimes useful in locating the uppermost feed zone, since the gas column can only extend down the well to this level, after which further accumulation of gas will leak off to the formation. Thus, the gas-liquid interface in the wellbore can identify both

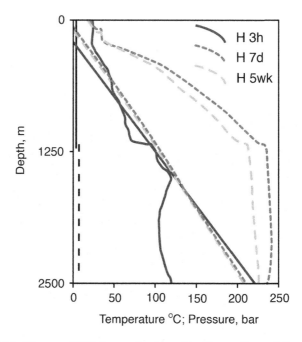

FIGURE 7.14   Warmup of PK5. *Source: Rotokawa Joint Venture, Personal Communication.*

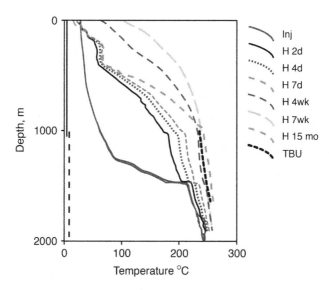

FIGURE 7.15   Warmup of PK1. *Source: Rotokawa Joint Venture, Personal Communication.*

the depth and pressure at this zone. When there are no leaks of gas at the wellhead while the well is heating or during the downhole temperature-pressure survey, reliable measurements of formation temperature can be made in the gas column. Sometimes the temperatures measured in both liquid and gas columns are identical and provide assurance that "real" temperatures of the formation outside the wellbore are being measured.

Vapor Wells

Where wells are completed into a vapor zone and steam rapidly fills the casing on termination of drilling, true formation temperatures cannot be determined. Heat transfer within the wellbore is much stronger than lateral heat conduction. Linear, "conductive" profiles have been observed in wells where adjacent shallow wells show much structure to the shallow temperatures. If data on true formation conditions above the steam zone are required, then the well must be completed in stages. After the production casing has been cemented in place, the cement can be drilled out to just above the casing shoe and then left filled with water and allowed to heat. True formation temperatures can be obtained from temperature buildup measurements made over two or three days or by waiting longer for conditions to approach equilibrium.

Convection Cells

Convection cells are largely controlled by the configuration of casing and liner in the well (Allis & James, 1981). Continuous temperature profiles and flow-meter measurements show that where lengths of plain liner are included in a string of otherwise perforated liner, convection cells will become established around the lengths of plain liner. Fluid flows in one direction within the liner and returns in the annulus. This is often observed at the top of the perforated liner, where one or two lengths of plain liner are usually installed (for example, see Figure 7.13). Configurations with sections of plain liner between otherwise perforated liner should be avoided if possible, and where they are used, inter-pretation of the temperature data must make due allowance for the circulation cells that are generated. Variation in diameter of the wellbore outside the liner and even the perforation density can also result in the development of circu-lation cells with apparent isothermal temperatures as measured inside the liner.

7.5. INJECTION PERFORMANCE

The preceding discussion generally focuses on the potential for geothermal wells to produce fluid. But equally, and perhaps more, critical to developing a geothermal resource is injection well performance, since for almost every resource today, the bulk of the geothermal process fluids remaining after extraction of the high-quality energy need to be injected back into the ground. Once feed depth, formation pressure, and injectivity have been measured, it is possible to calculate the relation between injection flow rate and wellhead pressure for an assumed fluid temperature.

The pressure gradient in the well is given by:

$$\frac{dp}{dz} = \left(\frac{dp}{dz}\right)_{static} + \left(\frac{dp}{dz}\right)_{fric} \tag{7.11}$$

The static gradient is:

$$\left(\frac{dp}{dz}\right)_{static} = \rho_W g \tag{7.12}$$

The frictional loss is empirically defined as:

$$\left(\frac{dp}{dl}\right)_{fric} = \frac{1}{2}f_M \rho_w V^2 / D = \frac{1}{2}f_M W^2 / \pi^2 \rho_w D r^4 = 8 f_M W^2 / \pi^2 \rho_w D^5 \tag{7.13}$$

where D is the diameter of the wellbore, V is the fluid velocity, l is distance measured along the wellbore ($= z$ for a vertical well), and f_M is a dimensionless number known as the Moody friction factor. (Note that there is an alternative Fanning friction factor that is one-quarter of the Moody factor.) The friction factor depends on the relative roughness, ε/D, of the pipe, where ε is the roughness of the pipe, and upon the Reynolds number, Re, of the flow:

$$Re = \frac{VD}{v_w} \tag{7.14}$$

For turbulent flow ($Re > 4000$), the friction factor is given by the implicit Colebrook equation (Colebrook, 1939):

$$\frac{1}{\sqrt{f_M}} = -2 \log_{10}\left(\frac{\varepsilon/D}{3.7} + \frac{2.51}{Re\sqrt{f_M}}\right) \tag{7.15}$$

Equation (7.14) can be approximately solved by iteration; take an initial guess for f_M, use that value on the right-hand side of Eq. (7.14) to compute a revised value, and iterate two or three times. At a large Reynolds number, which is usually the case, Eq. (7.14) is independent of the Reynolds number and becomes:

$$f_M = 1/[1.14 - 0.86 \ln(\varepsilon/D)]^2 \tag{7.16}$$

In typical geothermal operating conditions, the friction factor lies in the range 0.008−0.025. Given that there is often uncertainty in the pipe roughness due to the possibility of some surface corrosion or deposition, the friction factor often has to be fitted to observed results. In practice, the friction factor can be determined from pressure-temperature profiles measured during production or injection testing at several flow rates for known casing config-urations. For real injection wells, the friction factor can change dramatically as a result of buildup of silica in the wellbore. After determining the appropriate friction factor and knowing the proposed injection fluid temperature, feed zone details, and casing configuration, the pressure at the wellhead of an injection well can be calculated. The flowing pressure P_{wf} in the well at the feed depth is:

$$P_{wf} = P_r + W/II \tag{7.17}$$

where P_r is the reservoir pressure. For each section of casing or liner, the pressure drop in the wellbore is given by:

$$\Delta P = \rho_w g \Delta z + 8 \Delta l f_M W^2 / \pi^2 \rho_w D^5 \tag{7.18}$$

Adding together the contributions from the different sections of casing:

$$WHP = P_r - \rho_w g Z + W/II + C W^2 \tag{7.19}$$

where:

$$C = \frac{8}{\pi^2 \rho_w} \sum l_i f_{Mi} / D_i^5 \tag{7.20}$$

and the sum is over the different lengths of casing and liner.

For an example, consider a well with vertical $9\frac{5}{8}$ -inch casing to 1000 m, then with a 7-inch perforated liner deviated to reach permeability at 1900 m MD, 1700 m TVD. Reservoir pressure at this depth is 140 bar, and the well has injectivity 10 kg/s.bar. For the casing, internal diameter is 0.221 m, and roughness of 0.000046 m gives $f_M = 0.014$. For the liner, internal diameter is 0.16 m, and roughness of 0.0003 m gives $f_M = 0.023$. The injected water is at 85°C with density 968 kg/m³.

$$C = [8/(\pi^2\ 968)] \times [1000 \times 0.014/0.221^5 + 900 \times 0.023/0.16^5]$$

$$= 188\ \text{Pa (kg/s)}^{-2} = 1.88 \times 10^{-3}\ \text{bar (kg/s)}^{-2}$$

$$\rho_w g Z = 968 \times 9.81 \times 1700 = 1.61 \times 10^7\ \text{Pa} = 161\ \text{bar}$$

Then

$$WHP = 140 - 161 + W/10 + 1.88 \times 10^{-3} \times W^2 - 1\ \text{bar gauge}$$

$$= -22 + W/10 + 1.88 \times 10^{-3} \times W^2\ \text{bar gauge}$$

Figure 7.16 shows this calculated injection performance. Where the computed wellhead pressure is negative, the actual wellhead pressure will be zero, and there is a water level in the well with injected water falling as two-phase from the wellhead down to the water level. The roughness values used in this calculation are for new pipe and will increase as the casing and liner become rougher due to deposition or corrosion. For wells of good permeability,

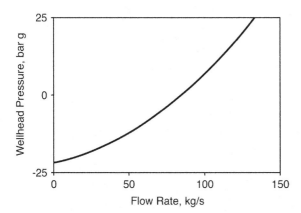

FIGURE 7.16 Calculated injection performance for model well.

casing friction limits the flow rate. This means that the maximum flow of injection wells depends on the casing diameter: for $9^5/_8$ -inch casing of, say, 1 km, flow is effectively limited to a maximum of 125−150 l/s.

7.6. VAPOR-DOMINATED SYSTEMS

A vapor-dominated reservoir contains fewer contrasts than a liquid-dominated one because there are few features to observe. The exceptions are the presence of superheat or of water entries into what is expected to be a steam-filled well.

Well behavior during drilling, completion, and heating depends on the drilling method. If the well has been drilled with water or aerated foam, it is cold (or cold in part of the wellbore) at the end of drilling, and the changes during heating are qualitatively similar to a well in a liquid-dominated reservoir. If the well was drilled with air only, then it is not cooled. Once permeability is found, drilling continues with a controlled discharge of steam. This provides a direct observation of permeability, since the gain in flow with each entry gives its flow. Combined with downhole flowing pressure, either from measurement or computation, productivity of the feed can be computed.

When drilled with water or foam, the well is cold at the end of drilling, and a column of water is present in at least part of the well. Injectivity can be measured by pumping cold water as in liquid-dominated reservoirs. As the well warms, the water column usually falls and steam enters the well. The more permeable the well, the faster the fall in the water level. The water column may disappear or may stabilize.

In some peripheral parts of Larderello, water levels in the wells appear to be the expression of a continuous water phase in that section of the reservoir (Barelli et al., 1977). At the feed zones the pressure of the water in the well must fall below that of the steam reservoir to allow steam entry into the well. Such wells may remain for years with a passive column of water until the water

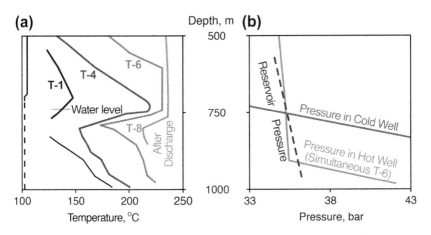

FIGURE 7.17 Heating temperature and pressure profiles in KMJ14. *Source: Pertamina Personal Communication.*

pressure falls enough to allow discharge. In other parts of the Larderello, reservoir wells sometimes contain a stable water column at the bottom of the well with the elevation of the steam-water interface randomly distributed. This is not the expression of a water surface in the reservoir. Rather, in such wells the pressure at the feed zone beneath the water level is balanced against mobile steam at that depth, and the location of the water level is controlled by the elevation of the feed zone, not by a water level in the reservoir (Barelli et al., 1977; Celati et al., 1978; Truesdell et al., 1981).

Figure 7.17 shows the heating of well KMJ14 at Kamojang, Indonesia. Water loss testing showed primary permeability at 700−740 m, and secondary permeability near 900 m. When the water level was above 740 m, pressure in the water column balanced reservoir pressure at this primary feed. After the well heated up, the water level fell to a stable position at 900 m. In this state, steam pressure in the well balanced the reservoir pressure at 700−740 m, and water in the well balanced reservoir pressure at 910 m. Temperatures show circulation effects, with heating being most rapid at the main feed due to steam flowing in at 740 m and exiting at 700 m. In contrast, the bottom of the well heated conductively much more slowly. Discharging wells can have inflows of saturated or superheated steam or even of water, and pressure-temperature profiles may deviate noticeably from simple isenthalpic expansion of steam up the wellbore.

Production Testing

8.1. INTRODUCTION

This chapter describes the discharge testing of wells, and the methods commonly used to measure the flow rate.

8.1.1. Basic Equations

For most conditions found in geothermal production testing, it can be assumed that steam and water properties are those for saturation conditions at the local pressure. Where significant amounts of noncondensable gases are present, their effect must also be allowed for. (This is covered in Appendix 2.) In some wells that produce "dry" steam, superheat conditions may exist; steam properties must be evaluated at the temperature and pressure of the steam. The following quantities are required to fully describe the flow characteristics of a well:

W_s separated steam flow [kg/s]
W_w separated water flow [kg/s]
W total mass flow [kg/s]
H fluid enthalpy [J/kg]
Q heat flow [MW]
X dryness
f noncondensable gas, weight fraction

Geothermal Reservoir Engineering. ISBN: 978-0-12-383880-3 DOI: 10.1016/B978-0-12-383880-3.10008-3

At saturation conditions, if any two of the preceding variables (excluding noncondensable gas) are known, the remainder can be calculated. The variables are related by the following formulae:

$$Q = W \times H \tag{8.1}$$

At separation pressure, P_{sep}

$$W = W_w + W_s \tag{8.2}$$

$$W_s = W \times X \tag{8.3}$$

$$X = \frac{H - H_w}{H_{sw}} \tag{8.4}$$

Example 1: Well EX12 produces two-phase fluid into a separator. The separated steam and water flows are 7.5 kg/s of steam and 48 kg/s water. The separation pressure is 9.2 bar g. Atmospheric pressure is 1 bar absolute. Calculate the mass flow rate and fluid enthalpy.

Procedure: In this case the separated steam and water flows are both measured at the same pressure, 9.2 bar g (10.2 bar abs), so they can be added together to obtain the total mass flow rate:

1. Calculate mass flow rate using Eq. (8.2):

$$W = W_w(10.2) + W_s(10.2) = 55.5 \, \text{kg/s}$$

2. To calculate the fluid enthalpy, first calculate the dryness using Eq. (8.3):

$$X(10.2) = \frac{7.5}{55.5} = 0.136$$

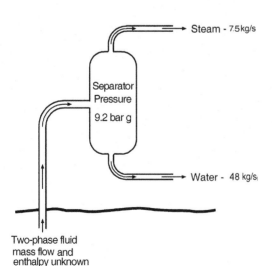

Steam - 7.5 kg/s

Separator Pressure

9.2 bar g

Water - 48 kg/s

Two-phase fluid mass flow and enthalpy unknown

FIGURE 8.1 Schematic diagram of a steam-water separator with flows and pressures used in example 1.

Then rearrange Eq. (8.4) and using the steam table values at 10.2 bar abs,

$$H = 0.136 \times 2018 + 766 = 1040 \text{ kJ/kg}$$

3. Calculate heat flow using Eq. (8.1):

$$Q = 1.04 \times 10^6 \times 55.5 = 57 \text{ MW (above 0}^\circ\text{C)}$$

8.1.2. Flash Correction Factor

This calculation is frequently used in flow-testing using separators to correct water flows measured at atmospheric pressure to some higher separation pressure. That is, after separating steam and water at some higher pressure (P_{ref}), the separated water is flashed again to atmosphere, and the water flow from this second flash is measured in a weir. If the water flow measured at atmospheric pressure (W'_w) needs to be corrected for the steam flashed off between the separator and atmospheric pressure, then

$$W'_w = W_{wsep} \times (1 - X')$$

where X' is the dryness of this second flash and is given by

$$X' = \frac{H_w(P_{ref}) - H_w(atm)}{H_{sw}(atm)}$$

since $H_w(P_{ref})$ is the enthalpy of the water separated at P_{ref}.
  The Flash Correction Factor (FCF) is:

$$W_{wsep} = W'_w \times FCF \tag{8.5}$$

$$FCF = \frac{1}{1 - X'} = \frac{H'_{sw}}{H'_s - H_w(P_{sep})} \tag{8.6}$$

At $P' = 1$ bar absolute:

$$FCF = \frac{2258}{2675 - H_w(P_{sep})} \tag{8.7}$$

Example 2: The most common application of the flash correction factor is in separator tests where the steam flow is measured at separation pressure and the water flow is measured after being flashed to atmospheric pressure. In this case the separated water flow must be recalculated to separation pressure before the mass flow and enthalpy calculations are made. In the example shown in Figure 8.2, the water flow at atmospheric pressure is 46.5 kg/s. Calculate the water flow at separation pressure; atmospheric pressure is 1 bar abs.
  Procedure: Calculate the flash correction factor using Eq. (8.5) or (8.7).

$$FCF = 1.182$$

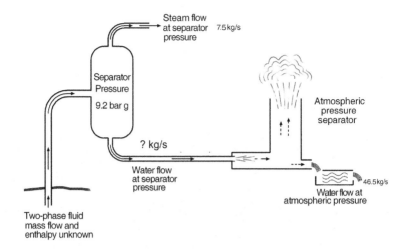

FIGURE 8.2 Schematic diagram of pressurised separator together with atmospheric separator with flows used in example 2.

Therefore, water flow from separator is

$$W_{wsep} = 46.5 \times 1.182 = 55 \text{ kg/s}$$

8.2. STARTING DISCHARGE

The first step in flow testing is starting the well discharge. For most wells this is not difficult, since they naturally develop sufficient pressure, either cold gas or steam, that opening the control valve will initiate flow. But in some wells it can be difficult to start flow. Even after waiting for weeks for the well to heat up following drilling, no pressure develops at the wellhead, and when the control valve is opened, there is no flow. This problem is most common in fields that are underpressured or where there is a cold section in the upper part of the wellbore.

The reason for difficulties in starting flow can be seen by considering the fluid column in the well. Figure 8.3 shows pressure profiles in a well while discharging and when shut. When the well is flowing, it contains a boiling column in the upper part of the wellbore and perhaps the entire depth. This dynamic fluid column has a lower pressure gradient than static liquid, so even though there is drawdown at the feed zone, there is still pressure on the wellhead.

If a well does not spontaneously discharge, it will have a water level some distance below the wellhead, and the upper part of the water column will be "cool" (below boiling point). Discharge requires boiling fluid in the wellbore, so to start a discharge, the cool water must be removed and replaced by hot

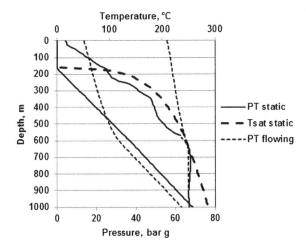

FIGURE 8.3 Discharging pressure profile in well with liquid feed and flashing in the wellbore.

water. Sometimes a well may be dormant, apparently "dead," with a water level not far below the wellhead but containing a water column at boiling point. In this case a small disturbance such as dropping an object into the well may be sufficient to initiate boiling and start discharge, analogous to starting a geyser. However, this is unusual, and for most dormant wells the alternative ways of starting discharge are as follows:

1. Pressurizing the well
2. Gas lift
3. Steam or two-phase injection
4. Workover

In high-temperature fields where wells are difficult to start, it is usually the practice, once the well has been heated and has discharged, to keep it "live" by maintaining a very small discharge (bleed) of either steam or two-phase fluid.

Minimizing the stresses on the casing caused by a sudden change in temperature should also be taken into account when selecting a suitable method to initiate flow. This is particularly important for high-temperature reservoirs (~280°C) and where long fully cemented casing strings are used (~1000 meters). For wells in both of these categories gradual heating using two-phase or steam injection is the preferred method of initiating flow.

8.2.1. Pressurizing the Well

This method is used where the gas-water interface can be depressed to a level such that the BPD profile from the depressed water level intersects the stable downhole temperature profile. Figure 8.4 shows profiles in a well being

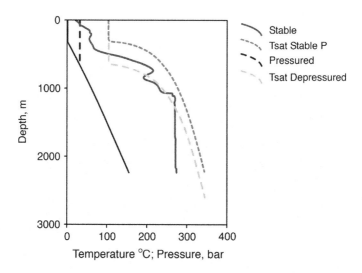

FIGURE 8.4 Profiles in BR65 under pressurization. *Source: Contact Energy, Personal Communication.*

pressurized. The well can be pressurized by pumping air or gas such as nitrogen. Some wells pressurize themselves by the buildup of gas pressure when the wellhead valves are shut.

The well is pressurized for several hours to allow temperatures in the water column to recover, and then the wellhead valves are opened rapidly. If this operation is successful, fluid in the well boils and discharge starts. Figure 8.4 shows the mechanism. The stable shut temperature profile is everywhere below boiling point for the pressure. When the well is pressurized, the water level is pushed down the well and water is displaced into the formation at the feed zone. After the well has been pressurized, it is necessary to wait for some time for the water in the wellbore to heat up again, since the cooler water from above has been pushed down into hotter formations. When the pressure is released, the boiling point profile (shown as Tsat Depressed in figure 8.4) drops to hydrostatic from the water level, and any interval where temperatures are above the boiling point for this lower pressure will boil. In the example illustrated in Figure 8.4, water will boil over the interval 600−800 and 1000−1100 m. If the excess temperature is sufficient or the boiling section is long enough, the boiling fluid expands and lifts the water column up to the wellhead and expels some of the boiling fluid, and then discharge can start. As the wellbore fluid is removed, new (hot) fluid can flow into the well at the feed zone and sustain the process. The minimum requirement for this method is that the water level can be depressed far enough that a significant section of the water column will boil when the pressure is released. In practice, a certain amount of upward fluid momentum is induced when the pressure is suddenly released, and this assists with ejecting the fluid column from the wellbore. It may not be possible to pressurize the well sufficiently to initiate flow. The available pressure is limited by the pump capacity, and usually the water level cannot be depressed very far below the casing depth or at most to just below the shallowest feed zone.

As with any sudden initiation of hot discharge, there is a thermal shock to the casing from the sudden heating, and this method is not favored for high-temperature wells (>280°C). This method is usually successful where the standing water level is less than 200 meters below the surface. For deeper water levels, the heat loss from the wellbore fluid to cool surrounding formations is excessive, and other methods to initiate flow are more suitable.

8.2.2. Gas Lift

Gas lift is a long-established method for flowing liquid from water- and oil-producing wells. Geothermal wells generally require a lower gas:water ratio than cold water wells because of the additional lift created by steam from boiling. A workover rig or coiled tubing unit will normally be needed to insert the gas injection tubing and then remove it under pressure once flow starts.

The result of a successful gas lift initially produces a gas-water flow at the wellhead. As cold fluid is removed from the wellbore by the gas lift, it is replaced by hot water flowing at the feed zone. Provided this fluid is hot enough, it will boil as it flows up the wellbore and create a sustained discharge. Self-sustained discharge may not occur if there is a feed of cold water in the upper part of the production zone since the airlift may simply bring more of this cold water into the well without disturbing the deeper, hotter fluid.

8.2.3. Steam Injection

If there is an extensive cold section at the top of the wellbore (>500 meters), the discharge initiated by pressurization (either natural or artificial) may not be sustained because too much heat is lost from the boiling fluid inside the wellbore to the cold surrounding formations, or the water level may be too deep to pressurize or gaslift. In these cases discharge may be best initiated by injecting hot fluid (Brodie et al., 1981; Siega et al., 2005). The heating fluid may be steam or two-phase fluid from a steam generator or from other wells. In developed fields the steam or two-phase production pipelines are commonly used to "backfeed" wells prior to starting discharge.

In either case the effect is the same. The casing is heated, and hot fluid is added to the well. In addition, the well is pressurized to the pressure of the boiler or two-phase supply. When injection is stopped and the flow control valves are opened, suddenly reducing the wellhead pressure (WHP), the heated water in the wellbore will boil, and, if successful, fluid is ejected at the wellhead, more hot fluid is drawn in at the deep feed zones, and sustained flow is initiated. This method has the great advantage that the casing can be heated in a controlled way, avoiding severe thermal shocks associated with sudden discharge using high-pressure gas.

Because a well is difficult to start, it does not mean that it is likely to be a poor producer. Well OK-5 at Palinpinon could not be discharged either by pressurization or airlift, and it was finally started by steam injection, after which it flowed up to 30 kg/s and 2000 kJ/kg. Menzies and colleagues (1995) modeled deep wells at Cerro Prieto with a very long (1.5—2 km) cold interval at the top. Some wells required stimulation by gas lift to a flow rate of 300 kg/s for several days before the wellbore was sufficiently heated that discharge was sustained. Gaslifting at a lower rate would not start the well, even though there was good permeability and a very high temperature at the feed zone.

8.2.4. Workover

Some wells defy all attempts to start discharge even though there is hot fluid and reasonable permeability deep in the well. There may be undesired permeability with a cool feed zone at some point in the wellbore. Where the well has a cold zone in the upper part of the open hole, it may be impossible to discharge despite the presence of hot fluid and good permeability deeper in the well. When gaslifted, there is only a flow of cool water from the shallow feed and deeper hot feed zones do not flow. To allow this type of well to sustain flow, the shallow cool feed zones must be cased off. If such cold zones can be predicted by association with particular formations or perhaps particular fault zones, the casing program in later wells can be redesigned to exclude these zones.

8.3. PRODUCTION TESTING METHODS

Selecting the type and size of equipment needed to test a geothermal well depends on the expected production rate, pressures, and the fluid type. The available equipment, the duration, and the accuracy required for the test will also be factors in selecting the test equipment. The expected fluid enthalpy will usually be known from testing of nearby wells and downhole survey information. The expected production rate can be estimated from the results of permeability tests made on completion of a well.

Where environmental conditions permit, a brief vertical discharge direct to the atmosphere can be used with the lip pressure method to get a first estimate of the longer-term production potential and to determine the most suitable equipment for longer-term testing. During a vertical discharge, the fluid enthalpy can be estimated by observation of the discharge plume, the fluid chemistry, and downhole conditions at the feed zone prior to discharge. The vertical discharge also helps to clear debris from the well.

These are the main methods used for flow-testing geothermal wells:

- Single-phase measurements (standard orifice or weir flows for single-phase steam or liquid)
- Total-flow calorimeter
- Separator
- Lip pressure (James) method
- Tracer dilution method

8.4. SINGLE-PHASE FLUID

When a well flows single-phase fluid, either liquid water or dry steam, testing becomes a simpler process as ambiguity about the fluid enthalpy is removed.

8.4.1. Low Enthalpy Wells

Where the temperature of the flowing geothermal fluid can be measured in liquid conditions, the fluid enthalpy can be obtained directly from the steam tables. The enthalpy of the fluid entering the well may be obtained by downhole temperature measurements in liquid conditions below the depth where produced fluid starts to boil. Depending on the fluid temperature, this may be below the flash point in the wellbore, at the wellhead, or after discharge to atmosphere if the enthalpy is very low (less than boiling point at the local atmospheric pressure). This information may be available only at smaller flow rates where well pressures are high enough to prevent boiling. Downhole temperature measurements are particularly useful in multiple-feed wells where the feed temperature of the different zones is variable. If the downhole temperature of a flowing well is used to obtain the discharge enthalpy, it is assumed that there is no heat loss from the wellbore or that the heat loss is less than the accuracy of the temperature measurement. For practical conditions, the accuracy of temperature/enthalpy measurement will be in the order of ± 10 kJ/kg. Modeling of flows for normal production rates—greater than 5 kg/s—shows that heat loss from the wellbore is somewhat less than this value and can be ignored.

Where no boiling occurs, the mass flow rate can be obtained directly using a standard orifice plate [ISO 5167] or weir [ISO 1438/1] and the enthalpy from the wellhead temperature. Where the fluid enthalpy is great enough to cause boiling, the total mass flow rate can be calculated from the liquid temperature (measured downhole at the feed zone), together with the water flow separated at atmospheric pressure using the Flash Correction Factor (Eq. (8.6)), where H, the discharge enthalpy based on the flowing wellhead or inflow zone temperature, is substituted for $H_w(P_{sep})$. Having measured the water flow at atmospheric pressure after separation, the total mass flowrate is then

$$W = W'_w \times \frac{H'_{sw}}{(H^-_s - H)}$$

and the heat flow is $Q = W \times H$

Example 3: From downhole temperature and pressure measurements, a geothermal well is known to have a liquid feed temperature of 200°C. While the well is flowing, the separated water flow at atmospheric pressure measured after flashing off the steam is 10 kg/s. Atmospheric pressure is 1 bar. What are the discharge enthalpy, the total mass, and the heat flow rates?

Procedure: Using the steam tables, at 200°C the liquid enthalpy is 852 kJ/kg, and the mass flow is

$$W_{wsep} = W'_w \times FCF = 10 \times \frac{2258}{2675 - 852} = 12.4 \, \text{kg/s}$$

and the heat flow is

$$Q = 12.4 \times 852 \times 10^3 \, \text{W} = 10.5 \, \text{MW}$$

8.4.2. High Enthalpy (Steam) Wells

At the other extreme from the low enthalpy "hot water" wells are those that produce saturated, or superheated, steam. Determining the actual flow rate and enthalpy is usually simple, requiring the measurement of flow using standard orifice (ISO 5167) or other devices such as pitot tube types (e.g., Annubar, Flobar, etc.) and temperature measurement techniques. Where the noncondensable gas content of the steam is small (less than ~2% by weight) and there is a slight superheat, the flowing enthalpy can be obtained by plotting the pressure-temperature conditions for the steam on the Mollier chart. Where the steam is slightly "wet," enthalpy can be measured using a throttling calorimeter to reduce pressure to saturation conditions (ASME PTC). This method assumes that a representative sample of the fluid can be obtained from the pipeline. If the steam is genuinely dry (i.e., superheated), it will be transparent as it exits the discharge pipe. Both pressure and temperature must be measured to determine enthalpy.

8.5. TWO-PHASE FLOW MEASUREMENT METHODS

The normal condition for geothermal flow is that there is two-phase fluid at the wellhead. This requires that the enthalpy of the steam-water mixture, or the ratio of steam to water, be measured as well as the total mass flow. There are a number of methods to do this. None of these methods is perfect and each has its limitations.

8.5.1. Total Flow Calorimeter

Calorimeters are excellent devices for flow and enthalpy determination, but their application is restricted to relatively small flow rates. Portable calorimeters with a capacity of up to 1.5 m3 can be easily moved between test sites on a lightweight trailer and have a maximum test capacity of about 10 kg/s total mass flow rate, depending on enthalpy of the discharge (800–2000 kJ/kg). Larger-capacity calorimeters are more difficult to use because they are not so portable and can have significant heat loss and environmental problems during operation.

To use a calorimeter for geothermal well testing, the total well discharge is directed into the calorimeter tank over a short time period. Before and after the test period, the well flow must be diverted elsewhere (such as a normal atmospheric pressure separator/silencer). In the calorimeter, the two-phase well discharge is condensed into cold water already in the calorimeter, and the change in volume and temperature of the liquid in the calorimeter is used to calculate the mass flow rate and enthalpy over the test period. This method has the disadvantage that continuous measurements cannot be made.

During the early days of field exploration at Wairakei, a sampling calorimeter (sampling part of the total well discharge) was developed to test the large two-phase wells. However, this technique was found to be less reliable than either the lip pressure or separator methods (Bixley et al. 1998), largely due to the difficulty in obtaining representative samples from the two-phase discharge line.

The flow rate is measured by determining the change of mass and the heat flow from the change in heat content in the sampling vessel over the sampling period.

$$W = \frac{\rho_2 V_2 - \rho_1 V_1}{\Delta t} = \frac{V_2/v_2 - V_1/v_2}{\Delta t} \qquad (8.8)$$

$$Q = \frac{\rho_2 V_2 H_{w2} - \rho_1 V_1 H_{w1}}{\Delta t} \qquad (8.9)$$

The flowing enthalpy is $H = Q / W$, as given in Eq. (8.1).

Example 4: At WHP 4 bar g, well EX3 gives the following results when tested with a calorimeter (volumes are in liters, l).

Run 1a $V_1 = 400\,l, T_1 = 21°C, V_2 = 511\,l, T_2 = 53°C, \text{Time} = 20\ \text{sec}$.

The procedure to calculate the total mass flow rate and fluid enthalpy is:

1. For data in Run 1a: specific volume of water at $21°C$, v_1 is 1.0021 l /kg and at $53°C$, v_2 is 1.0135 l /kg. Using Eq. (8.8) the mass flow rate is

TABLE 8.1 Calorimeter Test Data for Well EX3

Run	WHP bg	V_1 l	T_1 °C	V_2 l	T_2 °C	Δt sec	H kJ/kg	W kg/s
1a	4.0	400	21	511	53	20	731.3	5.25
1b	4.0	400	21	510	53	20	736.1	5.20
1c	4.0	400	20	512	53	20	742.3	5.29
1	**4.0**	**Average**					**737**	**5.25**
2	**4.0**	**Average**					**692**	**2.22**
3	**3.0**	**Average**					**688.5**	**0.68**

$$W = \frac{511/1.0135 - 400/1.0021}{20} = 5.25 \text{ kg/s}$$

2. At 21°C $h_w = 88$ kJ/kg and at 53°C $h_w = 222$ kJ/kg; using Eq. (8.9),

$$Q = \frac{511 \times 222/1.0135 - 400 \times 88/1.0021}{20} = 3840 \text{ kJ/s} = 3.84 \text{ MW}$$

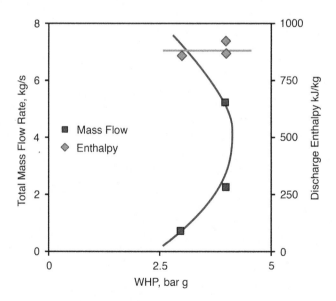

FIGURE 8.5   Mass and enthalpy plots for test data using calorimeter from well RX1 in example 4. In this case the well is being tested close to its maximum discharge pressure, and part of the deliverability curve is on the "bottom" side of the curve.

3. Flowing enthalpy using Eq. (1):

$$H = Q/W = 3840/5.25 = 731 \text{ kJ/Kg}$$

4. Repeating the calculations for the first set of three measurements (runs 1a–c) at the same throttle conditions and for other sets of data measured at different throttle conditions gives the results shown in Table 8.1. The average flow and enthalpy values can then be plotted against WHP to define the well's characteristic output (deliverability) curve.

8.5.2. Steam-Water Separator

The most accurate method for measuring two-phase flows from geothermal wells is to use an efficient cyclone separator (Bangma, 1961). This separates the steam (plus noncondensable gas) and water phases, allowing their individual measurement by conventional methods. Where a properly designed and adjusted separator is used, the separation efficiency can easily be better than 99.9%, so that accuracy depends on the individual steam and water flow metering systems, and the stability of the well flows. At some flow conditions there may be either unstable well flow or separation of the steam and water components, resulting in "slugging" of the two-phase flow. The overall accuracy will not normally be better than ±2% of the separated steam and water flows, but there may be cross-checks available, such as for liquid wells where the feed zone temperature may be known which might improve the reliability of the overall results.

Where the water phase is measured by an orifice plate, care needs to be taken to avoid excessive flashing as the water passes through the orifice because the water is at saturation temperature-pressure conditions as it exits the separator. This can be achieved by providing sufficient elevation difference between the separator water level and the orifice or by cooling the water before measurement. The separator water flow control valve must be downstream from the flow metering device. The presence of steam in the separated water line, due either to flashing in the line or to steam carryover from the separator, is readily detected by large and rapid fluctuations in the differential pressure at the metering orifice.

The general formula for flow through an orifice is:

$$W = C \times \varepsilon \times \sqrt{\frac{\Delta P}{v_{fluid}}} \tag{8.10}$$

where the expansibility ε for steam is:

$$\varepsilon = 1 - (0.41 + 0.35 \times \beta^4)\frac{\Delta P}{1.3 \times P} \tag{8.11}$$

where C is the orifice constant, which depends on the pipe and orifice geometry and pressure tapping configuration; ε is the expansibility factor for compressible gases; ΔP is the differential pressure across the orifice; v_{fluid} is the specific volume of the fluid passing through the orifice; and β is the diameter ratio (d/D) of the measuring orifice and steam pipe. For water, $\varepsilon = 1$. For steam flows, at the pressures generally used during geothermal well testing (5–15 bar g), if the differential pressure across the orifice is less than 0.3 bar, the correction for expansibility is small enough to ignore (less than 4%). However, the expansibility factor becomes significant in cases of large differential pressure ($\triangle P > 0.3$ bar) or low line pressure (P < 3 bar g). For simplicity, in Example 5 below ε is assumed to be 1.0. Calculation of well flow and enthalpy is made by using Eqs. (8.1–8.4).

Example 5: At well EX12 the following observations are made: At WHP 20.9 bar g the differential pressure at the separated steam flow orifice, $\triangle P = 42–49$ mbar, and pressure is 10.2 bar g; the differential pressure at the water orifice is $\triangle P = 63–71$ mbar, and the water temperature is 182°C. The separator pressure is 10.2 bar g, and the atmospheric pressure is 1.0 bar absolute.

1. Steam flow:

$$W_s = 10.525 \times \sqrt{\frac{\triangle P}{v_s}}$$

where $\triangle P$ is mbar and v_s is in cm^3/gm:

$$= 10.525 \times \sqrt{\frac{45}{174}} = 5.4 \text{ kg/s (variation } 5.17 - 5.59, \text{ or } \pm 4\%)$$

2. Water flow:

$$W_w = 5.69 \times \sqrt{\frac{\triangle P}{v_w}} = 5.69 \times \sqrt{\frac{67}{1.130}}$$

$$= 44 \text{ kg/s (variation } 42.5 - 45.1, \text{ or } \pm 3\%)$$

3. Mass flow:

$$W = W_w + W_s = 5.4 + 44 = 49 \text{ kg/s}$$

4. Enthalpy:

$$H = (H_w \times W_w + H_s \times W_s)/W = (785 \times 44 + 2781 \times 5.4)/49$$

$$= 1002 \text{ kJ/kg}$$

5. Heat flow:

$$Q = W \times \frac{H}{1000} = 1002 \times \frac{49}{1000} = 49 \text{ MW}$$

In summary, well EX12 produces a total mass flow of 49 kg/s at WHP 20.9 bar gauge. Discharge enthalpy is 1002 kJ/kg and heat flow 49 MW.

TABLE 8.2 Separator Test on Well WK207

Date	WHP bg	Separator Pressure bg	Steam Pressure bg	Steam ΔP mbar	Water Temperature	Water ΔP mbar
26 Mar	7.0	6.6	6.2	1688	168	536
19 Mar	7.2	6.6	6.2	1546	168	519
19 Mar	8.7	8.6	8.3	674	178	416
20 Mar	10.7	10.5	10.5	290	186	275
23 Mar	12.7	11.9	11.9	161	191	202
24 Mar	14.1	13.8	13.7	69	197	111
24 Mar	15.4	15.2	15.2	28	202	40

Source: Contact Energy, Personal Communication.

When using a separator to test two-phase wells, the steam flow is measured at separator pressure and the water flow may be measured using a sharp-edged weir after flashing to atmospheric pressure (rather than by using an orifice near the separator). In this case the water flow must be corrected for the steam flashed off between separator pressure and atmospheric pressure using the Flash Correction Factor (see Example 3 under "Flash Correction Factor").

Example 6: An output test is done on well WK207 using a separator, with the results shown in Table 8.2. Calculate the discharge enthalpy and total mass flow. Plot the enthalpy and mass flows against the WHP, and calculate the steam flow at separation pressure 7.0 bar g over the full test range (7−15 bar g). The equations for the separated steam and water flows are

Steam:

$$W_s = 4.89 \times \varepsilon \times \sqrt{\frac{\Delta P}{v_s}}$$

In this case the diameter ratio ($\beta = $ d/D) of the measuring orifice and steam pipe is 0.7, therefore the expansibility factor (Eq. (8.10)) for the steam flow calculation is:

$$\varepsilon = 1 - \frac{0.49 \times \Delta P}{1.3 \times P}$$

Water:

$$W_s = 2.23 \times \sqrt{\frac{\Delta P}{v_w}}$$

Procedure: Using the data for 19 March, at WHP $= 8.7$ bar g:
Steam flow:

$$\varepsilon = 1 - \frac{0.49 \times 0.67}{1.3 \times 9.3} = 0.97$$

$$W_s = 4.89 \times \varepsilon \times \sqrt{\frac{\Delta P}{v_s}} = 4.89 \times 0.97 \times \sqrt{\frac{674}{209}} = 8.5\,\text{kg/s}$$

In this case the steam expansibility correction is of the same order as the accuracy of the field data (i.e., $\varepsilon = 0.97$, and data accuracy is probably ~5%). Where there is a combination of large pressure differential and low static pressure, this correction becomes more significant.

Water flow: The expansibility is 1 for noncompressible fluids is therefore

$$W_w = 2.23 \times \sqrt{\frac{\Delta P}{v_w}} = 2.23 \times \sqrt{\frac{419}{1.125}} = 43\,\text{kg/s}$$

Mass flow:

$$W = 8.5 + 43 = 51.5\,\text{kg/s}$$

Enthalpy: Calculate the dryness at separation pressure using the steam and mass flows using Eq. (8.3), and then using the steam table values at separation pressure, calculate the enthalpy of the two-phase fluid using Eq. (8.4):

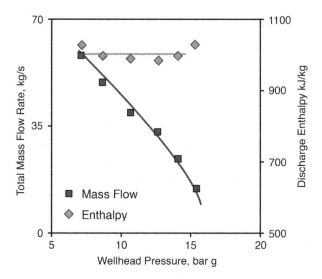

FIGURE 8.6 Deliverability curve plot for WK207 test data shown in Table 8.2. *Source: Contact Energy, Personal Communication.*

$$X = W_s/(W_w + W_s) = 8.5/51.5 = 0.165$$
$$H = X \times H_{ws} = 0.165 \times 2022 + 755 = 1089 \text{ kJ/kg}$$

Repeating these calculations for all the other test data gives the output or deliverability curve for mass flow and enthalpy (Figure 8.6).

For most wells the mass flow and enthalpy values will fall on smooth curves, and deviations from the smooth curve can be used as an indication of the reliability of the test data and the well stability. In this case the well has separate liquid and steam feeds and as the well is throttled the contribution from the steam feed increases relative to the liquid feed with the result that the total enthalpy increases at higher pressures (see Section 8.9.4 for further discussion). For some wells, particularly lower permeability and two-phase wells, flowrates and flowing enthalpy may stabilize one or two days after changing throttle conditions.

Sometimes well performance is plotted as flow rate against downhole pressure rather than WHP. This curve is known as an inflow performance curve. See Aragón et al. (2009) for an example.

8.5.3. James Lip Pressure Method

This method is based on an empirical formula developed by James (1966). For highly productive two-phase geothermal wells, the James method is the most versatile and economical method of testing. Although not quite as accurate as the separator method, the lip pressure method has the advantages of simplicity in both hardware and instrumentation and the ability to accommodate relatively large flows with less expensive test equipment. For many wells, the mass flow rate (and enthalpy) may not fully stabilize during the initial test period (in a flow test of a few days or even weeks). When a prediction of long-term production trends is required, the extra precision of the separator method is usually of no advantage over the lip pressure method, and longer-term enthalpy changes can be better estimated from a good knowledge of reservoir conditions. For wells that demonstrate short-term fluctuation or cycling flow, the separator is no more reliable than the lip pressure method.

Test results using both the separator and lip pressure methods have been frequently checked against each other (sometimes with simultaneous use of both techniques) and enthalpy derived from the surface measurements compared with the results of downhole data. Comparing these different methods, results are usually within 5% for mass flow and enthalpy. A theoretical study by Karamaraker and Cheng (1980) showed that the lip pressure method gives results within 8% of those predicted from one-component, two-phase critical flow theory.

To apply the lip pressure method, the steam-water mixture produced by the test well is discharged into an atmospheric pressure separator (or "silencer" designed to reduce the noise level produced by the discharge). The "lip pressure" is measured at the extreme end of the discharge pipeline as it enters the silencer, and the separated water flow exiting from the silencer is measured over a sharp-edged weir while the steam is discharged to the atmosphere. From these two observations, the flowing enthalpy and total mass flow rate can be calculated using the James equation.

Derivation of Formula: The James formula relates mass flow, enthalpy, discharge pipe area, and "lip pressure" as follows:

$$\frac{G \times H^{1.102}}{P^{0.96}} = 184 \qquad (8.12^*)$$

$$G = W/A$$

where W is the mass flow in kg/s, H is the enthalpy in kJ/kg, A is the cross-sectional area of the pipe in cm^2, and P is the lip pressure in bar (absolute). Note that the constant on the right-hand side of Eq. (8.12) is not dimensionless. When the flow is measured in tonnes per hour and enthalpy in kJ/kg, the relation is:

$$\frac{G \times H^{1.102}}{P^{0.96}} = 663 \qquad (8.13^*)$$

(The asterisk is used to denote equations where non-SI units are used.)

The separated water flow W_w is the water separated at atmospheric pressure from the total well flow with enthalpy H, so:

$$W'_W = \frac{W \times H'_{SW}}{H'_S - H} \qquad (8.14)$$

where the enthalpies with a dash are evaluated at atmospheric pressure. Substituting Eq. (8.14) into (8.12) gives:

$$\frac{W'_W}{AP^{0.96}_{lip}} = \frac{194}{H^{1.102}} \times \frac{H'_s - H}{H'_{SW}} \qquad (8.15)$$

In the particular case of separation at an atmospheric pressure of 1 bar, where $H'_s = 2675$ and $H'_{SW} = 2258$:

$$\frac{W'_w}{AP^{0.96}_{lip}} = Y = \frac{0.0815(2675 - H)}{H^{1.102}} \qquad (8.16)$$

or with the flow in t/h:

$$\frac{W'_W}{AP^{0.96}_{lip}} = Y = \frac{0.293(2675 - H)}{H^{1.102}} \qquad (8.16^*)$$

Figure 8.7 graphs the relation between Y and enthalpy.

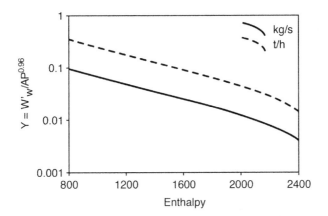

FIGURE 8.7 Relation between discharge enthalpy and lip pressure, water flow, and pipe area at atmospheric pressure of 1 bar.

The relation between Y and enthalpy is given with an accuracy of 1.5% for enthalpy in the range 800−2200 kJ/kg by:

$$H = \frac{2675 + 3329\,Y}{1 + 28.3\,Y} \tag{8.17}$$

or with flow in t/h:

$$H = \frac{2675 + 925\,Y}{1 + 7.85\,Y} \tag{8.17*}$$

Having obtained the flowing enthalpy, the mass flow rate can be calculated using the flash correction factor:

$$W = \frac{W'_w \times H'_{ws}}{H'_s - H} \tag{8.18}$$

where the atmospheric pressure is 1 bar absolute:

$$W = \frac{W'_w \times 2258}{2675 - H} \tag{8.18*}$$

Thus, the procedure for calculating well output from lip pressure and water flow measurements is a simple process:

1. Calculate Y value using Eq. (8.16).
2. Calculate enthalpy H using Eq. (8.17).
3. Calculate mass flow rate W using Eq. (8.18).

Example 7: At WHP 5.5 bar g, lip pressure is 3.0 bar g and water flow at atmospheric pressure measured is 210 t/h. Calculate the enthalpy and mass flow rate. Atmospheric pressure is 1 bar absolute and lip pressure pipe diameter is 200mm.

Method:

1. Calculate Y using Eq. (8.16*):

$$Y = \frac{W'_w}{AP_{lip}^{0.96}} = \frac{210}{\pi \times (200/20)^2 \times 4.0^{0.96}} = 0.177$$

2. Calculate H using Eq. (8.17*):

$$H = \frac{2675 + 925\,Y}{1 + 7.85\,Y} = \frac{2675 + 925 \times 0.177}{1 + 7.85 \times 0.177} = 1188\,\text{kJ/kg}$$

3. Calculate mass flow rate using Eq. (8.18):

$$W = \frac{W'_W \times 2258}{2675 - H} = \frac{210 \times 2258}{2675 - 1188} = 319\,\text{t/h}$$

Example 8: Well BR20 is tested using the James lip pressure method, with the results given in Table 8.3. The well is flowed through a lip pressure pipe into a silencer/atmospheric separator. The separated water flow is measured over a sharp-edged weir as it discharges from the silencer. Calculate total mass flow rate, enthalpy, and plot results. Calculate steam flow if separated at 7.0 bg. The calculated test results are shown in Figure 8.8.

8.5.4. Vertical Discharge

This is a variation of the James lip pressure method. Where environmental conditions permit, an initial estimate of well production can be obtained at small cost and with minimal hardware by conducting a vertical discharge test. This initial flow test is often a brief vertical discharge made soon after completion of drilling and the heating period. Although properly stabilized flows are not usually obtained from this test, it is useful for giving an initial estimate of well potential and to determine what test equipment will be necessary to carry out a longer-term test.

From the James formula in Eq. (8.12):

$$\frac{W \times H^{1.102}}{A \times P^{0.96}} = 184$$

$$Q = H \times W = \frac{0.184 \times A \times p^{0.96}}{H^{0.102}}\,\text{MW} \qquad (8.19^*)$$

Equation 8.19* is correct when the enthalpy is in kJ/kg and the heat flow in MW - this scaling converts the factor of 184 to 0.184. Over the range of well enthalpies normally encountered, 800 to 2800 kJ/kg, $H^{0.102}$ varies little, so the heat flow Q can be obtained from an estimated flowing enthalpy. In practice, this estimated enthalpy can be within ±300 kJ/kg using a combination of the downhole temperature at the feed zone before discharge, together with

TABLE 8.3 Output Test Data for Example 8, Well BR20 (Maximum Possible Flowing WHP (MDP) was 38.6 bg)

Date	WHP, bar g	Lip Press Diameter, mm	Lip Press, bar g	Water Flow t/h
10 Sep	10.5	203	3.81	233
11 Sep	12.4	203	3.71	227
12 Sep	14.5	203	3.57	216
13 Sep	16.9	203	3.34	207
14 Sep	19.7	203	3.1	198
15 Sep	21.7	203	2.95	188
16 Sep	24.5	203	2.74	180
17 Sep	28.3	203	2.40	161
18 Sep	32.1	203	1.81	142
19 Sep	33.4	203	1.41	124
20 Sep	36.6	203	0.68	89
	37.9	150	0.79	53

Source: Contact Energy, Personal Communication.

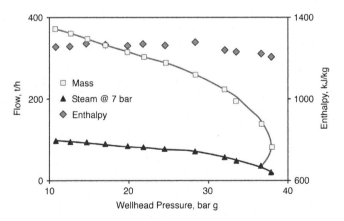

FIGURE 8.8 Enthalpy and mass flow plot for well BR20 calculated from data in Table 8.3. This is a typical output test plot for highly productive liquid-fed wells ($8^5/_8$-inch casing). *Source: Contact Energy, Personal Communication.*

experience in the particular field and a visual estimate of the plume of steam and water produced by the discharge.

Example 9: Using the data from example 7, at WHP 5.5 bar g, the lip pressure is 3.0 bar g, the lip pressure pipe diameter is 200mm and water flow is unknown. Additional information is that the temperature at the major feed zone before the well was discharged was 275°C.

Method: An estimate of likely flowing enthalpy is obtained from the steam tables; at 275°C the sensible heat for water is 1211 kJ/kg. This is rounded off to 1200 kJ/kg for the calculation. Using Eq. (8.19)

$$Q = \frac{0.184 \times \pi \times (200/20)^2 \times 4^{0.96}}{1200^{0.102}} = 106 \text{ MW (thermal above 0°C)}$$

Using the results of the previous calculations for example 7, mass flow 319 t/h and enthalpy 1188 kJ/kg, so the heat flow rate is

$$Q = M \times H = 319/3.6 \times 1188 = 105 \text{ MW}$$

Within the accuracy of the field data this is the same result as obtained by the "approximate" formula: Eq. (8.19).

8.5.5. Tracer Dilution Method

With this method the enthalpy and mass flow rate of a two-phase flow stream is determined by measuring the dilution of separate vapor and liquid phase tracers injected into a two-phase flow. Separated steam and water samples are taken downstream from the tracer injection point after the tracer has become mixed with the two-phase flow. The mass flow rates of the vapor and liquid phases can then be calculated, and knowing the pipeline pressure, the fluid enthalpy can also be obtained. The dilution method was refined for use in the Coso field (Hirtz et al., 1993; Hirtz and Lovekin, 1995) but is applicable to most two-phase systems. Special injection and sampling points are needed in the two-phase pipelines, together with pumps for injecting the tracer at precise rates and facilities for collection and accurate analyses of the samples. The method can be used on normal production two-phase pipelines and lends itself to routine testing without loss of steam and where other methods cannot be used—for example, where multiple wells feed single separators.

The technique requires that separate tracers for the liquid and vapor phases be injected at precise rates into the flow stream. Samples of each phase are collected downstream of the injection point. After chemical analysis the mass flow rates of the vapor and liquid phases are determined from the measured concentrations and injection rates of each tracer. Sampling points must be far enough downstream from injection in order that the tracer is fully mixed with the two-phase flow.

The liquid-phase mass flow rate and vapor phase flow rate are given by:

$$W_w = \frac{W_T}{(C_{Tw} - C_{Bw})} \qquad (8.20)$$

$$W_v = \frac{W_T}{(C_{Tv} - C_{Bv})} \qquad (8.21)$$

where:

W_w = Mass flow rate of liquid phase
W_T = Tracer injection mass flow rate
W_v = Mass flow rate of vapor phase
C_{Tw} = Concentration of tracer in liquid phase by weight
C_{Tv} = Concentration of tracer in vapor phase by weight
C_{Bw} = Background concentration of tracer in liquid phase
C_{Bv} = Background concentration of tracer in vapor phase

Knowing the pressure at the sampling point, the dryness fraction of the two-phase fluid at the sampling pressure can be calculated using:

$$X = \frac{W_v}{W_v + W_w}$$

Then the flowing enthalpy can be calculated using the steam tables, with all thermodynamic variables evaluated at the sampling pressure:

$$H = \left[\frac{W_v}{W_v + W_w} \times H_{WS} \right] + H_w$$

Tracer Selection

For use with this application, tracers that satisfy the following criteria are preferred:

- The liquid and vapor tracers should partition completely into their respective phases, or the distribution between phases should be accurately known.
- The tracer must be chemically and thermally stable under the injection/sampling conditions.
- Precise analytical methods over wide ranges of concentrations must be available.
- The natural background levels must be low and constant.
- The equipment needed for injection and sampling should be simple and robust.

Example 10: Liquid and vapor phase tracers are injected into a two-phase flow pipeline. The liquid tracer background concentration is 4.6 grams/tonne; the liquid tracer injection rate is 710 grams/hour; and the downstream concentration of tracer is 9.8 grams/tonne. For the vapor phase tracer, the background concentration is zero; the tracer injection rate is 1750 grams/hour; and the downstream concentration is 75 grams/tonne. The pipeline pressure is 6.6 bar abs. Calculate the total mass flow rate and enthalpy for the two-phase fluid.

For the liquid phase:

$$W_w = \frac{W_T}{C_{Tw} - C_{Bw}} = \frac{710[g.hr^{-1}]}{(9.8 - 4.6)[g.t^{-1}]} = 137 \text{ t/h at 6.6 bar}$$

For the gas phase:

$$W_v = \frac{W_T}{(C_{Tv} - C_{Bv})} = \frac{1750[g.h^{-1}]}{75[g.t^{-1}]} = 23 \text{ t/h at 6.6 bar}$$

Total mass flow rate is steam rate plus liquid rate, since separation pressure is the same for both phases:

$$W = 137 + 23 = 160 \text{ t/h}$$

Enthalpy of two-phase fluid:

$$H = \left(\frac{W_v}{W_v + W_w} \times H_{ws}\right) + H_W = \left(\frac{23}{137 + 23} \times 2074\right) + 687$$
$$= 985 \text{ kJ/kg}$$

This technique is now becoming the preferred method for monitoring production where multiple two-phase wells feed common steam separation stations.

8.5.6. Other Methods

Where there is a large pressure drop across a valve or orifice, the change in dryness fraction can be determined using the change in concentration of noncondensable gas (in the steam-gas phase) or chloride (in the liquid phase). Assuming that there is no heat loss between the fluid sampling points (and for the gas method that all gas is partitioned into the steam-gas phase), the enthalpy can be calculated using the change in concentration and sampling pressures (Ellis and Mahon, 1977; Blair and Harrison, 1980; Marini and Cioni, 1985).

The mole fraction of noncondensable gas in the steam-gas phase is

$$n = \frac{no.moles\ gas}{no.moles\ steam}$$

The mole fraction of noncondensable gas in the water-steam-gas mixture at the upstream sample point is

$$n_t = n_{s1} \times X_1$$

And at the downstream sample point is

$$n_t = n_{s2} \times X_2$$

The total amount of gas flowing along the pipeline is the same at both sample points. Assuming there is no change in fluid enthalpy between the sample points, the concentration of gas in the steam phase has changed as the dryness has changed in accord with the pressure. Therefore

$$n_{s1} \times X_1 = n_{s2} \times X_2$$

$$n_{s1} \times \frac{H - H_{w1}}{H_{ws1}} = n_{s2} \times \frac{H - H_{w2}}{H_{ws2}}$$

If:

$$R = \frac{n_{s1}}{n_{s2}} \text{ and, } r = \frac{H_{ws1}}{H_{ws2}}$$

then substitute and rearrange:

$$H = \frac{R \times H_{w1} - r \times H_{w2}}{R - r} \tag{8.22}$$

Example 11: A geothermal well is discharging through a two-phase pipeline with the flow rate controlled by a throttling orifice. The upstream pressure is 20 bar with a noncondensable gas content of 25.0 mM/M (millimoles gas per mole condensed steam). Downstream of the control orifice, the line pressure is 15 bar, and a second gas sample gives 22.1 mM/M noncondensable gas in steam. Calculate the enthalpy of the two-phase fluid.

At the upstream sample point, the line pressure is 20 bar, the partial pressure of noncondensable gas is $20.0 \times 0.025 = 0.5$ bar, and the vapor pressure 19.5 bar. The steam table values at 19.5 bar are $H_w = 902$ kJ/kg, $H_{ws} = 1896$ kJ/kg. At the downstream sample point, the line pressure is 15 bar and partial pressure of gas is 0.3 bar. Vapor pressure is 14.7 bar, $h_f = 841$ kJ/kg and $h_{fg} = 1951$.

$$R = \frac{n_1}{n_2} = \frac{25.0}{22.1} = 1.131$$

$$r = \frac{H_{ws1}}{H_{ws2}} = \frac{1896}{1951} = 0.972$$

$$H = \frac{R \times H_{w1} - r \times H_{w2}}{R - r} = \frac{1131 \times 902 - 0.972 \times 841}{1.131 - 0.972} = 1275 \text{ kJ/kg}$$

This method of estimating the well enthalpy can be considered as an approximation (± 50 kJ/kg at best) for most conditions, since the accuracy of analyses and pressure measurement is normally not sufficiently precise.

8.5.7. Superheat

When testing "dry" steam wells, it is important to determine if the flow is in fact dry or saturated. Temperature and pressure at the metering point should be measured simultaneously and the temperature compared with the saturation temperature after allowing for the effect of noncondensable gas. Superheat may be identified by observing the discharge from a sampling point. Saturated steam flowing out of the sample point will be initially transparent, and after a few centimeters it expands, cools and a mist of water droplets forms, producing normal "steam clouds." Superheated steam is transparent for some distance from the discharge point, and in many conditions a cloud of condensed steam will not be observed. Where the "steam" is wet, the flow will have no transparent portion and the plume is translucent at the exit point.

8.6. CYCLING WELLS

Some wells will not flow under stable conditions. They may have a regular or irregular cycle in which mass flow, enthalpy, and WHP vary periodically, while the flow control valves are all at constant setting. There may be a smooth cycle, or the well may operate steadily for most of the time, interrupted by surges. This variable flow is commonly referred to as "cycling" and can result in difficulty with well control and flow measurement. Some wells can also cycle when shut, although this is not a production management problem.

Cycling is normally caused by the presence of two significant feeds of different enthalpy and permeability. One of the feeds has poor permeability such that it cannot sustain continuous flow. Figure 8.9 illustrates one such case where cycling is common. In this case the reservoir has a steam cap, a

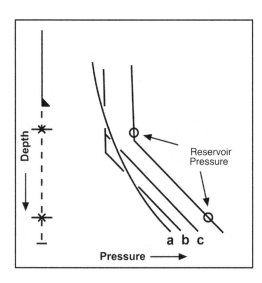

FIGURE 8.9 Pressure profiles in cycling well.

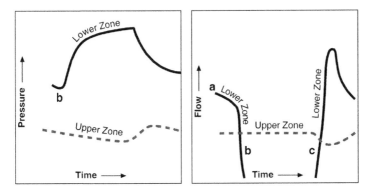

FIGURE 8.10 Pressure and flow profiles and upper and lower feeds on cycling well.

vapor-dominated zone overlying a liquid-dominated zone, and the well has a feed in each zone. The liquid feed has relatively poor permeability. Figure 8.9 shows pressure profiles in the well at three phases of the cycle, and Figure 8.10 shows the history of pressure and flow of each zone.

In phase (a) of the cycle, both zones are flowing. The lower zone draws down and the two-phase column between the two zones collapses, starting phase (b), where only the upper zone is discharging and the well discharge is dry steam. Pressure at the lower zone recovers until the water level rises to the upper feed zone at (c). Water is then entrained into the discharge, initiating

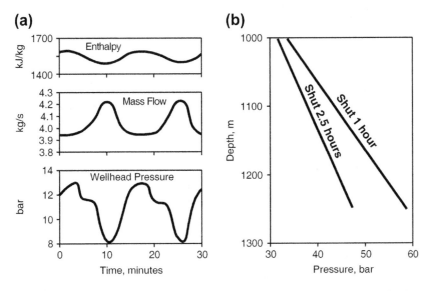

FIGURE 8.11 Well BR14, Ohaaki. (a) Output cycle. (b) Downhole profiles after shutting. *Source: Contact Energy, Personal Communication.*

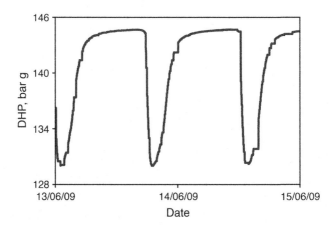

FIGURE 8.12 Cycling downhole pressure in shut well. *Source: Rotokawa Joint Venture, Personal Communication.*

discharge of the lower zone. The liquid column in the wellbore flashes, the wellbore unloads, and the cycle continues. The discharge of the well is dry steam during part of the cycle, with water added during the period when the lower feed flows, so the discharge is steam with periodic slugs of water. The mechanism driving the cycle is the oscillation in the pressure gradient in the wellbore between a liquid column and a flashing two-phase column. This is coupled with oscillatory flows at each feed. Low permeability in one feed is needed so that flow stops. If both feeds have high permeability, there is simply a sustained discharge from both feeds. The form of the cycle varies with the well control.

Cycling is also possible without the presence of a steam zone—for example, a well in a liquid-dominated reservoir with a deep, high enthalpy, low permeability feed that discharges intermittently. Figure 8.11 illustrates such a well, which cycles while flowing and for a period of time after shutting. The figure shows the form of the cycle and pressure profiles measured after shutting that show two different phases of the cycle. When shut 1 hour, there is water in the well, but at $2\frac{1}{2}$ hours there is a two-phase column. Cycling has also been observed in a well with nearly dry discharge. Figure 8.12 illustrates a well cycling when shut, with probably intermittent upflow from a deep, high enthalpy, poor permeability feed. Note how the pressure record looks like a succession of pressure buildups and drawdowns because a feed zone is successively stopping and starting.

8.7. ACCURACY OF FLOW MEASUREMENTS

Testing geothermal wells to obtain accurate and reliable data remains more of an art rather than a standard engineering exercise. For example, a test may be performed where accurate data are obtained, but if the well flow conditions are not stable, the test data may be of little value in evaluating well performance. The accuracy and reliability of flow measurements are affected by many factors, the most important being test method, test equipment design, instrumentation, test procedures, and well characteristics. Some of these factors can be evaluated in an engineering sense, such as calibration of the instrumentation and design of primary metering devices, whereas other factors are difficult to quantify. To establish a traceable set of measurement data, a field check sheet is required that lists all the pressure-temperature-flow measurement points and instrument calibration records, with their condition before (possibly during) and after a well test. Then during the test, all times of flow and test equipment adjustments must be noted, together with any instrument changes in order that any discrepancies found in the data after the test may be checked.

8.7.1. Instrumentation

In the environment found around most geothermal fields where there is significant gas, particularly H_2S, moisture from steam condensation and chloride water, high-quality instrumentation and good maintenance procedures are essential to obtain reliable test data. Most instrumentation will consist of pressure and differential pressure gauges (and temperature for steam-producing wells). Mechanical pressure gauges should be maintained at least to standard ANSI B40.1M, class 2A ($\pm0.5\%$) and preferably to class 4A ($\pm0.1\%$). Differential pressure sensors should have an accuracy of better than $\pm1\%$. Greater accuracy is not normally needed, since there is nearly always some degree of short-term fluctuation in pressures due to the two-phase fluids being measured. All instrumentation should have calibration checks before and after each major well test. Calibration records for each instrument must be maintained so the basic field data can be reviewed should problems be found once a test has been completed.

8.7.2. Test Procedures

The design of well test procedures can have a big influence on the reliability of test data. As a general rule, the longer the test period, the more reliable the test data. However, with tight environmental and time constraints, this is often not possible, so the well characteristics must often be determined from a brief flow test lasting only a few days or even hours. Geothermal wells generally fall into two categories:

- Liquid-fed, highly permeable wells that stabilize within hours of changing their flow conditions
- Wells with lower permeability and those producing from two-phase and steam reservoirs, which may never reach constant flow conditions

As a general rule, the throttle conditions at the wellhead are adjusted as little as possible during the test to allow stable (constant flow or constant rundown)

TABLE 8.4 Accuracy of Flow Measurements

Method	Good Test Control		Average Test Control	
	H error, kJ/kg	W %	H	W %
Single-phase orifice		3	10	
Sharp-edged weir		3	10	
Lip pressure method	20	5	50	10
Separator	10	3	30	5
Calorimeter	10	3	30	5
Tracer	20	5		

conditions to develop but at the same time providing sufficient spread of flow-pressure data at four or five different throttle settings. Other essential data from fluid chemistry sampling, noncondensable gas, and downhole measurements also need to be incorporated into the test program to fully test the well and the local reservoir. If the time available to maintain continuous well flow is very limited (because of environmental or other time constraints), it is best to obtain a set of reliable production data at one reasonably stable wellhead pressure, together with a flowing PTS profile, and use wellbore simulation to compute performance at other flow rates.

8.7.3. Accuracy of Individual Test Methods

Where care is taken to design and properly operate a well test, results of surprising apparent accuracy can be obtained. Often the test data is self-checking to some degree. For example, in liquid-fed wells, downhole temperature and pressure data with the well flowing will identify the inflow temperature and the pressure at which boiling commences. The flowing enthalpy at the wellhead can then be checked against these values. Chemical data can also be used to cross-check against feed zone temperature and flowing enthalpy, especially for liquid-fed wells. Where a deliverability/output test is done, the flow-WHP values will normally fall on a smooth curve (for example see Figure 8.8), and any significant deviations from this are an indication of unstable conditions (usually insufficient time being allowed for the wells to stabilize between adjusting throttle conditions). This is also the case with flowing enthalpy values, which should be almost constant for liquid-fed wells. In fact, the variation for enthalpy for liquid-fed wells over a range of flow-pressure combinations provides a practical indication of the accuracy and reliability of the test data; this is not true for two-phase and multiple feed wells. Based on experience with the various well testing methods in New Zealand, accuracies are assessed as shown in Table 8.4.

8.8. CALCULATING WELL PERFORMANCE

If the location, reservoir pressure, fluid enthalpy, and productivity of each production zone are all known, the production well performance can be calculated. There are two simple cases where the fluid in the well is single-phase only and the normal and more complex case where there is two-phase fluid in part or all of the wellbore during flow. In the following two sections, explicit formulae are given for pressure drop in single-phase wells with a single feed. Multiple feeds can also be calculated, but it is usually simpler to use a wellbore simulator.

8.8.1. Single-Phase Liquid

If the fluid in the well is liquid and does not reach boiling point, the pressure drop calculation is the same as for an injection well, with appropriate sign changes. This arises in low-temperature artesian wells or in pumped wells where there is liquid at wellhead. For pumped wells, the pump is set deep enough to prevent any boiling when the well is flowing. The pressure at the pump for a well with a single feed is given by:

$$P = P_r - \rho_w g Z - W/PI - C W^2 \tag{8.23}$$

where P is the pressure at the pump, Z is the depth from the pump to the feed, and:

$$C = \frac{8}{\pi^2 \rho_w} \sum l_i f_{Mi}/D_i^5 \tag{8.24}$$

where the sum is over the different casing sections, starting at the depth of the pump.

8.8.2. Single-Phase Steam

This case is very similar to single-phase water except that allowance must be made for the variation of steam density with pressure. Note that it is only required that there be single-phase steam in the wellbore. Reservoir fluid may be two-phase as long as there is no significant liquid in the discharge. The equation for the pressure gradient is:

$$\frac{dp}{dz} = \left(\frac{dp}{dz}\right)_{static} + \left(\frac{dp}{dz}\right)_{fric} \tag{8.25}$$

$$= \rho_s g + 8 f_M W^2 / \pi^2 \rho_s D^5 \tag{8.26}$$

The density of steam is given by the imperfect gas law, which can be written as:

$$P = K\rho_s$$
$$K = ZR(T + 273)/M_s \qquad (8.27)$$

and K is approximately constant for saturated steam and is 2×10^5 Pa.m³/kg at 240°C. Multiplying Eq. (8.26) by P gives a differential equation for P^2, which integrates to give:

$$P_{wf}^{\,2} + A = (P_{WH}^2 + A)e^{\alpha z} \qquad (8.28)$$
$$A = 8Kf_M W^2/\pi^2 D^5 \qquad (8.29)$$
$$\alpha = 2g/K = 2gM_s/(ZR(T + 273)) \approx 10^{-4} \text{ m}^{-1} \qquad (8.30)$$

where P_{wf} is the downhole pressure at the steam entry. If there is no flow, the static pressure profile is given by:

$$P = P_{WH}e^{\alpha z/2} \qquad (8.31)$$

The effect of drawdown can be included by the method of Acuña (2008), Acuña and Pasaribu (2010), and Peter and Acuña (2010). Assuming a steady-state drawdown, described in terms of pressure-squared, the drawdown is:

$$W = PI'(P_f^2 - P_{wf}^{\,2}) \qquad (8.32)$$

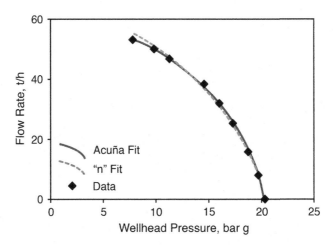

FIGURE 8.13 Flow of well WK216, Wairakei. *Source: Contact Energy, Personal Communication.*

where $PI' = 2PI/P_f$. Then, combining these equations and observing that αz is small, we get:

$$P_{WHO}{}^2 - P_{WH}{}^2 = W/PI' + C'W^2 \tag{8.33}$$

$$C' = 16g \sum l_i f_{Mi}/\pi^2 D_i^5 \tag{8.34}$$

where the sum is over the different lengths of casing. This is similar in form to the equation for single-phase pressure drop, (Eq. (8.23)), with the use of pressure-squared rather than pressure.

Performance of gas wells is often represented by the empirical formula (Sanyal et al., 2000):

$$W = C(P_{WHO}{}^2 - P_{WH}{}^2)^n \tag{8.35}$$

where n is a constant usually between 0.5 and 1. Eq. (8.33), when productivity is low, corresponds to $n = 1$, and when productivity is high and wellbore friction dominates, it corresponds to $n = 0.5$. In practice, the two methods produce fits that are equally good. Figure 8.13 shows the flow of well WK216 at Wairakei, which produced dry steam from the steam cap, fitted to both formulae. The respective fits are

$$P_{WH}{}^2 = 453 - 2.40 \times W^2 - 0.087 \times W$$
$$W = 0.85 \times (453 - P_{WH}{}^2)^{0.704}$$

Because the two fits are equally good at describing the actual performance, Eq. (8.33) is preferable since the coefficients relate to measurable physical parameters. The inflow is in fact not entirely steady but does change slowly with time. The transient changes in long-term flow can be used to infer transmissivity (Enedy, 1987).

8.8.3. Two-Phase Fluid in Wellbore

Flowing high temperature, geothermal wells contain a two-phase mixture of steam, gas and water in all or part of the wellbore. Two-phase flow through pipes occurs in many situations, from nuclear reactors and boilers to petroleum and geothermal wells. There is no precise mathematical model of two-phase flow in pipelines, but a range of empirical relationships has been developed.

The fluid mixture distributes itself differently in different "flow regimes," depending on the ratio of the vapor and liquid fluxes and the velocities. With only a small amount of vapor, bubbles of steam are present through the fluid column. With an increased fraction of steam, there are slugs of steam and liquid, then a transition to the final regime, which is mist or annular flow, where the center of the pipe contains vapor with a mist of liquid droplets and there is a continuous water phase coating the walls.

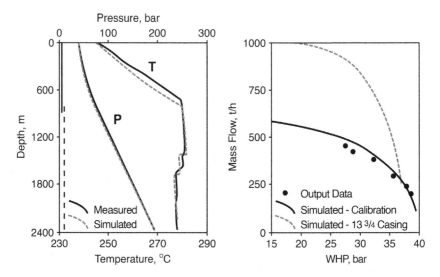

FIGURE 8.14 Flowing profile and output curve match by wellbore simulator. *Source: Aunzo, Z.,; Rotokawa Joint Venture, Personal Communication.*

One of the consequences of this range of regimes is that there is a minimum flow needed to sustain discharge in an underpressured reservoir. If the well is steadily throttled from a large flow, WHP rises and flow rate falls until a maximum discharging pressure (MDP) is reached. Further throttling of the well results in a lower WHP and possible collapse of the discharge. Basically, in this state, the flux of steam up the wellbore is barely sufficient to keep the steam and water mixed. If the flow is reduced any further the flow regime changes to a continuous column of water with steam bubbles dispersed in it, the pressure gradient also changes to that of liquid water increasing the overall density of the fluid column and fluid no longer reaches the surface. Figure 8.5 illustrates this situation on well RX1, where the "lower branch" of the output curve is shown.

Another consequence of the form of two-phase flow in the well is that well performance is sensitive to discharge enthalpy. Increased enthalpy means a higher steam fraction in the two-phase column in the wellbore and thus a lower-pressure gradient up the well and higher WHP. Gas content of the fluid can also significantly affect well performance because the gas causes boiling at a higher pressure—that is, deeper in the well—and gives a higher vapor fraction in the two-phase fluid, which is essentially some gas-lift to the discharge.

Calculating discharging pressure profiles with two-phase fluid is best done using a wellbore simulator, which contains empirical approximations for the pressure gradient. A number of these are available, such as HOLA & GWELL (Aunzo et al., 1991); WELLSIM (Gunn & Freeston, 1991); GEOFLOW (Acuña, 2008); WELBOR (Pritchett, 1985); and SIMU000 (Upton, 2000).

A wellbore simulator should be able to represent a well with a number of specified feed depths with inflow or outflow, with fluid temperature or quality specified for each inflow. Usually it also calculates wellbore heat transfer. Figure 8.14 shows a typical match. The well has a $9^5/_8$ inch casing and a $7^5/_8$ inch liner. Well performance is matched with match to a flowing profile and to an output test. There are four feed zones, three of which are visible in the temperature profile. The fourth zone at the casing shoe is very small. Note that the temperature profile is not quite isothermal between zones, but the water cools a little as it rises. This is because of the adiabatic (isenthalpic) expansion of the liquid as it rises up the wellbore. There is no wellbore heat transfer cooling the flow because downhole temperature is very close to reservoir temperature. Having made the match, the wellbore simulation was used to estimate the well performance if a a larger-diameter well had been drilled. This is a common use of wellbore simulators to calculate the benefits of different well completion designs.

8.9. INTERPRETATION OF OUTPUT DATA

8.9.1. Introduction

At the completion of the flow testing, the stable or quasi-stable well performance has been measured at a range of WHPs and perhaps some transient changes in flow. From the way that mass flow, enthalpy, and perhaps chemistry vary with WHP, some inferences can be made about the reservoir. In a reservoir of liquid water at constant temperature and gas content, variations in output are controlled by only variations in pressure at wellhead and in the reservoir, and similarly for a dry steam well. In all other cases the performance of the well is sensitive to the enthalpy, and variations in mass flow and enthalpy must be considered together.

8.9.2. Maximum Discharging Pressure

The well flow depends on pressure and enthalpy. At low flow rates, the resistance to flow in the reservoir and the wellbore can be relatively unimportant, and the flow depends only on the pressure and enthalpy of the fluid entering the well. If the reservoir is normally pressured, temperature is the only variable.

The MDP of the well is the maximum pressure that can be attained at wellhead with the well flowing. If the well is progressively throttled from a large flow, WHP increases until it reaches a maximum and a nonzero flow. Further throttling causes both pressure and mass flow to decrease. There is a minimum flow needed to keep the two-phase column in the wellbore sufficiently well mixed. Observation shows that there is a simple correlation between MDP and the temperature of the source fluid for reservoirs where the pressures are approximately hydrostatic from ground level (James, 1970, 1980a, c):

$$T = 100\,P^{0.283} \tag{8.36}$$

where the temperature is in degrees Celsius and the pressure is in bar absolute.

8.9.3. Mass Flow

Figure 8.15 illustrates some of the common variations in the form of the output curve of mass flow against WHP. It is assumed in these cases that enthalpy and gas content do not vary greatly with WHP. Curve A is the basic curve, representing the results from a water-fed well of high permeability. The MDP is determined by the feed temperature, and the maximum flow is controlled by the well design—by friction in the perforated liner and casing. Curve B shows the effect of a fall in reservoir pressure, and curve C shows the effect of a rise in reservoir pressure. Curve D shows the effect of scaling in the wellbore, and curve E shows the effect of lower permeability. In both cases there is little effect at small flow (high WHP) but additional resistance at higher flow. The additional resistance due to scaling is proportional to the square of the flow, whereas the resistance due to lower permeability is linear. Curve F results when the feed is two-phase (at the same pressure as for curve A). The higher enthalpy results in higher pressure at low flows, but greater specific volume gives more friction at larger flows.

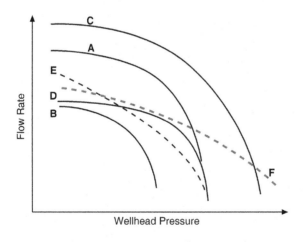

FIGURE 8.15 Output curves: form of the variation of mass flow with wellhead pressure.

8.9.4. Enthalpy Variations

The discharge enthalpy, and any variation with flow rate or WHP, defines the type of fluid feeding the well. Figure 8.16 shows the main possibilities for the variation of enthalpy with WHP. The enthalpy is plotted with respect to a scale between saturated water and saturated steam and the main feedpoint.

Curve A is the least complicated. The enthalpy is that of liquid water, which enters the well at constant temperature at all flow rates. This corresponds to

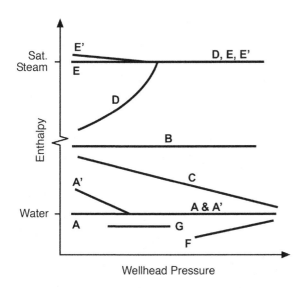

FIGURE 8.16 Output curves: form of the variation of enthalpy with wellhead pressure.

a feed of liquid water at constant temperature. Curve A' also corresponds to a liquid water feed, but at some flow rate there is enough drawdown to start flashing in the formation and a little additional enthalpy is gained.

Curve E is also straightforward. Dry saturated steam enters the well at all flow rates. The reservoir contains steam and immobile water. Curve E' is obtained when the well reaches dry-out. Below dry-out the steam entering the well is superheated. To definitely identify superheated conditions, it is necessary to allow for the wellbore heat transfer and the expansion up the wellbore.

Curve B corresponds to a two-phase reservoir of high permeability. The fluid entering the well is a two-phase mixture at all flow rates, with an enthalpy above liquid water at the measured reservoir conditions. Curve C is more common. It corresponds to a two-phase reservoir of lower permeability. At higher flow rates there is more drawdown, and this results in some heat transfer from rock to fluid.

Curve D represents a well with two entries: a steam entry in a vapor-dominated zone and a deeper liquid entry. At high WHP the well discharges only steam from the upper zone. As the wellhead is lowered and flow rate is increased, there is more drawdown and water from the lower zone is entrained into the discharge. Such a well is often unstable near the point where water first enters the discharge.

Curves F and G illustrate hidden cold entries. In curve G the downhole temperature has been wrongly identified. The water produced comes from a cooler zone, so perhaps a flow in the well has disguised the true reservoir temperature. Curve F is the case where the well produces water at the expected temperature at low flow rates, but at higher rates increasing amounts of cooler water are mixed into the discharge, so there is a hidden cold feed in addition to the expected hot feed.

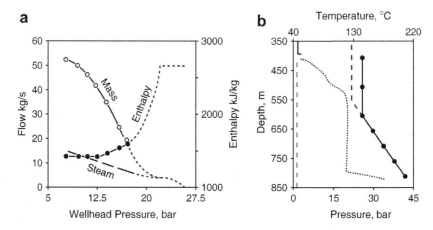

FIGURE 8.17   Well WK215, Wairakei. (a) Output test. (b) Downhole profiles. *Source: Contact Energy, Personal Communication.*

Example 12: Figure 8.17 shows measurements in well WK215. It is a well of type D, producing steam only at WHPs above 22 bar. The mass flow was not measured above 18 bar, but it was observed that it was dry. Note that the mass flow varies smoothly only within each type of discharge. Some downhole measurements are shown. Pressure profile P2 (solid line with dots) is a static profile and approximates the reservoir pressure. A steam cap is present. P3-(dashed line) is a constructed profile with the well flowing only steam at WHP 22 bar. The water level is at 530 m. Lowering the WHP brings water into the discharge, so this identifies 530 m as the bottom of the steam feed. Temperatures T1 (dotted line) under injection in the completion test show permeability at 425−550 m and at 800 m. Thus, both downhole data and output data show that the well produces from two zones, The upper zone at 425−530 m is in vapor-dominated conditions and the lower zone at 800 m is in liquid.

Case Study: A History of Well BR2, Ohaaki

9.1. INTRODUCTION

Well BR2 has been chosen for this case study of the drilling, measurement, discharge, and data analysis/interpretation of a geothermal well. It has exhibited many of the possible behavior modes of a geothermal well, and the comprehensive discharge and downhole measurements obtained during its lifetime make it a particularly good example. Naturally, not all possible behavior modes are displayed by this single example, nor is it intended that BR2 be presented as typical. The case history illustrates the practical application of the techniques of well measurement and interpretation that have been presented in the previous chapters.

Ohaaki geothermal field is located 20 km north of Wairakei, in the Taupo Volcanic Zone of New Zealand. Its history is described by Clotworthy and colleagues (1989, 1995) and Newson and O'Sullivan (2001). The first well, BR1, was drilled in the center of the resistivity anomaly, and although it encountered high temperatures, the permeability was insufficient to support a sustained flow. BR2, drilled in 1966, was the discovery well. When first discharged, it had an electric power potential of about 14 MW and at that time was the largest producer in New Zealand. The success of BR2 led to further exploration drilling and well testing in the Ohaaki area, and from 1966 until

Geothermal Reservoir Engineering. ISBN: 978-0-12-383880-3 DOI: 10.1016/B978-0-12-383880-3.10009-5
169

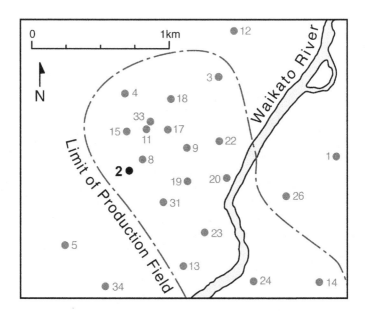

FIGURE 9.1 Ohaaki geothermal field. *Source: Contact Energy, Personal Communication.*

1971, 23 more wells were completed. In 1971, natural gas from the Maui field became the favored energy source for new electricity generation, all geothermal development in New Zealand was terminated, and the Ohaaki wells were "mothballed."

BR2 and other production wells had been discharged as they were completed with a sustained flow test of the western part of the field equivalent to about 30 MWe from 1968 to 1971 (Hitchcock & Bixley, 1975). This production testing resulted in a decline in reservoir pressure in the western part of the reservoir by up to 17 bar by the time of field shutdown.

With the energy crisis of the mid-1970s, geothermal development resumed. At Ohaaki, well drilling and testing restarted in 1974, and flow testing continued at a lower rate through 1988, by which time deep reservoir pressures had recovered to about 7 bar below the undisturbed values (Figure 9.2). In 1988, the Ohaaki 110 MWe power station was commissioned. BR2 was connected to the steam-gathering system as a production well from 1988 until 1996, when it ceased production due to cooling and was abandoned. Figure 9.1 shows the West Bank production area at Ohaaki with BR2 and its neighboring wells.

9.2. THE DRILLING AND TESTING PERIOD: MAY–AUGUST 1966

Figure 9.3 shows the heatup temperature-pressure logs of BR2. After cementing the anchor casing at 301 m, the well was drilled to 560 m using mud

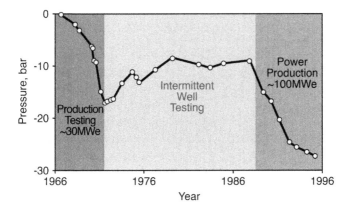

FIGURE 9.2 Pressure change in the deep liquid reservoir at Ohaaki 1966—1995. *Source: Contact Energy, Personal Communication.*

with gel plugs to seal off circulation loss zones. In the interval between 300 and 560 m, several such losses occurred: 430 m temporary loss, gel-plugged, and cemented; 440 m temporary circulation loss and drilling break of 0.3 m; 500 m temporary loss; and at 510 m partial loss of circulation. At this stage it was decided to set the production casing above these loss zones, and the wellbore was then filled with gel, a cement plug was placed at 420 m, and the $8^5/_8$ -inch production casing was cemented at 417 m. While drilling through the gelled section between the casing shoe at 417 m and the previous drilled depth of 560 m, there were no further signs of permeability, and the remainder of the hole was drilled, using mud, to the total depth of 1034 m. Below 560 m there were further partial circulation losses at 660 m and at 780 and 810 m, where the mud returns were heavily flocculated and gas-cut. After setting the perforated liner on the bottom, the mud was circulated out of the hole, and during this process the well came under pressure of 5—6 bar g while circulating with water.

9.2.1. Completion Tests and Measurements

The completion test indicated only moderate permeability. While injecting 24 kg/s, the wellhead pressure increased by 13 bar, indicating an injectivity of 1.8 kg/s.b. The completion test and heatup temperature-pressure profiles are shown in Figure 9.3. The two temperature profiles measured while injecting cold water at 12 kg/s, T1 and T2, indicate the major permeable zones are above 800 m and there is little permeability below 900 m, since the temperatures from T1 and T2 (measured 3 hours apart) show minimal change below this depth. Heating over the first 22 days after completion (runs T3, T4, and T5) shows a well-developed heating peak over the 450—600 m interval, and T5 indicates an interzonal flow is present between 420 and 610 m.

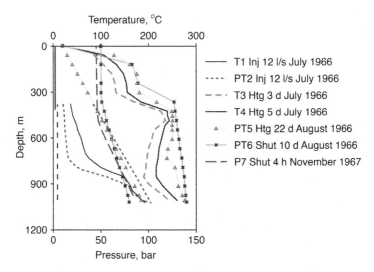

FIGURE 9.3 Pressure-temperature profiles in BR2 1966–1967. *Source: Contact Energy, Personal Communication.*

During the cold water injection test made on well completion, the wellhead pressure had stabilized at 4–5 bar g, and the P2 pressure profile plotted in Figure 9.3 was calculated from the stable wellhead pressure and measured temperatures in T2. During the initial heatup period, the wellbore remained completely filled with liquid, with wellhead pressure around 10 bar g, as shown on run P5. After one month of heating, the wellhead pressure suddenly increased over two days from the previous "stable" value about 10 bar g to about 50 bar g. No downhole runs were made with the well in this condition, and the well was opened for the first time a few days later.

9.2.2. Initial Discharge

This initial discharge was large. On full-open vertical discharge, a wellhead pressure of over 10 bar g was continuously sustained. It was estimated that more than 5 m³ of rock particles were ejected during the two-hour discharge. Normally, less than 1 m³ of drilling debris is ejected on first discharge. The ejecta contained quartz and platey calcite, which petrological examination suggested was derived from fissured rock between 420 and 510 m. These ejecta helped explain the poor permeability indicated by the injection test. At that time the fissures in the main feed zones were blocked with drilling waste, and the completion test injectivity of 1.8 kg/s.b reflects the performance of the well in this damaged state.

After the first discharge, the shut-in wellhead pressure stabilized at about 50 bar g. Downhole measurements 10 days later (PT6) showed that the wellbore now contained an interzonal two-phase flow up the wellbore. This is indicated by the low-density fluid from 380 m to bottomhole, with gas above 380 m (since temperatures are far below saturation for the measured pressures in this interval, the fluid in the wellbore must be gas rather than steam). This rapid accumulation of gas in the cased part of the wellbore indicates that that the upflowing, boiling fluid below 380 m contained significant noncondensable gas. In spite of the apparently poor permeability below 800 m, the low-density column extended to bottomhole. This implies that a small inflow was present near bottomhole, with the bulk of the interzonal flow deriving from above 800 m.

9.2.3. Interpreting the Early Measurements

The sudden rise in wellhead pressure as the well heated can be explained by considering the changes in the deep wellbore fluid. While the well temperatures were relatively cool, the pressure gradient in the fluid column was above the formation pressure gradient and the water could only flow downward. Temperature profiles T3 and T4, measured prior to discharge, indicate inflow at 450 m, flowing down the well to about 850m. When the fluids had heated sufficiently and the fluid pressures in the wellbore were balanced or slightly underbalanced with respect to the formation pressures in the lower part of the well and an upflow could initiate as shown by PT6. Once this upflow began, hotter fluids were produced from the deeper feed zones, and the fluid could boil as it flowed up the wellbore with decreasing pressures. This boiling further decreased the fluid density in the upflowing column, increasing the pressure difference between wellbore and formation to create the strong upflow of the interzonal flow. The boiling upflow then disguised reservoir temperatures in the complete wellbore (except to say that the true formation temperatures below 1000 m must be—slightly—greater than the measured value of 280°C). Profile T5 after 22 days of heating, the last temperature measurement obtained before the interzonal flow began, shows a maximum of 252°C at 450 m. The stabilized temperatures at this upper feed zone are likely to be close to this value of 252°C.

So far, the measurements in the well indicate permeability at 450–600 m and near 800 m. Pressure profile P5 is probably close to the reservoir profile (between 400 and 1000 m depth). Pressure profile P5 is the first to reflect actual formation pressures. From this time (August 1966) until P11 in May 1971 (Figure 9.8), the interzonal flow disguises both the true formation pressures and temperatures. Many other pressure-temperature profiles were measured over this period, and almost all showed two-phase boiling upflow conditions. These are not plotted here. Comparing P5 when the wellbore is liquid-filled with P6 when the interzonal flow present, it can be seen that the drawdown at the deeper feed zone is about the same as the overpressure at the upper zone. That is, there is about the same pressure drop drawing fluid into the wellbore as there is overpressure injecting it from the wellbore into the upper feed zone, so the two zones are of similar permeability. The interzonal flow identifies two-phase fluid in the reservoir at the lower feed. The fluid in the upper feed is, so far, unidentified, although the maximum measured temperature of 252°C at 450 m is also close to boiling for the reservoir pressure at that time. Formation temperatures at 1000 m must be at least 280°C.

9.2.4. Output Tests

The performance of the well was measured by two output tests in August and October 1966 using a sampling calorimeter (Bixley et al., 1998). Flow and enthalpy data from the two tests agreed within the measurement error (and are plotted as a single test on Figure 9.4). A maximum flow of over 100 kg/s was obtained, and the well stabilized quickly after changes in the throttle condition. The enthalpy of the discharge was 1200–1270 kJ/kg, equivalent to that of liquid water at 270–285°C. These temperatures are attained within the hole, but only below the upper permeable zones where temperatures are about 250°C. The lower zone is already known to be two-phase, and the upper zone is very probably also two-phase, since the enthalpy is always significantly above liquid at the upper zone temperature.

With the average enthalpy measured during this test of 1230 kJ/kg and flow of 100 kg/s, the separated steam flow would be almost 30 kg/s, capable of generating about 14 MW of electricity. The quick stabilization of both the well pressure and flow rate despite two-phase feeds indicates high permeability, as does the large flow. In these tests, neither the enthalpy nor the well cycle varied with flow rate or showed any instability. The absence of enthalpy variation indicates that excess enthalpy is due to two-phase conditions in the undisturbed reservoir rather than being produced by the boiling of water near the well due to

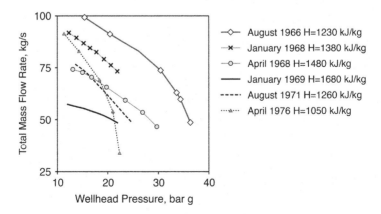

FIGURE 9.4 Output tests 1966–1976. *Source: Contact Energy, Personal Communication.*

the drawdown. In that case the excess enthalpy is expected to increase with increasing flow/drawdown at the feed zone.

9.3. THE DISCHARGE PERIOD: 1966–1971

At the time BR2 was being tested, the Wairakei field had been in production for several years, and a significant increase in the discharge enthalpy had been observed in some groups of production wells. This change had rendered unworkable the "hot water scheme" designed to recover additional steam from waste-separated water (Thain and Carey, 2009). There was also concern with the relatively high noncondensable gas content at BR2 and the potential effect of this on power generation and of possible calcite scaling on wellbore performance. While wellbore scaling had not been a serious issue at Wairakei, severe scaling had been evident at another field, Waiotapu, being explored at that time. It was recognized that reliable prediction of enthalpy would be critical in design of a multiple flash steam gathering system for Ohaaki, and this led to the sustained production test carried out from 1967 until 1971 using BR2 and other wells as they became available. The total production rate was equivalent to about 30MWe from 1968 until 1971.

For modern geothermal developments, reliable prediction of production enthalpy is very important for the design of efficient steamfield plant and separated water/reinjection systems, since relatively small changes in the average enthalpy of the produced fluid leads to large changes in the quantities of separated water that must be injected. For example, if the average production enthalpy changes from 1200 kJ/kg to 1600 kJ/kg, then assuming a separation pressure of 5 bar g, the fraction of separated water available for injection changes from 75% to 55% for the same separated steam flow. This impacts on the injection system design in terms of the number of wells required, pipeline sizes, and associated holdup time in the injection system to minimize silica scaling.

Figure 9.5 shows the discharge history of BR2 from 1966 to 1971. From August 1966 until June 1971, almost all of the pressure profiles measured in BR2 showed two-phase conditions with a significant interzonal flow similar to PT6 (Figure 9.3). Pressure changes in the deep reservoir over this period (see Figure 9.2) are derived from nearby wells where only liquid conditions were present (usually in the few weeks after well completion, since many of the wells developed interzonal flows when they had fully heated after drilling).

9.3.1. Further Output Tests

In order to obtain reliable and accurate production data from BR2, a separator was installed in 1967, and in January 1968 another output test was carried out. This test showed that the enthalpy had now increased nearly 1400 kJ/kg, significantly above that of liquid water. The enthalpy again varied little with flow rate. The enthalpy increase over the previous tests must therefore be due to

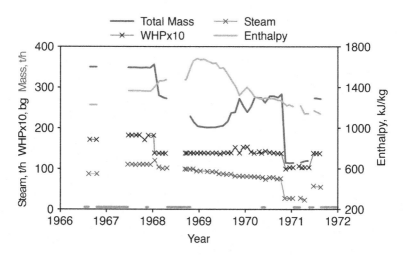

FIGURE 9.5   Discharge history of BR2, 1966−1971.

increased steam fraction in an extended area of the reservoir, rather than changes in this steam fraction only in the immediate vicinity of the well. The mass flow has reduced mainly because of the enthalpy rise and consequent greater specific volume of the discharge fluid. This behavior has now been observed in many production wells that feed from reservoir fluids at or close to boiling point conditions. The separated steam flow remains constant or sometimes increases slightly, while the separated water flows decline as the enthalpy of the total produced fluid increases. Reservoir pressure showed little change at this time (see Figure 9.2), but the flowing enthalpy of other nearby wells was also increasing.

In early 1968, a series of discharge tests and downhole surveys on BR2 and surrounding wells was carried out. The April 1968 output test on BR2 showed a continuing rise in enthalpy to 1480 kJ/kg and a consequent further decline in mass flow. During this test, some pressure transients were also recorded downhole at 1021 m, near bottomhole (rather than closer to the main feed zones at 450m). Figure 9.6 shows some of the interpretative results: (a) an MDH plot of the drawdown when opened to a flow of 46 kg/s and (b) a Horner plot of the recovery when shut from a flow of 61 kg/s. The pressure chart record was fairly clean, with some small oscillations present when the well was opened or closed but not at increases in flow rate.

The recovery section of the record (Figure 9.6b) shows severe humping. Downhole profiles show why this happens. Although a low-density two-phase column in the wellbore is the stable state when the well is shut for long periods and during discharge, a liquid column may be present for a short period after being shut. This is clearly indicated by the pressure profile P7 in Figure 9.3, which shows liquid density gradient below 500 m, in contrast to the more normal "shut-in" two-phase profiles being measured about that time—for

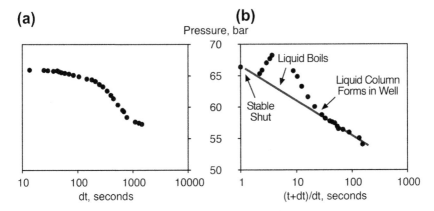

FIGURE 9.6 (a) Drawdown when opened to 46 kg/s, (b) recovery from 61 kg/s. *Source: Contact Energy, Personal Communication.*

example, P6. There is an initial linear segment on the Horner recovery plot that projects toward the stable shut pressure at this depth in the well. This is assumed to be a usable straight line, and it has a slope of 5.3 bar/cycle. Allowing for the changes in density of the normal fluid column in the well, this is a slope of 4.3 bar/cycle at 500 m depth, implying

$$kh/v_t = 2.6 \times 10^{-5} \text{ m·s}$$

The enthalpy measured at all flow rates in the output test was 1480 kJ/kg. If the produced fluid comes from the shallow feed, where an undisturbed temperature of about 250°C applies,

$$\rho_t = 80 \text{ kg/m}^3$$

and so

$$kh/\mu_t = 3.3 \times 10^{-7} \text{ m}^3/\text{Pa·s}$$

Assuming that fracture flow relative permeabilities apply,

$$v_t = 0.31 \times 10^{-6} \text{ m}^2/\text{s}$$
$$kh = 8 \text{ d-m}$$

If Corey relative permeabilities are taken,

$$v_t = 1.3 \times 10^{-6} \text{ m}^2/\text{s}$$
$$kh = 34 \text{ d-m}$$

It is not possible to make a clear interpretation of the drawdown results because of the difficulty of distinguishing features of the response due to density changes in the wellbore from changes caused by permeability or reservoir structure.

9.3.2. Interference Test

During 1968, well BR8 was completed 145 m away from well BR2. BR2 was shut simultaneously with the opening of BR8, and downhole pressures measured in BR2 remained steady. When BR8 was shut, there was a response in BR2. Figure 9.7b shows an MDH plot of the recovery of BR2 at 500 m and 800 m.

The feed zone depths in BR8 are similar to BR2: It has a primary feed close to 500 m, and the two wells probably communicate at this depth. It is thus the recovery at this 500 m depth that best indicates the reservoir pressure change as transmitted to BR2. The changes at other depths in BR2 reflect the pressure changes transmitted from the primary feed down the fluid column in the wellbore. Observation of this recovery terminated after some 80 days when BR11, another neighboring well, was opened. Pressures in BR2 declined in response to the BR8 discharge, but the measurements were not sufficiently accurate to interpret quantitatively.

Taking the recovery at 500 m as the best indicator of the reservoir response, a straight line with a slope of 2.7 bar/cycle can be drawn on the plot (see Figure 9.7b). Given the error between repeat measurements, there is considerable latitude for drawing different lines. The flow rate of BR8 stabilized at about 50 kg/s, with a discharge enthalpy of about 1600 kJ/kg, although this varied with the flow rate. Using the same reservoir parameters as for the test on BR2, the slope of 2.3 bar/cycle gives

$$kh/v_t = 3.9 \times 10^{-5} \text{ m} \cdot \text{s}$$

$$kh/\mu_t = 5.7 \times 10^{-7} \text{ m}^3/\text{Pa} \cdot \text{s}$$

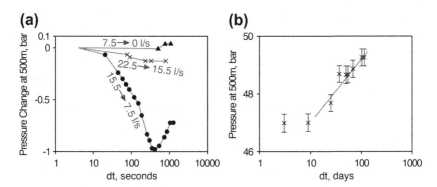

FIGURE 9.7   Pressure transients in BR2 measured at 500 m depth. (a) During injection, at three changes of flow rate. (b) During recovery, after shutdown of BR8. Error bars reflect a measurement error of 0.3 bar. *Source: Contact Energy, Personal Communication.*

Depending on the relative permeability functions, the permeability-thickness is 50 dm (Corey) or 15 dm (fractured system). Given the uncertainty in drawing the straight line and the variation in the enthalpy of the discharge from BR8, these results are a good match with the earlier analysis of BR2 tests. From the position of the straight line, the storativity can be obtained:

$$\varphi c_t h = 3.5 \times 10^{-5} \text{ m/Pa}$$

This value is sensitive to the uncertainty in the correct straight line. The value of the total compressibility, c_t, depends on both the composition and the purity of the two-phase mixture, and since BR2 and BR8 both discharge significant noncondensable gas, this dependence should be taken into account. As with BR2, the actual reservoir pressure and temperature in BR8 at that time were disguised by an interzonal flow. Based on the measured downhole pressure-temperature conditions, the partial pressure of CO_2 present in the shallow feed of BR8 could have been as much as 10 bar. Assuming a partial pressure of 5 bar and a 250°C feed temperature

$$c_t = 5.5 \times 10^{-7} \text{ Pa}^{-1}$$

which gives

$$\varphi h = 60 \text{ m}$$

Core data gave porosities in the range 0.2—0.4, the average being perhaps 0.25. Given the uncertainty in φ and φch, these results indicate that the aquifer joining BR2 and BR8 has a "hydraulic" thickness of about 200—300 m. Thus, although the wells feed through distinct fractures, they have, through these fractures and their extensions, access to the fluid stored throughout an aquifer 200—300 m thick. Despite the fracturing, as fluid is withdrawn by well discharge, this aquifer behaves on the gross scale like a homogeneous medium. In contrast, in 1977 during an injection trial, tracer tests showed rapid preferential returns from the shallow injection well BR33 to BR8 and BR11 (McCabe and Barry, 1977). These well locations are shown in Figure 9.1. By this time the fluid in fractures linking the shallow injection aquifer and the deeper production aquifer had cooled and filled with liquid.

9.3.3. The Period 1968—1971

From 1968 to 1971, development drilling continued and there was extensive discharge from the new wells. Over this time, reservoir pressures declined by up to 17 bar in the area near BR2, as shown in Figure 9.2, with resultant boiling where subsurface conditions were already close to boiling point. There were substantial changes in the discharge enthalpy of BR2. This rose and peaked in 1968, but thereafter declined steadily, as shown in Figure 9.5. This "enthalpy hump" is common in wells producing from two-phase reservoirs. For typical wells there is an initial enthalpy rise as discharge lowers pressures and causes increased boiling around the well. Ultimately, however, the well must draw on recharge at greater distance, and the discharge enthalpy then declines toward that recharge enthalpy. During the January 1969 output test, BR2 was discharging at its peak enthalpy. The trend of the tests up to that time continued: Mass flow was reduced because of higher enthalpy and reduced reservoir pressures, but high wellhead pressures were maintained because of the low density fluid in the well.

Gas content of the discharge was measured regularly from 1968 onward. It was relatively high and changed with time, generally following the trends in the enthalpy, and reached a peak value of 4% by weight in the total discharge in January 1969. The production history (see Figure 9.5) shows a sudden steep decline in mass flow in October 1970, with no corresponding changes and wellhead pressure or enthalpy. Go-devils run at this time showed the wellbore was blocked at 510 m, and a workover to remove the scale buildup was performed in May 1971. Scale samples recovered during the workover confirmed that calcite scale was the reason for the sudden decline in flow rate.

Following the workover, the well was reopened and the subsequent output test results showed the discharge enthalpy had declined to 1260 kJ/kg (see Figure 9.4, August 1971) and was now similar to that obtained in the first tests in 1966 (1230 kJ/kg). Because of this lower enthalpy, the maximum discharging pressure had declined, while the maximum flow had risen compared to the previous tests, due to the decreased enthalpy and associated higher-density fluid in the flowing two-phase fluid column. Even though the flowing enthalpy in 1971 was similar to the original flow tests, the total flow rate was smaller because the deep pressures were around 17 bar less than those when BR2 was first tested in 1966.

After 50 days of discharge, the well was shut, and a routine downhole survey later found the wellbore to be totally blocked at 550 m. A second workover to remove scale was performed in September 1971, and a completion test was performed immediately after. The injectivity now was about 30 kg/s per bar, and the temperature profile (T10, Figure 9.8) showed a step profile indicating an inflow at 420−500 m, with isothermal temperatures below this indicating flow all the way to a deeper feed zone near bottomhole. This analysis was confirmed by an experimental spinner profile measured during the completion test.

9.4. SHUTDOWN AND PRESSURE RECOVERY: 1971−1988

In August 1971, all wells at Ohaaki were shut. The subsequent downhole pressure profiles in BR2 and other nearby wells showed a gradual pressure recovery over the next few years, averaging about 2 bar per year. During this pressure recovery period, several changes, both qualitative and quantitative, were measured in BR2. The pressure profiles immediately before the 1971 field

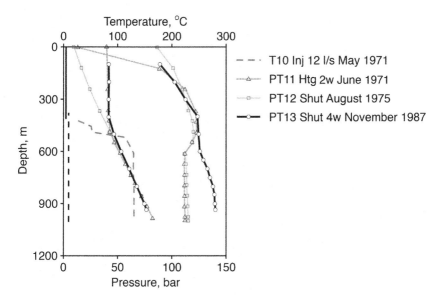

FIGURE 9.8 Downhole profiles 1971–1987. *Source: Contact Energy, Personal Communication.*

shutdown showed the well standing shut with a column of steam over a column of water (Figure 9.8, PT11). The steam-water interface is below the 420 m feed point, indicating the 420 m feed zone is steam-dominated at this time. This "steam cap" profile indicates subhydrostatic reservoir pressures with, in this case, a vapor zone present or forming in part of the reservoir connected to the 420 m feed. This single measurement in BR2 would not be significant were it not that other wells nearby also showed the same behavior at this time, and one of these wells, BR19, discharged a large flow of dry steam for a short period. Thus, a vapor zone was beginning to form in this part of the field and was extending to BR2. All subsequent pressures in BR2 show the section of well below the cemented casing full of water, indicating that the developing vapor zone quickly disappeared as the deep liquid pressures recovered from discharge, and an influx of cooler water into the boiling zone quenched it.

A second change now affected the well, which was caused by a "new" feed zone not previously apparent. The temperature profile (Figure 9.8, T11) has an isothermal section from 600 to 1000 m, indicating that the well had now developed a downflow. Water at about 220°C entered the well at 600 m flowing down to a deeper feed zone below 1000m. As just discussed, higher temperatures with two-phase conditions persisted at the 420 m feed zone, unaffected by the entry of the cooler water at 600 m. This was the first appearance of cooler fluid in the deep reservoir.

In 1976, a further output test was performed. The enthalpy was constant at 1050 kJ/kg, corresponding to that of the 250°C liquid measured at the 420 m

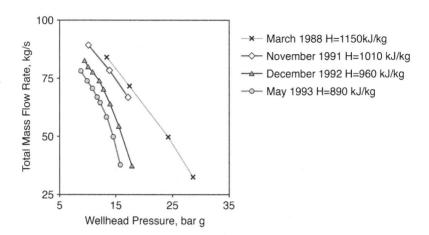

FIGURE 9.9 Output tests 1988–1997. *Source: Contact Energy, Personal Communication.*

feed and significantly higher than that of the 220°C water entering at 600 m. The fact that the 600 m feed did not contribute significantly to the discharge indicates minor permeability at this level, despite the major effect of this flow on the downhole temperature profiles. In 1979, downhole samples showed that while water at 590 and 900 m was chemically similar to that from previous samples taken while the well was flowing, samples from 500 m depth showed the well contained water with normal chloride but very high sulfate (Henley, 1979). This high sulfate content reflects an inflow of steam-heated surface water. The downflow persisted from 1971 until early 1979, and spinner profiles measured at that time confirmed the interpretation of the isothermal temperature profile as a downflow from 600 to 1000 m at 5 kg/s. In August 1979 this downflow disappeared and was replaced by an upflow and the temperature profile reverted to a typical boiling-point upflow profile again, with two-phase conditions persisting down to the bottom of the wellbore (Figure 9.8, P13). Although P13 was measured in November 1987, the 1979 profile was almost identical. The small pressure differences between the two-phase pressure gradient (P13) and liquid pressure gradient (P12) indicates that the upflow that developed in 1979 was not as vigorous as the interzonal flow that was present from 1966 to 1970.

9.5. PRODUCTION: 1988–1997

From 1988 to 1996, BR2 supplied steam to the Ohaaki power station. Few downhole surveys were done in this period as tubing to inject calcite scale inhibitor was inserted in the wellbore and no continuous flow measurements are available as BR2 fed two-phase fluid combined with other wells to a common separator. Prior to commencing production in 1988, a "baseline" output test was carried out, followed by further tests over the next few years. This production period is characterized by gradual cooling of a liquid feed zone indicated by declining enthalpy from 1150 (263°C) in 1988 to 890 kJ/kg (209°C) in 1993 (Figure 9.10). At the same time, declining chloride and increasing sulfate values indicate that the source of the cooling fluids was dilute, shallow steam-heated water. Output curves for the production period 1988−1994 are plotted in Figure 9.9. The output curves are controlled by both declining feedwater temperature and declining reservoir pressure.

At the start of the production period in 1988, the feed zone temperature and discharge enthalpy had recovered to 1150 kJ/kg, only 80 kJ/kg below that measured when first tested in 1966. At this time the two-phase interzonal upflow had reestablished itself as a result of pressure recovery in the deep reservoir (T13, Figure 9.8). However, once fieldwide production commenced, deep reservoir pressures quickly declined, and by 1988, temperature-pressure conditions at 400 m were no longer close to boiling. Despite the initial rapid pressure decline in the reservoir (see Figure 9.2), a shallow two-phase zone could not develop. It had been assumed that the generally low-permeability "caprock" (Huka Falls formation mudstone, siltstone, and tuffs) overlying the top of the reservoir would isolate the deeper high-temperature fluids from cool, shallow groundwaters, with limited interconnection via known faults and fractures. However, once the deep pressure started to decline, it became apparent that cooler fluids overlying this part of the reservoir were not fully isolated from the deeper hot resource, and a cool intrusion developed in the top of the reservoir, affecting BR2. Note that in the case of BR2, tracer tests had shown minimal returns from brine injection wells, and the feed zone cooling was entirely due to influx of shallow cool fluids overlying this part of the reservoir.

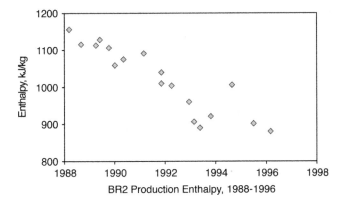

FIGURE 9.10 Production enthalpy 1988−1997. *Source: Contact Energy, Personal Communication.*

Geothermal Reservoir Engineering

Control of calcite scaling by downhole injection of inhibitor was not fully effective, and the well suffered periods of rapid rundown, requiring several workovers to remove calcite scale and repair inhibitor tubing installations. In 1991, following a workover to repair the well and remove calcite scale, a further completion test, heatup and discharge test, was carried out. By this time, more detailed downhole data were available from high-temperature electronic logging equipment. Profiles from these tests are plotted in Figure 9.11. The high permeability is confirmed by small pressure changes at the main feed zone during injection and production, with injectivity now less than 50 kg/s per bar, and by the step temperature profile during injection (T15, Figure 9.10) characteristic of wells with high permeability and multiple feed zones. The temperature steps locate the hot inflow zones at 410−420 and 500−510 m, with a deep zone outflow zone near 1000 m. (There will still be a feed zone at 700−800 m, as shown in the early tests, but this is not evident in the 1991 injection test.) The one-hour heating temperatures show a heating peak of 178°C at 430 m and a downflow at 152°C from the 500 m feed zone. Above 500 m the fluid column appears to be stagnant, judging by the cool interval between 430 and 500 m. After four days of heating, there appears to be a 220°C downflow originating from the 500 m feed zone and flowing down to 1000 m (T17). Then a few days later, with the well flowing, the 500 m feed zone is producing water at 238°C (T18). These changes are probably best explained if the four-day heating profile (T17) is a mixing profile, with 197°C inflow at 430 m, flowing down the well to 500 m, where a hotter inflow—about 238°C—enters the wellbore to

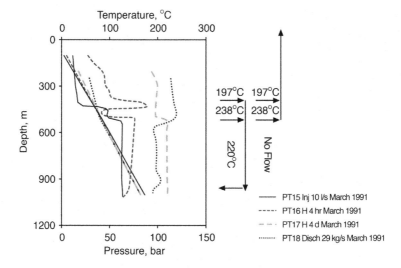

FIGURE 9.11 BR2 Pressure-Temperature profiles 1991. The schematic flow diagrams to the right of the figure illustrate likely mixing models to explain the temperatures profiels for T17 (well shut) and T18 (well discharging) *Source: Contact Energy, Personal Communication.*

make a 220°C mixture. Interpretation of these temperature profiles is a little uncertain, and without spinner data, it is difficult to develop a consistent flow model to explain all the details—a case where more detailed information does not necessarily lead to better understanding. By 1996, the discharge enthalpy in BR2 had declined to less than 900 kJ/kg, and the well was unable to sustain production wellhead pressures. BR2 was finally abandoned in 1998.

9.6. CONCLUSIONS

This case history of one well in the two-phase, liquid-dominated Ohaaki field illustrates several points: the application of the various well testing techniques that have been used to some effect in the New Zealand fields; the changes that can take place in the production characteristics of a single well over a period of time as a result of changing subsurface conditions; and the use of a single well to give some sort of representative picture of the behavior and changes taking place in the reservoir in the area. It is not in any way the intention here to suggest that the history of BR2 is either typical or representative of production well behavior. Rather, the extent of the data available, its changing nature, and the range of tests that have been carried out make BR2 ideal as a demonstration to apply various interpretive methods to understand the changes that can occur in the deep reservoir as a result of disturbance by fluid production.

Conceptual Modeling and Simple Inferences

10.1. INTRODUCTION

A conceptual model is a concise graphic and verbal description of the reservoir that combines the relevant structures and processes that determine its existence and response to exploitation. It is most commonly represented as a cross section or map, or both together. Figure 10.1 shows a common example for a high-temperature system where there is significant topographic relief. The characteristic pattern of steam-heated features at higher altitude and chloride springs at lower altitude identifies the point where the liquid upflow reaches the ground surface at elevations lower than the steam discharges. The pattern of isotherms, alteration, and surface activity is a consequence of the natural flow. As a contrast, Figure 10.2 shows a conceptual model of the low-temperature system at Landau in Germany. This example says that the reservoir is defined as the aquifer, and the only significant information is the geology that defines its depth and the increase of temperature with depth.

A conceptual model integrates the data available from several geoscientific disciplines, bringing together a consistent interpretation of all this data. Such a conceptual model provides a clearer rationale for well targeting and field development and, furthermore, one that can be tested. Depending on the exploration results, the model may be supported, modified, or refuted. A good conceptual model has a number of qualities:

1. It should not be unnecessarily complicated. The simplest model that fits the available data is the best. Any additional complexity merely introduces

Geothermal Reservoir Engineering. ISBN: 978-0-12-383880-3 DOI: 10.1016/B978-0-12-383880-3.10010-1
187

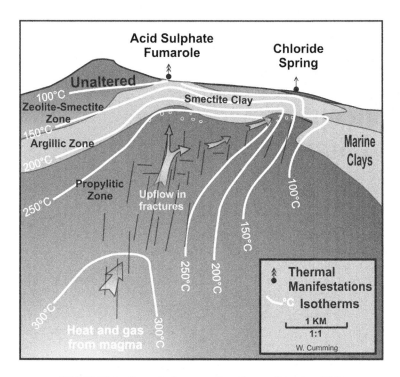

FIGURE 10.1 Conceptual cross section. *Source: Cumming, 2009.*

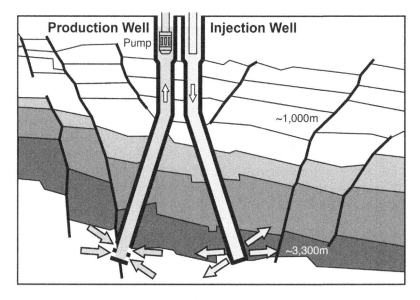

FIGURE 10.2 Geological cross section. *Source: Geox, Personal Communication.*

spurious detail, since there is no data on which to base further specification. Elaboration can always be added later if new data do not fit the simpler form. This new data may indicate the nature of the changes that should be made.

2. It should not be so simple that essential characteristics of the system are dropped from consideration.

3. It should not be biased—for example, by fitting one specific data set with great accuracy while ignoring other information. For a flow model, chemical or enthalpy variation is just as significant as pressure variation. A conceptual model that describes one body of data at the expense of others is unlikely to be totally valid.

4. It should, where possible, fit observed rather than interpreted data. The latter may have already assumed some particular model and thus bias conceptualisation toward that model.

Not all of these requirements can be met with every model. Each geothermal field has its own individual characteristics, and its model has to be viewed in that light. Conceptualizations can depend on unrecognized assumptions as well as observed information and will ultimately be checked only by long-term observation of the field. To the field developer, the model that in the long term most closely matches the performance of the reservoir is best.

10.2. MAPPING THE RESERVOIR

The first conceptual models are constructed by the exploration geoscientists: geologists, geophysicists, and geochemists. Reservoir engineering has little to contribute until wells have been drilled and tested, although there have been simulations of undrilled fields to verify that a proposed conceptual model could produce the observed surface signature.

The first step in conceptual modeling is to assemble the available data in a readily usable form. Typically this comes as maps, cross sections, or 3D computer visualizations of the data. Early in the exploration stages, there will be geological and geophysical maps and maps of surface thermal activity. Once drilling starts, there is additional data from the wells and well tests. This data must be collated into a convenient form to show any patterns that may exist (for example, by plotting the permeability onto the geology), and any anomalies that appear need to be checked or confirmed. Because these early descriptions can exert a strong influence on later interpretations, the mapping process is a very important phase of the reservoir analysis.

The following are some of the properties that might be mapped:

1. The reservoir geology
2. Surface and downhole geophysics
3. Reservoir temperature

4. Reservoir pressure
5. Permeability distribution
6. Zonation within the reservoir: liquid, two-phase, or steam zones
7. Fluid chemistry
8. Natural discharges
9. Hydrothermal alteration
10. Well discharges
11. Surface deformation

This list is not complete, and not all these points are used in any particular field. Deformation of the ground surface, for example, is relevant only in an exploited field. These maps are normally not only contours of actual measurements, but will also use interpreted or summarized data. This implies some sifting of the data, usually based on judgments as to what is relevant. These judgments will change with time, and revision is necessary as new information becomes available. Occasionally a new interpretation will cause a search through old records to check or reinterpret old data.

The reservoir engineer assembles data from the wells: reservoir temperature, pressure, and permeability. The temperature is probably the most important of these data, and when sufficient wells have been drilled, isothermal maps of the reservoir can be prepared. Isothermal maps are useful in understanding the subsurface processes observed by different disciplines. The geologist may look for correlations with structures and alteration, the geophysicist to see if resistivity correlates with temperature, and the chemist wants to locate boiling, deposition, or mixing processes. Because the isotherms imply the direction and location of the natural flow and the location of economic reserves, they have a very strong impact on possible reservoir models.

The reservoir engineer therefore must provide the best estimates of reservoir temperature. If interzonal flows within a well mean that reservoir temperatures over some depth interval are not known, that lack of information must be reported." If the well is not fully heated so that only a minimum estimate of reservoir temperature is available, that must be reported. Temperatures outside the reservoir are as important as those within in it. These peripheral temperatures help define the field boundaries, will be important in the later simulation, and they can directly imply permeability or its absence.

The location of significant permeability within a well is probably the next most important factor. The geologist and other specialists will be looking for any pattern in the location of permeability or a correlation with geological structure in order to guide future drilling. Pressure is of interest to the reservoir engineer and the modeler. If there is a measurable pressure differential horizontally or differential from static vertically and the natural flow is known, there is a direct measurement of permeability times the cross-sectional area of the flow.

10.3. TEMPERATURE PROFILES

Temperatures are easily measured and, with care, readily interpreted. A reservoir temperature profile may be determined by measurements in any well as it warms up after drilling and is discharged. Plotting these temperatures in plan view or section produces isothermal maps, perhaps the most basic representation of any aspect of a geothermal reservoir.

These maps immediately indicate the convective or conductive nature of the reservoir. Some inferences about the natural flow patterns are immediate. A natural hot recharge inflow must enter the reservoir at its hottest point and move toward cooler zones. A temperature reversal similarly implies a flow of colder water entering the reservoir. A zone of constant temperature implies convective mixing of fluid. In all cases, maximum or minimum zonal temperatures imply some flow and thus some permeability. However, a thermal feature need not imply a large permeability. Convection so dominates conduction as a means of heat transport that even a small flow in a zone of small permeability can produce a major thermal feature. Generally, once exploitation starts, the subsurface flow pattern generated by exploitation overwhelms the natural flow. Changes in reservoir temperatures reflect changes in the reservoir as water boils or cooler water sweeps in. Note that there is no inference that recharge under exploitation enters the reservoir at its hottest point.

In an undisturbed field, the vertical pressure gradient can be inferred from the pattern of the natural flow. For the effect on well profiles, there are three significant cases (and many more complicated special cases):

1. The upflow area of the field where the reservoir pressure gradient exceeds hydrostatic and the wells often contain upflows from a deep zone to shallower zones.
2. The outflow area where the fluid flows horizontally or nearly so, and wellbore pressures are close to equilibrium with the formation. In this part of the reservoir, there is often little flow within the wellbore, and consequently wellbore temperatures often reflect true formation conditions over a substantial depth interval.
3. The outflow area where hot fluid moves out over colder zones. Usually the hotter fluid is slightly overpressured, and wells typically have downflows.

In an exploited field, the past pattern of fluid withdrawal and injection determines the pressure distribution, which can be far from static.

10.3.1. Upflow Conditions

Figure 4.4 shows an example of a well in an upflow area. The shaded area represents the reservoir temperature profile, as determined from measurements on this well and adjacent wells during drilling and warmup before upflow started in the well. Note that there is some structure, so the reservoir does not have uniform upflow. There are crossflows present also, but on average there is a net gradient upward. WK24 was drilled in two stages: to 578 m and then to 832 m. Casing is at 345 m. The stable profiles for each stage are shown.

Before deepening there is a boiling-point temperature profile. Boiling fluid enters the well near bottomhole and flows upward. Steam separated off from the upflow rises into the cased section, heating this part of the wellbore also. After deepening, the well shows an isothermal section with another boiling point profile above it. Liquid water enters near bottomhole, flows up the well, and boils as it does so. In both cases, the boiling-point segments of the downhole profile obscure the reservoir temperatures in the upper part of the well and in the casing. The deepened well never shows the highest temperatures measured in the initial well, since the flow of liquid up the well completely obscures this detail.

Although reservoir temperature data are lost because of these internal circulations, some other data are gained from interpreting the flows in the well. The boiling-point profile identifies boiling conditions and, when present down to the inflow, identifies boiling conditions at that feed. Similarly the isothermal profile identifies liquid inflows and thus liquid conditions in the reservoir at that depth. The boiling-point profile is often the most sensitive means of identifying two-phase conditions. It may first appear late in the heating period, after which the upflow of steam rapidly heats the upper part of the open hole and the casing.

A second and more striking type of internal flow is the boiling crossflow, when the well contains fluid with a pressure gradient intermediate between water and steam. The profile is like that of a discharging well, and for the same reason. Steam and water are flowing up the wellbore sufficiently rapidly to mix and produce the typical intermediate pressure gradient. The flow must be sufficient to move the flow regime in the well from bubble to slug. This is a more vigorous form of upflow than the boiling point profile, which is a flow in bubble regime.

Figure 10.3 shows a good example: well Y-13 at Yellowstone National Park. Temperatures were measured during drilling, and the corresponding saturation pressures were plotted. Note that the reservoir pressure gradient exceeds hydrostatic. The stable downhole pressures are also shown, both shut and bleeding. There is an upflow of boiling fluid in the well with pressure gradient less than hydrostatic, which exits into the formation at a shallow feed zone. Above the feed zone the pressure gradient is close to hydrostatic, so there is apparently water balanced on top of lower-density fluid. Such a profile is only possible because it is dynamic, not static. (Similar profiles are discussed in Chapters 4 and 9.) With the vigorous boiling upflow, steam rises into the casing. The casing is heated extremely rapidly, and gas pressure (left after the steam condenses) builds up if the well is shut. The gas pressure will rise until it is bled off or the water level is depressed to the shallowest feed so any further gas is lost into the formation. The gas content of the reservoir need not be high to create a high gas pressure at wellhead; gas is concentrated in the casing by this process of distillation.

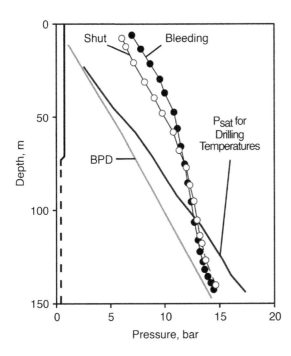

FIGURE   10.3 Pressure profiles in well Y-13, Yellowstone National Park. *Source: White et al. 1975*

The presence of a boiling interzonal flow normally indicates that the well discharge will have significant excess enthalpy. Note that the low-pressure gradient in the well does not indicate the same gradient in the reservoir. The reservoir gradient will usually be near hydrostatic in an undisturbed reservoir. The reservoir gradient can only be determined by determining reservoir pressure at different depths, in different wells.

10.3.2. Static Conditions

In the peripheral or outflow regions of the field, temperature profiles may show an absence of marked convective effects in the well. Figure 10.4 shows profiles in two wells from Tongonan, which penetrate the Malitbog outflow from the reservoir. The profile in MB-1 is similar to a boiling-point profile but colder. The reservoir fluid here derives from boiling fluid, which has flown laterally outward, cooling somewhat as it does so. The absence of marked convective effects in a well with good permeability indicates genuine equilibrium between well and reservoir; the reservoir is vertically hydrostatic. MB-7 shows an outflow with a temperature reversal.

Sometimes most or all of a field development takes place on an outflow. Examples are Ahuachapan, Yangbajan, Rebeira Grande, El Tatio, and the early developments of Wairakei and Kawerau. Exploratory drilling found excellent production on an outflow, and development was based on this productive area.

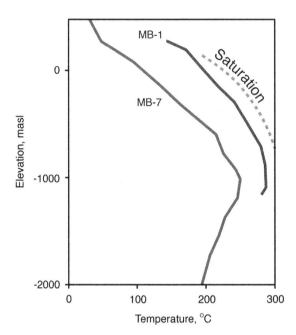

FIGURE   10.4   Stable
temperatures in wells MB-1
and   MB-7,   Tongonan.
*Source: EDC, Personal Com-
munication.*

Only later is the upflow drilled, and sometimes it is less permeable than the outflow region.

10.3.3. Downflow Conditions

Like an upflow, a downflow is recognized by a near-isothermal profile. The water flowing in the well may gain or lose some heat conductively, so it heats or cools a little as it flows. Sometimes it is possible to recognize the existence of a flow in a well from temperature data alone but be unable to decide whether the flow is up or down. In this case spinner profiles with the well shut can identify flow direction and flow rate. Note that interzonal liquid flow provides an opportunity to sample the deep reservoir fluids without discharging a well.

10.3.4. Conductive or Cold Water Layers

In many places, the upper layers above the high-temperature reservoir may be quite cold, while in the area of surface discharge hot or boiling conditions may extend to the surface. Such activity normally occurs over only part of the field. Away from the surface activity, varying thicknesses of cold rock may be drilled.

Two forms of temperature profiles are often found in such regions: a linear conductive gradient indicating poor permeability and roughly isothermal cold

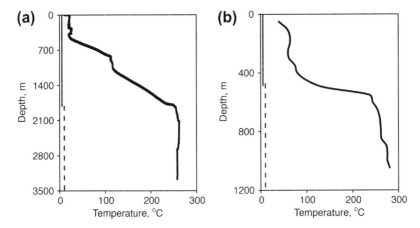

FIGURE 10.5   (a) Stable temperatures in NM6. (b) Stable temperatures in BR31. *Source: (a) Rotokawa Joint Venture, Personal Communication; (b) Contact Energy, Personal Communication.*

temperatures, or large temperature inversions indicating cold aquifers; such cool aquifers usually have strong correlation to specific geological formations.

Figure 10.5a shows an example of the first type: well NM6 at Ngatamariki. The reservoir top is at 1800 m, below which there is little temperature change in the convecting reservoir. Above this depth, there is a region of roughly linear gradient to 500 m, interrupted by an aquifer at 1000 m. Above 500 m are cold aquifers. Figure 10.5b shows an example of the second type: well BR31 at Ohaaki. This well lies toward the edge of the reservoir, as shown in Figure 9.1. The reservoir top is at 550 m. Above this is a very steep temperature gradient to cold aquifers above 400 m. The steep gradient between 400 and 550 m implies low permeability at this depth, whereas above 400 m there must be fluid movement, and thus permeability, in relatively cool rock. This interpretation based on the temperature profile correlates with the well geology: In the 400-550 metre interval low permeability siltstones are present and above this is fractured very permeable rhyolite.

10.3.5. How Permeable Is Permeable?

Temperature profiles have been categorized here as convective or conductive, with convective profiles implying the presence of permeability. It should be noted that there may be sufficient permeability for convection but still insufficient for production. As a very rough guide, a permeability of 1 md is required for convection in fluids at 200−250°C (Straus & Schubert, 1977; Hanano, 2004). However, this is insufficient for a good production or injection well, which generally requires a transmissivity of several darcy-meters or more, and values of tens or hundreds of darcy-meters are commonly encountered. Thus, the presence of convective flows is a necessary, but not

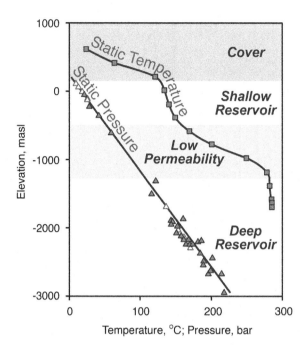

FIGURE 10.6 Pressure-temperature profile in Mt. Amiata system. *Source: Barelli, Cei et al., 2010a.*

sufficient, condition for the existence of a productive reservoir. As with drilling losses, convective temperature effects are encouraging but do not prove a productive reservoir. This explains the common observation of the "impermeable upflow."

The initial exploration of a field proves a productive outflow zone, but further drilling into the upflow zone finds little or no productive permeability. In the upflow zone there is only sufficient permeability for convection, not enough for production. The anomaly is actually the outflow, with much higher permeability.

Figure 10.6 shows an example of apparently contradictory indications. The Mt. Amiata system (Barelli et al., 2010a) hosts a shallow reservoir and a deep reservoir. The temperature profile shows that the two reservoirs are separated by a conductive zone between the two convective temperature zones. However, there is, within the data scatter, a common pressure gradient implying a hydrologic connection. These indications need not be contradictory. The temperature profile implies that vertical permeability in the conductive interval is less than some value, $k_v < k_1$. The lack of an observable pressure difference implies that vertical permeability exceeds some other value, $k_v < k_2$. It is possible that $k_2 > k_1$, so that both criteria can be satisfied. To be more precise would require modeling the system to determine how low the permeability must be to show the conductive gradient, and how high it must be to create the common pressure gradient. In the actual event, under exploitation no pressure interference has been observed between the two reservoirs.

10.4. PRESSURE

Pressure is a property directly reflecting fluid flow in the reservoir. It is readily measured and responsive to production and injection. Unfortunately, as discussed in Chapters 4–7, downhole pressures do not necessarily correspond simply to pressures in the reservoir at the same depth. Mapping the pressure distribution in the reservoir requires a number of measurements in each of a number of wells. If successful, the direction of any pressure gradient (in excess of fluid static) implies the direction of fluid flow in the reservoir.

Figure 11.2 shows the initial pressure distribution with depth in the Kawerau geothermal field. The pressures are feedpoint pressures in early wells, measured before much production had occurred. The wells are all located in one part of the field and do not include southern areas drilled later. Within the data scatter, there is a single linear trend of pressure with depth, which is plotted in Figure 2.6. This trend has a gradient of 0.085 bar/m. By comparison, the average reservoir temperature is 250°C, which would have a hydrostatic of is 0.078 bar/m. The difference between these values is 0.007 bar/m = 700 Pa/m, which drives the upflow through the reservoir. The natural discharge from Kawerau has been estimated at 80 MW. With a deep recharge temperature of 290°C, this represents a flow of 67 kg/s. If the area of upflow is A, and the vertical permeability k_v, applying Darcy's Law in the vertical direction gives

$$W = 67 = \frac{k_v A}{v}\left(\frac{\partial P}{\partial z} - \rho_w g\right) = \frac{k_v A}{0.13 \times 10^{-6}} 700$$

$$k_v A = 1.25 \times 10^{-8} \text{ m}^3 = 12.5 \text{ md.km}^2$$

If the upflow rises over an area of 1 km², the vertical permeability is 12.5 md. By contrast, interference testing among the wells shows kh values of the order of 100 dm or greater (Grant & Wilson, 2007). On the gross scale, horizontal permeability is much better than vertical. For another example, see Mak-Ban in Chapter 12, where the vertical pressure gradient was used to calibrate the vertical permeability in the simulation model.

Figure 10.7 shows initial reservoir pressures in Los Azufres, with the data being the identified pressure at the main feed of each well. A steam zone at a pressure over 40 bar is clearly identified above the deeper liquid-dominated region. In this case, the pressures of the different wells fall reasonably close to a single trend, indicating good connectivity across the reservoir. This profile is present over only part of the reservoir, the south zone. In the north there is a liquid-dominated reservoir with a normal gradient over its entire depth (Torres-Rodriguez et al., 2005).

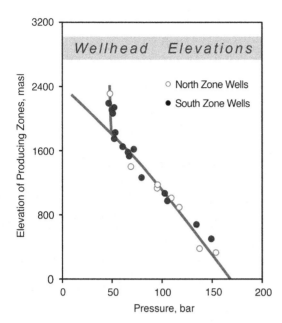

FIGURE 10.7 Pressures in Los Azufres. *Source: Arellano et al., 2005.*

Comparing pressures at different depths in different wells can be difficult. In a petroleum or groundwater reservoir, it is simple to correct to a standard elevation using static gradients. In a geothermal reservoir with varying temperatures and nonstatic gradients, there is no simple comparison. It is convenient to define a standard pressure profile for the reservoir by plotting reservoir pressure from different wells against elevation and finding the gradient. Having defined this "standard hydrostatic pressure" (Hitchcock & Bixley, 1975), pressure data can then be expressed as deviations from this trend. This approach is also useful in an exploited field. The standard initial pressure profile is used as a reference, and pressures in a newly drilled well can be expressed as a deviation from this trend. Figure 10.8 shows an example.

10.5. EXPLOITED FIELDS

In an exploited field, additional types of well behavior may appear due to the very dynamic pressure distribution. If part of the field was initially boiling or near boiling, the drop in pressure creates a general expansion of two-phase conditions. Interzonal flows in wells, both up and down, tend to become more prominent. If there are distinct aquifers in the field, differential drawdown may create a general pattern of fluid flow between them in wells that have feedzones in both aquifers.

If there is sufficient boiling, a steam zone normally forms at the top of the reservoir—or in some part of the reservoir where there is a laterally extensive

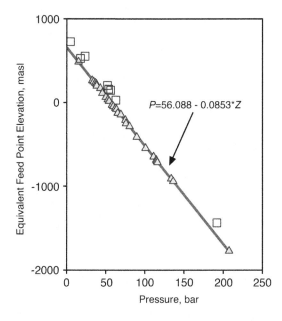

FIGURE 10.8 Initial pressure gradient in Oguni field. *Source: Garg & Nakanishi, 2000a.*

$P = 56.088 - 0.0853 \cdot Z$

low permeability capping layer that prevents vertical drainage of water—and expands as deep pressure declines. A well that is open to the steam zone and to the underlying liquid will usually show a steam cap profile: a column of steam above a column of water. The well equilibrates against reservoir pressure at two points, one in the steam zone and one in the water, since the downhole pressure in the steam and water columns can move independently, with the water level moving correspondingly (Note that the steam-liquid profile observed inside the wellbore does not mean there is a "water level" in the reservoir. Above the liquid saturated zone there will be a gradual transition of increasing steam saturation, with "free" steam only present in large fractures). Convection or circulation effects are possible within both the steam and water sections of the well.

10.6. SUMMARY

In this chapter, the first analyses of reservoir information has been discussed as opposed to information relevant to a single well. The natural state information is limited and often cannot be obtained once exploitation is under way. The natural flows, continuing for tens or hundreds of thousands of years, have established the heat and fluid distribution in the reservoir that is now a target of development.

In simple terms, it is the movement of fluid up and through the various types of geothermal systems that established the geothermal reservoirs. To the developer, this is probably only of scientific interest. It is the flow and the way in which it relates to the reservoir hydrogeologic structure that set the form of the reservoir. There are two distinct forms of geothermal reservoir: liquid-dominated and vapor-dominated, each with a different natural flow.

For both types of systems there are simple models of the natural state that can be useful in focussing the conceptualization of the system and for estimating some field parameters. What these simple models imply can be useful in the more complex real systems. This chapter has been qualitative. Later chapters become more quantitative and therefore much more detailed and using a wider range of data. Data from all disciplines need to be taken into account when forming a conceptual model.

Simulation

11.1. INTRODUCTION

This chapter outlines the use of simulation by the reservoir engineer. The approach assumes that the reservoir engineer provides data to and receives output from another person, who carries out the simulation. Sometimes one person performs both roles of reservoir engineer and modeler/simulator. In this discussion only the reservoir engineering input is considered, and the mechanics of carrying out a simulation is not discussed. For the simulator's viewpoint, see O'Sullivan and colleagues (2001). The focus here is on what data are needed, the procedures for calibrating the model, and what results may reasonably be expected.

The process is not a simple step of providing data and receiving results. One of the great advantages of a simulation is that it makes computations in accordance with physical and mathematical laws: it is logically and internally consistent. In contrast, when constructing a lumped-parameter model, it is possible to make simplifying assumptions that are not consistent with the system structure. Modeling, in theory, avoids this hazard by explicitly representing the known information. Often the process of model building shows that some data are inconsistent or very difficult to fit and require review. This sends the reservoir engineer back to check on actual measured data and its processing and interpretation. Perhaps it is possible to cross-check the data against other measurements. Perhaps the interpretation was wrong. Or if the data are solid, then more

Geothermal Reservoir Engineering. ISBN: 978-0-12-383880-3 DOI: 10.1016/B978-0-12-383880-3.10011-3
201

work is implied for the modeler to make the model fit what appeared to be inconsistent data. Or perhaps the conceptual model requires revision.

Simulation codes currently available, most commonly TOUGH2 (http://esd. lbl.gov/TOUGH2) or TETRAD, allow two-phase fluid with the addition of gas and salt, or tracer, and dual-porosity models. TETRAD requires a rectangular grid, while TOUGH2 does not. Automated parameter matching is available in iTOUGH2, but this is very computationally intensive and at present can be used on only a small number or parameters. One current shortcoming is the lack of a "front-tracking" option, where a sharp interface is moving across a block—the most obvious case being a free water surface near the model top. The interface is represented only as saturated blocks below with unsaturated above, and the water level can only move by increments of a block thickness. This means that surface effects can be represented only approximately or require very fine vertical gridding.

11.2. INPUT DATA

The basic inputs for modeling are as follows:

1. A conceptual model of the reservoir
2. Natural state pressure and temperature distribution
3. Well tests and interference tests
4. Production and injection history: mass flow, enthalpy/temperature, and possibly wellhead pressure (WHP)
5. Changes in reservoir pressure and temperature during production
6. Well specifications (locations of feed zones)

Other data that can be used include the following:

1. Gas and chemical distribution in the natural state
2. Changes in gas and chemistry during production and injection
3. Changes in gravity during production and injection
4. Changes in surface elevation during production and injection
5. Changes in surface activity
6. Tracer tests

Research work is underway to use changes in other observables, such as SP or magnetic field (Ishido & Pritchett, 2001; Nakanishi et al., 2001; Pritchett, 2007). In principle, any physical parameter that is changed with fluid change and is conveniently measurable could be used, provided that there is a computable physical model to relate its change to the reservoir changes.

New parameters introduce new properties to be set. For example, modeling subsidence requires specification of the elastic properties of the reservoir rock. Matching subsidence is primarily about fitting these elastic properties, but the pattern of subsidence also reflects pressure changes in and above the reservoir so there is some constraint on possible reservoir pressure and thus some additional constraint on reservoir permeability and porosity. For example, there is often a broad pattern of subsidence across the entire reservoir; this pattern essentially outlines the area over which pressures have fallen and so indicates the boundary of the area where pressure has fallen. Conversely, a localized area of subsidence implies compaction close to the surface, so the reservoir model must have pressure or temperature change close to the surface.

In general a model needs constraints: observational data of relevant physical parameters. More constraints improve the model by restricting possible models. In Chapter 12, the discussion of Wairakei lumped parameter models shows that different physical models produced nearly identical fits to one data history: the field pressure. The models were separated by adding additional data: the possible physical size of the reservoir. Introducing a different type of data usually provides a different constraint on the model, affecting different parameters. For example, the natural (steady) state of a reservoir depends only on rock permeability and not at all on porosity, while changes in enthalpy, gas, or chemistry under production or during well testing in a two-phase reservoir are highly sensitive to porosity.

A model can be constructed at the exploration (predevelopment) phase, using only natural-state plus well-testing (including interference testing) data. Such a model is partly calibrated. It provides some estimates of future reservoir performance under production but lacks the constraints provided by a production history. Porosity is little constrained, and the fitted permeabilities are dependent on measurements of the natural flow. If there is undetected subsurface discharge, the natural flow will be understated and the permeabilities underestimated. There is no constraint on possible stimulated flow from external aquifers which are in equilibrium in the natural state, since the strength of this connection is not tested by the natural state modeling. The following sections discuss how the information that is collected by the methods of the previous chapters is used in the model construction and calibration.

11.3. CONCEPTUAL MODEL

The conceptual model is the first guide to the numerical model. This incorporates the mental models of the geoscientists working on the reservoir: the construction of a pattern from the mixture of physical information available. The reservoir engineering data is important; temperatures and pressures provide an indication of where fluid is flowing. The geophysics provide an estimate of the boundary of the reservoir beyond the drilled area. Normally the model will start by assuming that the permeable reservoir extends across the area of geophysical anomaly, with permeability decreasing at the edges. The drilled wells or geophysics may or may not indicate a bottom to the reservoir. The geology provides the structure of the model. For a first assumption porosity and permeability values will be assigned to each geological unit. Any relevant structure such as a fault or formation contact can be incorporated as a region of higher permeability. If there is some regional trend or major structure, it is usually more convenient to orient the model grid parallel to that trend so structures can be conveniently represented as a line of blocks or by changing the grid detail.

11.4. NATURAL STATE

The first step in model calibration is to match the natural state. The relevant data are the natural temperatures and pressures and the amount of surface discharge as both heat and mass. The reservoir model is constructed with an input of mass and heat at the bottom, possible infiltration from the surface, and leaks at sites of surface or subsurface discharge. It is then run until a steady state is reached. The temperature and pressure distribution is then compared against the measured data. Then some parameters are changed, and the model is rerun until a steady state is obtained that is closer to the observations. The process of adjusting parameters to get a natural state match is usually slow. As well as structure within the reservoir model, the boundary conditions may be adjusted; a side boundary may be impermeable or may allow contact to a lateral aquifer, and the model bottom may be deepened to bring more of the region in which fluid flow is affected by production within the model.

It is important, at least in the initial stages of development, that the model structure be as simple as possible but still contain within it the mechanisms that affect reservoir processes. For example, if there is a complex pattern of deep inflows rather than one or two sources, this indicates that there is structure at greater depth influencing the flow pattern, which should be included within the model. If there are constant pressure boundaries (except at surface) that contribute significant flow under exploitation, this again indicates that an important assumed property of the reservoir has been located outside the model, and again the model should be enlarged so that nearly all flow is contained within the model.

It is important that the information used as input into the model is "real"— that is, the pressure and temperature measurements must be carefully interpreted to provide the best estimates of the reservoir temperature and pressure. Some data may turn out to be critical and this should be rechecked. Model structure is often sensitive to anomalous pressures, and it will need to be checked that an anomalous pressure genuinely is high or low.

Temperatures are best presented as well-by-well matches of interpreted well temperature profiles and model results. Sometimes there will be several wells within one model column. In this case it will be necessary to decide which well data is most representative of the reservoir, to compare with the modelled results. It is useful also to have isotherms in maps or sections, but because this involves some smoothing of the actual data, it is not as accurate a calibration. Contours on particular layers can be helpful in visualizing fluid flow in those layers.

Figure 11.1 shows a set of well temperature matches for the Kawerau model of White (2006), and Figure 11.2 shows a match of pressure-depth data from the model of Holt (Personal Communication). The pressure-depth data for

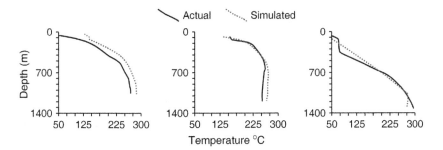

FIGURE 11.1 Downhole temperature matches. *Source: White, 2006. Reproduced with permission of Kawerau Geothermal Ltd., and Ngati Tuwharetoa Geothermal Assets Ltd.*

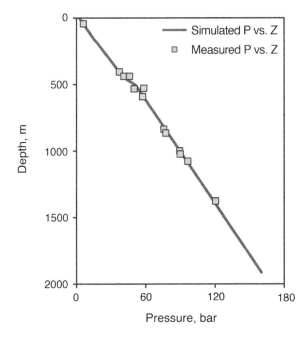

FIGURE 11.2 Pressure-depth profile in natural state of Kawerau *Source: Holt, Personal Communication. Reproduced with permission of Ngati Tuwharetoa Geothermal Assets Ltd., and Kawerau Geothermal Ltd.*

Kawerau were previously discussed in Chapters 2 and 10. The model incorporates a reduced permeability capping layer around 500 m depth that is visible in the simulated pressure profile, whereas the actual data are too coarse to discriminate this feature. Hoang and colleagues (2005) showed a pressure match in a vapor-dominated reservoir. Figure 11.12b shows the comparison of modeled against measured pressures at Ngatamariki. Each of these presents a comparison between simulated and observed data.

Sometimes an explicit goodness of fit measure is used, usually some variant of the sum of the deviations or squares of deviations in pressure and/or temperature data, normalized by reference to some pressure/temperature. This provides a more objective measure of how closely the data have been matched than visual inspection. Visual comparisons are still needed to ensure that the simulated results have the right pattern.

Pressure matches are sometimes presented as pressure-depth plots for each well. This does not present the data clearly, since in liquid-dominated systems, all the well profiles are normally hydrostatic and the differences between wells are small compared to the total pressure, making it difficult to judge if the simulation correctly represents the relatively small deviations from hydrostatic equilibrium. It is better to present the data as reservoir pressure at identified points and compare those data against simulated results. This is done in Figures 11.2 and 11.12b, where the reservoir pressure profile or point values from the simulation is compared against the reservoir pressure data.

It will normally take considerable time, adjusting permeabilities, to get an acceptable natural state match. Automated parameter matching can be done (using iTOUGH2), but only a few parameters can be matched at a time without greatly escalating computing time. There may be only a limited number of wells drilled before production, limiting the data defining the natural state. Often wells drilled after production has commenced are used, assuming that the temperature has changed little so the later profiles are representative of the natural state profile. A more rigorous approach is to match the data from later wells with simulated data for the time of measurement—that is, match the later profiles as a part of production matching.

Interference tests should be available at this phase of field development and can be used to provide additional calibration. The tests are normally analyzed using standard uniform aquifer models to obtain transmissivity and storativity, which are then used as initial estimates for the reservoir in the area around the wells involved. When the interference deviates from a uniform aquifer, it is usually simplest to simulate the interference test with the model and then adjust model parameters to obtain a best fit. The simulation will not get the fine details of the short-time interference correct due to the block size, but it should get the larger and longer-time response correct. Figure 11.13 shows such a match to the interference testing at Ngatamariki.

11.5. WELL SPECIFICATION

Each of the wells in the field should have been fully interpreted to create a well model. If the reservoir simulator is coupled to a wellbore simulator, then the specification of feed depth(s) and productivity/injectivity of each feed, together with the well casing and deviation provides a full specification of the well. The simulator provides the reservoir pressure and fluid quality at each feed, and from these the wellbore simulator can calculate the well flow at specified WHP.

Most current simulators do not include a coupled wellbore simulator, and it is necessary to make a simpler well specification. This may be an allocation of flow to different zones at fixed ratios, these ratios having been assessed at some flowing state that is representative of reservoir conditions for the simulation period, or use of a lookup table, which is based on the results of wellbore simulations, or some other computationally convenient simplification. Total flow of the well is calculated by the simulator using a simple formula or lookup table, again defined from wellbore simulations of a representative flowing state.

The full details of well specification may or may not be important to the simulation. If the reservoir is reasonably homogeneous and permeable, the details of flow allocation may not be particularly important because the reservoir response is more controlled by the total flow, and well performance is not much affected by the depth of fluid entry.

In other cases, detailed well specification is more important, particularly when there is significant stratification within the reservoir. One such case is when there is a two-phase zone over a deeper liquid zone. Well performance and discharge enthalpy are then strongly influenced by the proportion of fluid taken from the two-phase zone, and this in turn affects the power capacity of the production and design of the plant. A second case is when there are cool zones within an otherwise homogeneous liquid reservoir—for example, from injection returns. This again adversely affects well performance and discharge enthalpy and, in turn, power capacity and plant design.

11.6. HISTORY MATCHING

Once the natural state matching has been done, wells are specified, and the model is then used for production runs to simulate the changes under exploitation. The simulated changes are compared with actual measurements, and another cycle of parameter adjustments are made to generate a fit. The parameter changes also affect the natural state model, so it is necessary to rerun the natural state to make sure this match has not been degraded. Further iterations may be needed to get acceptable matches to both the natural state and production history.

Figures 11.3-11.5 show history matches from the Wairakei-Tauhara model of O'Sullivan and colleagues (2009), matching pressure in one well, enthalpy in one well, and surface heat flow. These are only a sample of the data matched, since many well pressures and enthalpies were matched, giving a match to data across the field and its history.

Earlier models of Wairakei tended to match the average field pressure, since pressure is fairly uniform across the field. Later models matched specific well histories (Figure 11.3), thereby ensuring a match to the distribution of pressure across the reservoir as well as with time. Figure 11.4 shows a match to the enthalpy of one well. Because Wairakei developed a steam zone with drawdown, well enthalpy depends on the depth of feed zones. To get an accurate match, it is necessary to have an accurate representation of the feed depths so

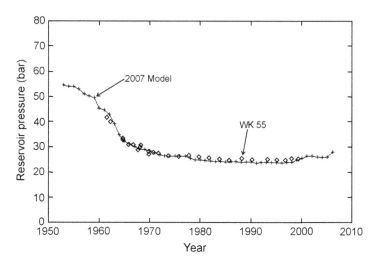

FIGURE 11.3 History match to pressure in the WK55. *Source: O'Sullivan et al., 2009. Copyright © Elsevier.*

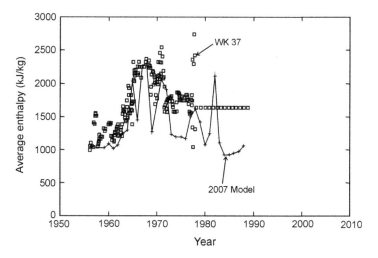

FIGURE 11.4 History match to WK37 production enthalpy. *Source: O'Sullivan et al., 2009. Copyright © Elsevier.*

there is the correct proportion of fluid taken from steam and water zones and an accurate simulation of the process of vertical drainage and segregation, which tests the vertical permeability. The match to surface heat flux, shown in Figure 11.5, requires fine structure in the shallow layers of the model.

With the development of the steam zone, there came a large increase in surface heat discharge that early models did not match. Making the model match required calibration of the shallow model layers and finer layer structure

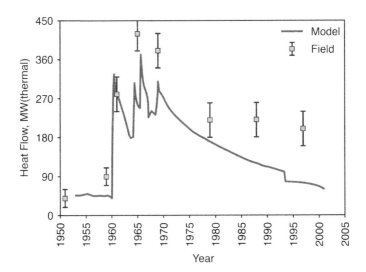

FIGURE 11.5 History match to surface heat flow at Karapiti. *Source: Mannington et al., 2004. Copyright © Elsevier.*

near the surface. These changes have little direct impact on reservoir performance, but they are critical for calibrating the model representation of surface and near-surface changes, which are increasingly important for assessments of environmental impact.

Figure 11.6 shows a history match from the model of Hatchobaru in Japan (Tokita et al., 2000). The model matches a range of data, tracer, pressure,

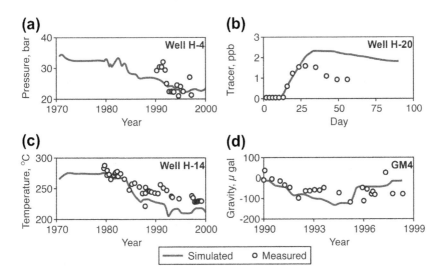

FIGURE 11.6 (a) Pressure, (b) Tracer return, (c) Temperature, (d) Gravity. *Source: Tokita et al., 2000.*

temperature, and gravity. Matches to such a wide range of data provide considerable constraint to the possible reservoir structure and provide increased confidence in simulated outcomes. In the case of Hatchobaru, thermal interference from injection returns is a management problem, and the simulation, by matching both temperature changes and tracer returns, is calibration of the parameters critical to this interference.

11.7. DUAL POROSITY

This discussion so far has implicitly assumed that the medium is homogeneous. Most simulators offer a dual-porosity option, with representation of the fractures and block matrix as connected media. This requires additional parameters to calibrate the model: fracture spacing, fracture and matrix permeability, and porosity. The interaction between the block matrix and the fractures varies from the simple model of Warren and Root, and Barenblatt, to more complicated and more accurate representation of the transient flow from matrix to fissure, such as the MINC formulation (Pruess et al., 1999).

Many simulations are perfectly satisfactory using single-porosity models. A fractured structure is needed when there is a marked difference between the fluid in the blocks and in the fractures. One case is a vapor-dominated reservoir, with dry steam in the fractures but liquid in the matrix. A more common case is when cold water is advancing through hotter rock—say, in an EGS or when injection fluid is returning to production. A third case is in two-phase fluid: single-porosity and dual-porosity models give different enthalpy histories under drawdown. A homogeneous porosity model produces overoptimistic predictions of reservoir longevity under injection returns. Figure 11.7 shows a typical example of a simulation, comparing the rate of cooling due to injection in

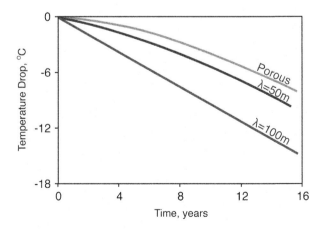

FIGURE 11.7 Change in temperature with time near reinjection area for porous medium and fracture spacings of 50 and 100m. *Source: Nakanishi et al., 1995.*

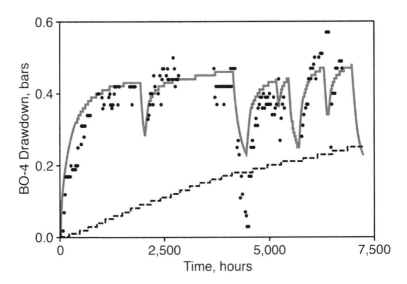

FIGURE 11.8 Interference—simulated pressure in fracture and matrix. Dots are measured data, the solid line is the fracture pressure, and the dashed line is the matrix pressure. *Source: Lopez et al., 2010.*

a homogeneous medium and a fractured medium with two different fracture spacings. Figure 11.8 shows an interference test modeled with TOUGH2 using the MINC formulation. The fracture and matrix pressures are shown. While the fracture pressure responds rapidly to the flow rate changes, drawing down with flow and recovering during shut, the matrix pressure changes much more slowly and simply draws down over the entire period. Figure 11.9 shows an extreme case where a fractured medium is essential, the temperature changes in a well used for injection and then returned to production. In this case it was essential to use a MINC model, since conventional Warren and Root models underestimated the time for the well to heat up (Acuña et al., 2008.). Kumamoto and colleagues (2009), simulating the Ogiri field in Japan, found it necessary to use MINC blocks in the two-phase production zone, the high-permeability Ginyu Fault.

While dual porosity simulations are essential to model injection returns, this additional sophistication is of little value until the model can be appropriately calibrated. This requires data to determine if a preferential flow pattern is present. The first such data are provided by tracer testing; the simulator can be used to match a tracer test. Better calibration is provided by observation of some measured chemical and thermal changes. A practical constraint is that tracer tests and injection returns are typically highly individual, with returns strongly influenced by details of fracture structure. It is common to observe one well with strong tracer returns, while a nearby well has much lower concentrations. Short of detailed mapping of individual fractures, it is not possible to represent such variations in the model.

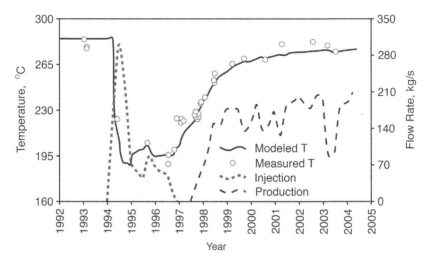

FIGURE 11.9   Computed and observed thermal recovery of well Awi 10-1 at Awibengkok after 3 years' injection. *Source: Acuña et al., 2008. Copyright © Elsevier.*

11.8. VALIDATION OF THE SIMULATION PROCESS

Ideally a model would be validated by producing a prediction of future field performance and then observing actual behavior and comparing the two. The traditional rule of thumb is that model predictions are valid for roughly the length of time that history has been matched. While there is substantial operating history with some simulations, very little has been published providing such a direct validation. There are three fields for which there are such published "post-audits": Olkaria, Nesjavellir, and Wairakei. In addition, a number of simulations in a range of fields (The Geysers, Cerro Prieto, Mammoth, Heber, Geo East Mesa, Salton Sea, Puna, and Steamboat Springs) are reported to have provided accurate forecasts for periods of 5 to 14 years (Sanyal, quoted in Kneafsey et al., 2002).

Industry practice provides further support. Simulations have been maintained since the 1980s in major geothermal fields, and these models have been refined, recalibrated, and upgraded. Normally the level of detail is increased as time goes by and computing power increases. The modeling of Wairakei (see Chapter 12) shows this process. That the operators continue to use and improve the models demonstrates a degree of confidence in their value.

11.8.1. Olkaria

Olkaria is a large and complex field. In the region first drilled, the reservoir contains a thin vapor-dominated zone overlying a liquid-dominated zone. This region of the field was modeled by Bodvarsson and colleagues (1987a, b), with the model calibrated against 6.5 years of production history. A "postaudit" (Bodvarsson et al., 1990b) showed that the model had provided a good prediction of field-wide steam decline rate and individual well predictions that were accurate in about 75% of cases. The model was then recalibrated.

11.8.2. Nesjavellir, Iceland

Nesjavellir is a high-temperature field in Iceland. A model was constructed in 1986, calibrated on production data for 1 to 3 years (Bodvarsson et al., 1988, 1990a, 1991). The results were evaluated by Bodvarsson and colleagues (1993) for the following 6 years. Decline rates, pressure decline, and enthalpy rise were all overestimated, and the model was recalibrated.

11.8.3. Wairakei

O'Sullivan and colleagues (2009) reviewed the history of Wairakei modeling, including a review of the 1994 model predictions compared to later performance. There are reservations because the production and injection histories do not correspond exactly to the actual flows. With this reservation, it was found that deep pressures were well predicted. The model was in error in one aspect: the predicted performance of a group of shallow steam wells. This error was mostly due to an incorrect well model, although there were also errors in predicted steam pressure.

11.8.4. Summary

The experience at Olkaria, Nesjavellir, and Wairakei and industry practice in continuing to maintain and develop simulation models provide support to the validity of the modeling process, with the limitation that predictions are generally valid for a time similar to the calibration period. The following section briefly presents one simulation. Other examples can be found in Chapters 12 and 13.

11.9. NGATAMARIKI

Ngatamariki in New Zealand was first explored in the 1980s and then left idle until new geophysical and geochemical surveys were done in 2004; exploration drilling resumed in 2008. It is a good example of the capabilities and limitations of simulation in a greenfield site—a reservoir with no production history. Following the previous section, specific predictions have a limited period of validity. However, the model can bring out important aspects of the reservoir performance that may affect development plans. The field and its exploration are described in Boseley and colleagues (2010), and the field layout is shown on Figure 11.10. The geophysical boundary shown is derived from MT-TDEM surveys and downhole data.

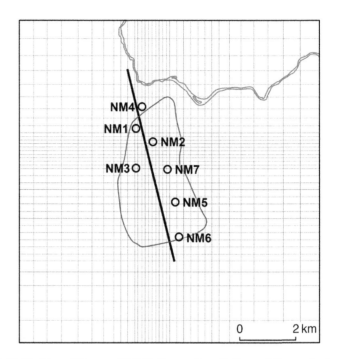

FIGURE 11.10 Map of Ngatamariki field. *Source: Burnell, Personal Communication. Reproduced with permission of Rotokwa Joint Venture.*

Figure 11.11 shows a cross section of the conceptual model along the line shown in Figure 11.10. A critical feature of the field that is likely to impact on reservoir management is communication between the deep high-temperature reservoir and shallower aquifers near wells NM2 and NM3. The reservoir top dips to the south from about 700 m in NM3 to 1500 m in NM6. The temperature profile of NM6, shown in Figure 10.5a, shows the top of the reservoir and also the disturbance created by shallow aquifers. Temperature profiles in the northern wells all show shallow temperature reversals associated with a highly permeable shallow, cooler aquifer at around 400 m depth. Geochemistry shows that geothermal fluid rises from the high-temperature reservoir into this aquifer, where it mixes with cool groundwater and then flows northward, feeding surface activity.

This conceptual model, with interconnected deep reservoir and shallow aquifers, was the basis of the simulation. The model has a deep high-temperature recharge and outflows (represented in the model as wells) at the springs. Reservoir temperatures in all wells were matched. Figure 11.12a shows the temperature match for two wells, which includes the shallow aquifer temperature structure. Figure 11.12b shows the natural pressure match; measured pressure at feed points is compared against simulated pressure. The pressure

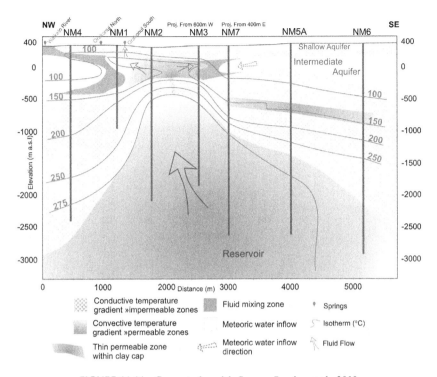

FIGURE 11.11 Conceptual model. *Source: Boseley et al., 2010.*

data include some shallow pressures measured during drilling or in shallow monitor wells. The pressure plots in Figure 11.12b show a tolerance of ±2 bar, which is the observed repeatability of measurements.

An interference test was conducted among the deep wells by discharging three wells for varying periods and monitoring pressure in well NM2. The results were fitted to line-source models, but showed significantly different reservoir properties in the response to the different source wells. Figure 11.13 shows the match to the interference test, using the simulation model. Simulators are limited for modeling interference tests, since they cannot resolve fine time details unless fine-gridded nested well blocks are used around each producing well. Without such fine gridding, the response is smoothed at short times by the storage capacity of the block containing the source well. However, the simulation has the great advantage that it can include structural detail when there is contrast in reservoir properties.

The simulation of production indicated that pressure drawdown in the reservoir produced significant downflow from the shallower, cooler aquifers. To provide more constraints on this part of the model, a series of shallow wells were drilled and interference tests were carried out. Aquifer parameters from these tests were used to refine the shallow structure and its interconnection to

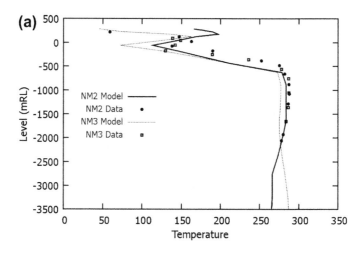

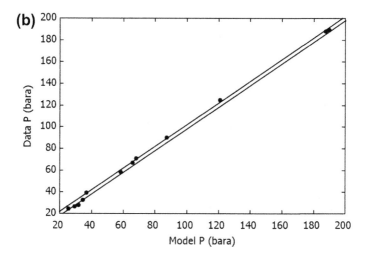

FIGURE 11.12 (a) Temperature matches in two wells, (b) Pressure matches. *Source: Burnell, Personal Communication. Reproduced with permission of Rotokawa Joint Venture.*

the deep aquifer. The model was then used to simulate the effects of production and injection over 50 years. Because there is no production history to provide calibration, the model is not fully constrained and these simulated results could be significantly in error. However, the model has highlighted the significant physical processes that might control long-term reservoir behavior. It identified the possibility of an influx of cool fluids from the shallow aquifers overlying the

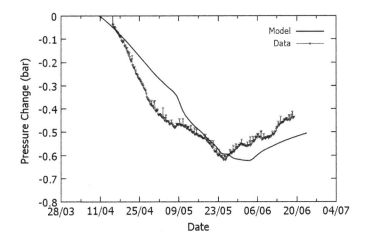

FIGURE 11.13 Interference test match. *Source: Burnell, Personal Communication. Reproduced with permission of Rotokawa Joint Venture.*

deep reservoir with consequent degradation of the production area. This issue constrains possible development options and field management plans emphasise the importance of pressure maintenance in the deep reservoir.

Field Examples

12.1. INTRODUCTION

This chapter presents case studies on a number of fields. Two fields, Wairakei and Svartsengi, are discussed at some length, since there is sufficient history to show not only the evolution of the reservoir under production but also the evolution of the understanding, modeling, and management of the field.

12.2. WAIRAKEI

Wairakei field, in the North Island of New Zealand, was the first high-temperature liquid-dominated field to be substantially exploited. Exploratory drilling began in 1949, and the first power generation was in 1958. Due to its long history, it has been the subject of much study. There have been different concepts of the field, different reservoir interpretations, and changing management priorities. The history as it is now understood is discussed here. For additional background and more detailed discussion, see O'Sullivan and colleagues (2009), Bixley and colleagues (2009), Carey (2000), Hunt (1995), and Bignall and colleagues (2010). Monitoring of deep liquid pressure in wells drilled in the adjacent Tauhara field has shown that there is a good connection between these two fields at depth.

Geothermal Reservoir Engineering. ISBN: 978-0-12-383880-3 DOI: 10.1016/B978-0-12-383880-3.10012-5

12.2.1. Natural State

Natural activity at Wairakei consisted of about 400 MW of water-fed and steam-heated features, plus a further 100 MW at the adjacent Tauhara field. The first exploratory wells at Wairakei were drilled to depths of around 300 m near areas of surface activity. The results from these shallow wells were sufficiently encouraging to prompt deeper drilling. This "deeper" drilling to depths of 1000–1200 m commenced in 1953 and soon proved an area of very high productivity at depths of 500–600 m. Downhole temperatures were measured in all wells after drilling and regularly thereafter. Downhole pressures were not measured until 1961 but can be calculated for earlier times for some wells where temperature profiles indicated that a well was liquid-filled and a well-head pressure (WHP) or water level was recorded.

Figure 2.6 shows the initial pressure distribution with depth. The pressure profile identifies the reservoir as liquid-dominated. The maximum temperatures approximate BPD down to a depth of 400 m and a maximum temperature of 260–265°C (Banwell, 1957). The measured pressures exceed hydrostatic gradient by about 7%. Using this excess pressure and the natural mass flow through the system of 400 kg/s, an average vertical permeability of 8 md can be estimated. No prominent break in the pressure gradient is apparent in the profile shown in Figure 2.6, indicating that the reservoir is uncapped on the large scale. The discharge enthalpies of most wells were initially equal to temperature at the feed zone, although higher enthalpies were obtained in some shallow-feeding wells. Thus, the reservoir in its natural state was liquid-dominated, with a base temperature of about 260°C. In the upflow region of the reservoir, liquid-dominated two-phase fluid was present at depths shallower than 400 m and in parts of the groundwater aquifers.

Mass flow of most wells was usually large, and flowing pressure measurements on some wells showed little drawdown. The permeable zones have very high horizontal permeability, with kh of productive wells being typically in the range 10–100 dm.

12.2.2. Exploited State

Under exploitation, pressures in the liquid part of the reservoir declined, and a two-phase zone formed, with a "water level" at about 400 m depth. As pressures declined, boiling conditions extended to greater depth within the reservoir. Over a few years, the two-phase zone segregated into a vapor-dominated zone, with vertical gradient less than half hydrostatic and a liquid-dominated two-phase zone with a pressure gradient near hydrostatic. In the vapor-dominated zone, well discharges consisted mainly of saturated steam, and in the liquid-dominated zone, discharge enthalpies matched the liquid temperatures. Wells with feed zones in both the liquid and steam zones had enthalpies intermediate between liquid and steam.

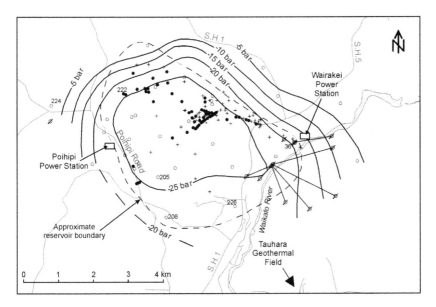

FIGURE 12.1   Pressure change from natural state to 1997. *Source: Bixley et al., 2009. Copyright © Elsevier.*

Although pressures in the deep liquid have declined by about 25 bar from the predevelopment values, within the productive area the drawdown is very even within 1 bar. Figure 12.1 shows the pressure distribution in 1997. Wells at the margins of the productive area had temperatures lower than observed in the productive wells and smaller pressure drawdown. The uniformity across the field of pressures in the liquid-dominated zone means that the history can be conveniently described by a single pressure. The more permeable production wells feeding from the liquid-dominated zone recover to the reservoir pressure within minutes of being shut. This observation, together with the horizontal pressure uniformity, indicates high horizontal permeability

12.2.3. Changes with Time

When the original production system at Wairakei was designed, there was no appreciation of the potential environmental effects of surface disposal of separated geothermal brine, which at that time was discharged directly into the nearby Waikato River. In the late 1970s, injection of used process fluids, separated brine, and condensed steam was investigated in several geothermal operations worldwide, and since that time, injection has gradually become the accepted method for disposal of these fluids and, in some cases, management of the resource. At Wairakei, significant injection did not begin until the mid-1990s.

Initially the field was liquid-dominated, with the overall discharge enthalpy generally corresponding to the liquid feed temperatures, but as just mentioned, some wells produced high enthalpy fluids before significant drawdown was observed. With exploitation, pressure fell everywhere, the boiling zone extended, and the steam fraction within the boiling zone increased. By 1962, there was a general trend toward rising enthalpy because many wells produced excess steam. After this time, well behavior tended to stabilize again, although some shallow feeding wells continued to rise in enthalpy and eventually produced "dry" steam. Steam cap profiles became common in downhole pressures, reflecting development of the steam zone, and the pressure profile in the reservoir at this time started to take on a consistent shape with distinct vapor- and liquid-dominated zones becoming evident.

Figure 12.2 shows temperature changes observed in the reservoir. Intrusions of cold water appeared in localized zones of the eastern field. Cool near-surface fluid flowed downward into the productive reservoir in the area of original surface discharge. When such cool flows impacted on production wells these were worked over (to instal deeper cemented casing), or abandoned. There was also a more general and gradual cooling of production temperatures in the east side of the field.

Over time, the overall field discharge enthalpy has not increased by more than 10%. The pressure drawdown in the deep reservoir is moderate by world standards, and excess enthalpy of the two-phase and steam wells is not a major

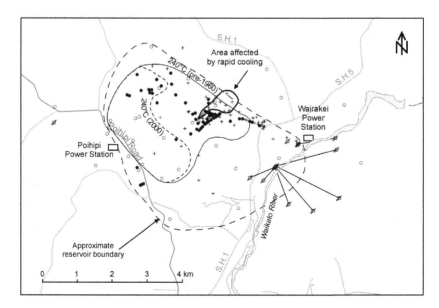

FIGURE 12.2 Temperature changes at −400 masl from initial state to 2000. *Source: Bixley et al., 2009. Copyright © Elsevier.*

feature of the field. Up to the mid-1980s, over 95% of the mass discharge from the field was derived from the liquid-dominated zone, and for the most part, it consisted of liquid water. In fact, changes in the overall field enthalpy since 1970 are the result of reservoir management options rather than deriving from changes in the resource. In the late 1980s, new production wells were drilled, targeting the shallow steam zone in the western part of the field, and eventually up to a third of the total steam production came from this source. The exploited pressure profile indicated that the liquid-dominated zone had developed a free surface. The discharge enthalpy and pressure profiles suggested that the field pressure would behave like that of an unconfined aquifer.

By 1980, the pressure in the liquid reservoir had stabilized at 25 bar below the original pressures (without any reinjection. Figure 11.3) and together with repeat gravity surveys (Hunt, 1995) indicated that 100% of the discharge was being replaced by recharge (both hot and cold). Thus, the high-temperature reservoir at Wairakei had very substantial recharge, increasing from the natural rate of about 400 kg/s to more than 1000 kg/s as a result of pressure drawdown.

12.2.4. Conceptual Model

Figure 2.9 shows a simple conceptual model, the "drainage model," describing the natural state of the field and changes with time. Similar figures can be found in Bolton (1970). In the natural state the reservoir conforms to the boiling-point-for-depth (BPD) model, with base temperature 260°C. In its exploited state, a vapor-dominated zone forms, and the temperatures immediately beneath this zone are controlled at boiling point by falling pressures. Recharge can come from two sources: deep recharge of 260°C water and recharge of cooler near-surface water from the edges and above the high-temperature reservoir.

12.2.5. Lumped-Parameter Models

James (1965a) was the first to attempt to calculate the behavior of a field like Wairakei. The first quantitative model was that of Whiting and Ramey (1969), who modeled the reservoir as a confined box of liquid water with transient recharge and obtained the reservoir storage coefficient. The assumption of a confined liquid reservoir resulted in a very large reservoir volume ($\varphi V = 310 \text{km}^3$). A good fit to the early pressure history was obtained. Subsequent lumped-parameter models are developments of this successful history match, assuming, however, an unconfined reservoir or a reservoir with liquid-dominated and vapor-dominated zones (Sorey & Fradkin, 1979; Zais & Bodvarsson, 1980; Wooding, 1981). Although the reservoir as a whole acted like an unconfined aquifer, the response of outlying wells fit a confined aquifer model.

All of these simple models obtain correlation coefficients of 98% or better to the pressure history and so cannot be discriminated by quality of fit. The interpretation of the storage coefficient indicates that the drainage model is a better fit over the first two decades, as does the mass loss determined from gravity surveys. In all of these models, the discharge enthalpy is assumed to be near liquid, and thus enthalpy provides no test of the model. Given the amount of recharge, as determined by fitting the drainage model, an energy balance on the system can be performed once the history of the steam zone is known. From this assessment it is found that the overall enthalpy of the recharge is significantly below 260°C water, so a significant proportion of the recharge is cold.

12.2.6. Simulations

Further work on lumped-parameter models of the field ceased once simulation became possible in the 1980s. Trend fitting continues to be used as the main management tool for short-term predictions, particularly of production well performance. The simulation history is presented by O'Sullivan and colleagues (2009) and Mannington and colleagues (2004). Figure 12.3 shows the evolution of the size of successive reservoir models. The first models were limited in the number of blocks, and the first model was a 1D vertical model with 14 layers. The number of model blocks has increased with time as computer capacity increased and expectations for field management became more detailed. All models to date have been single-porosity models.

The first models of the field included the Wairakei area only, with a recharge condition at the Tauhara boundary. Pressure measurements had shown that wells in the adjacent field, Tauhara, were also responding to production at Wairakei. In 1992, Tauhara was included in the reservoir model (at this stage becoming a system model), and in 1997, the fluid type included in the model was changed from steam/water to steam/water/air to allow for development of a groundwater surface in the model. Finer layers have also been added near the surface to enable better representation of shallow changes. Figure 12.4 shows the 2008 model grid.

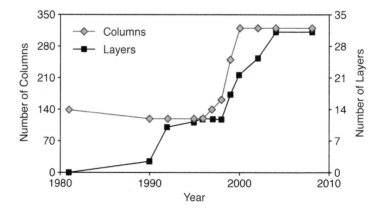

FIGURE 12.3   Number of layers and columns in Wairakei-Tauhara model.

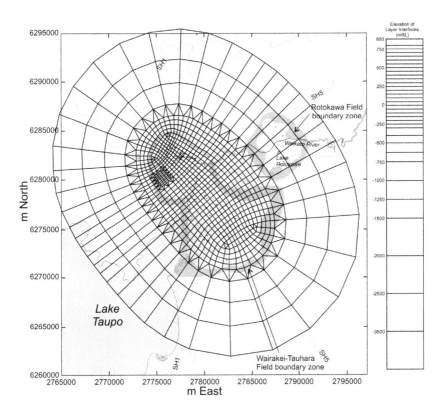

FIGURE 12.4   2008 Wairakei-Tauhara model. *Source: O'Sullivan, 2009. Copyright © Elsevier.*

In 2009, the reservoir simulation model was coupled to a finite-element subsidence package in order to model subsidence (O'Sullivan et al., 2009). Model calibration at this time included well-by-well matching of temperature in the initial state, or the year the well was drilled, well-by-well matching of pressure and enthalpy history, and matching of changes in surface activity. Some of the history matching at Wairakei is shown on Figures 11.3-11.5. The pattern of surface deformation at Wairakei includes a broad area of even subsidence similar to that where deep pressure decline is observed (Figure 12.1), plus a smaller area of accelerated subsidence toward the northeastern part of the field. The subsidence model reproduces reasonably well the broad pattern of surface deformation, while the localized areas of high subsidence are described by local models because they need much finer gridding, and the correspondence between finite-difference and finite-element formulation is not straightforward.

12.2.7. Summary

Wairakei and Tauhara are two linked liquid-dominated fields—Wairakei with a long production history and base temperature of about 260°C and Tauhara with no production until 50 years after Wairakei and a base temperature close to 300°C. Under exploitation, vapor-dominated zones have formed at different locations and depths in response to pressure drawdown in the deep liquid zone. Pressures in the liquid-dominated zone of Wairakei and parts of Tauhara follow similar trends with time, but with an offset of a few bars, indicating there is a deep connection between the fields. There is substantial recharge enhanced by the pressure drawdown, both of deep hot water and of cooler near-surface waters.

12.3. THE GEYSERS

The Geysers geothermal field in northern California is the largest electric power producing field in the world. It has been developed by a number of independent developers, although most of these independent developments have now been "combined" into a single operator. There is a large geothermal anomaly in The Geysers—Clear Lake area. Part of this area is hot but unproductive, and part contains the permeable vapor-dominated reservoir.

The reservoir is vapor-dominated, with an initial pressure at sea level of 35.4 bar (Barker et al., 1991). Prior to development, the steam was saturated and remained so until pressure and temperature had declined to dry-out conditions at 165—190°C. (Reyes et al., 2003). No deep-water level has yet been found beneath the steam reservoir. In the northwest, there is at depth the high temperature reservoir (HTR), where temperatures reach 345°C (Walters et al., 1991) in superheated steam. Figure 12.5 shows a conceptual model of the natural state of the reservoir. In the southeast, steam rises from a conjectured deep boiling zone of liquid water, and in the northwest, the water is absent and superheated steam rises from depth. Within the reservoir, there is an upflow of steam and return downflow of condensate.

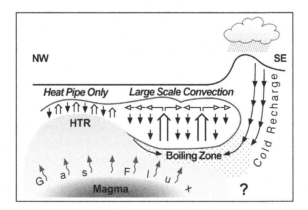

FIGURE 12.5 Conceptual model of The Geysers showing flow of steam and liquid. *Source: Truesdell et al., 1993. Copyright © Geothermal Resources Council.*

Development began in 1955, with the first 11 MW unit being installed in 1960. Expansion then continued with a number of independent developments adding capacity. Figure 12.6 shows the production history. Resource assessment during the period of expanding development was largely confidential due to the presence of competing developers. Three approaches to the management or prediction of reservoir performance are reported in published material (Sanyal et al., 2007, 1989; Koenig, 1991):

1. A well spacing of 1 well per 40 acres (6 wells per km^2)
2. Decline analysis of well flow to infer reserves as total cumulative production
3. P/Z analyses

The use of well spacing as a development guideline was taken from the petroleum industry. The choice of 40 acres as the spacing is a rule of thumb based on operating experience. With typical production rates of 45−70 t/h (Barker et al., 1991), it is equivalent to a power density of over 40 MW/km^2. Decline analyses using harmonic decline were used as a basis for development plans (Koenig, 1991). There has also been extensive use of decline analyses to extrapolate the performance of individual wells or groups of wells. P/Z analyses (see Chapter 3) are simply a plot of average P/Z against cumulative production by analogy with a natural gas field, and can be used after significant production has produced a significant change in pressure.

By the late 1980s, development was intense and decline rates rapidly increased due to interference between the increasing numbers of wells. Installed capacity reached a peak of 2043 MWe in 1989 (Koenig, 1991). Drilling ceased, and some of the power stations committed in the late 1980s

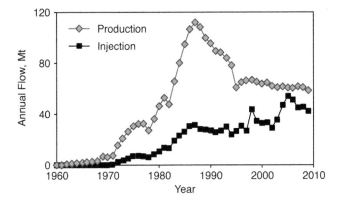

FIGURE 12.6 Steam production and injection. *Source: California Division of Oil, Gas, and Geothermal Resources.*

never generated. The rate of reservoir pressure decline increased with increasing total withdrawal, as shown in Figure 12.7. In the 1980s, the development of superheated areas of the reservoir showed operators that the reservoir was approaching dry-out conditions and that injection was needed to provide additional fluid in order to continue mining the heat stored in the reservoir rock.

A numerical dual-porosity model was developed by Unocal (Williamson, 1992) and later extended to include data from all operators in an Industry Consortium (Menzies & Pham, 1993, 1995). The model is shown in Figure 12.8 with six layers, for a total of 2880 matrix blocks and 2880 fracture blocks. Reservoir top was defined from shallowest steam entry, and reservoir bottom by deepest steam entry, location of microearthquakes induced by injection, and history matching. The upper part of the reservoir was initially saturated, while the two bottom layers were superheated. Initial liquid saturation of four upper layers in the fractures was set at 1−25% and in the matrix at 83%. The model was calibrated against the pressure history of the field. The primary parameters that were adjusted in this matching were fracture and matrix permeability, and initial matrix liquid saturation. This model showed that the declining production trend would continue into the future and that additional drilling was not economical (Sanyal, 2000). The model also indicated that injection should reduce the decline in production rates. Injection was initially only the surplus steam condensate, but the modeling had shown that additional injection would be beneficial.

Further injection was provided in 1997 and 2003 by pipelines bringing treated wastewater from Santa Rosa more than 40 km away (Goyal & Pingol, 2007). These injection projects reduced steam decline rates by increasing production above the prior declining trend. Figure 12.9a shows the simulated

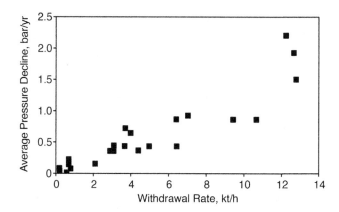

FIGURE 12.7 Reservoir pressure decline rate as a function of mass withdrawal. *Source: Barker et al., 1991. Copyright © Geothermal Resources Council.*

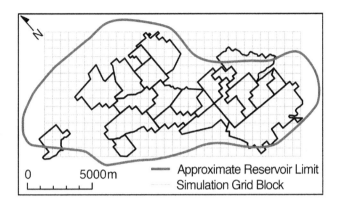

FIGURE 12.8 Simulation grid. *Source: Menzies & Pham, 1993. Copyright © Geothermal Resources Council.*

effect of SRGRP (Santa Rosa—Geysers Recharge Project), and Figure 12.9b shows the observed response by comparing actual production against extrapolation of the preceding trend. Both show a gain of about 80 MW. The gains from injection have decreased with time. Wright and Beall (2007) show decreasing returns of Freon tracer with successive tests and the development of a temperature reversal at depth in the southeast Geysers. Thus, with time, a cooler liquid-saturated region is developing, and less of the injected water is vaporized to make fresh steam.

Sanyal (2000) estimated from past production and extrapolated decline a reservoir capacity of 40,000 MW-yrs, or 1367 MW for 30 years, equivalent to a power density of 17 MW/km^2.

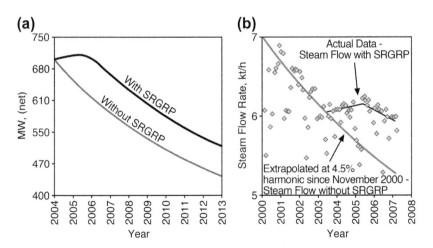

FIGURE 12.9 (a) Simulated production, (b) Actual production. *Source: Stark et al., 2005; Goyal & Pingol, 2007. Both Copyright Geothermal Resources Council.*

12.4. SVARTSENGI

Svartsengi in Iceland shows the progressive development of understanding the reservoir with time and production experience. At the end, a well-calibrated numerical model was produced that is somewhat simpler than some earlier versions.

The geothermal field of Svartsengi has been described by Thorhallson (1979) and Georgsson (1981). It is located on the Reykjanes Peninsula, the landward extension of the mid-Atlantic ridge in the southwestern part of Iceland. In this region geothermal fields are associated with fissures in the spreading zones. Energy from the Svartsengi reservoir is used for electricity generation and district heating, and the total plant capacity has expanded with time.

In the first stages of development, reservoir testing obtained some specific results, showing communication between wells in a compressed liquid aquifer, with some boundaries. Although there are some problems in interpretation of some parameter values, these are not important for the particular aspects of field behavior being considered. This first analysis was described in detail by Kjaran and colleagues (1979). A conceptual model of the natural flow in the earthquake zone is illustrated in Figure 12.10.

The fluid tapped in the wells at Svartsengi is derived from a mixture of seawater and freshwater concentrated by boiling, and it is at a temperature of about 240°C in the reservoir. The hydrology of the peninsula, including Svartsengi, was modeled conceptually by Kjaran and colleagues as shown in Figure 12.10a. A regional flow from Lake Kleirfarvatn to the sea mixes with intrusions of seawater while gaining heat as it moves through the system, with

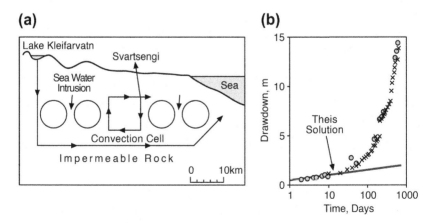

FIGURE 12.10   Svartsengi field. (a) Regional flow on the Reykjanes peninsula. (b) Drawdown in observation well H-5, semilog scale. *Source: Kjaran et al., 1979. Copyright © Geothermal Resources Council.*

some loss of steam to surface activity. The estimated heat input to the Svartsengi reservoir is 300 MWth.

In 1979, five wells had been drilled; of these, two were used for production, and one, H-5, was used as an observation well. H-5 responds to the two producers and to barometric pressure. The barometric efficiency of 75% implies a storativity of 3×10^{-7} m/Pa. The response of H-5 to the first 750 days of production is shown in Figure 12.10. For an initial period the drawdown is approximately linear with log (time), and this response implies a transmissivity of 170 dm and storativity 1.5×10^{-6} m/Pa. The drawdown later deviates from this semilog form and approaches an almost linear pressure-time trend. The entire response has been modeled as a system contained within a semi-infinite trench, with impermeable walls and one end at great distance. In this model the "box" is bounded by impermeable walls defined by earthquake activity. With these bounds the average storativity is reduced to the value implied by the barometric efficiency, whose average reflects the stiffer wall structures.

Kjaran and colleagues (1979) use the model to project future drawdown. This is important for well performance because the wells are subject to deposition above the flashpoint in the wellbore. The future pressures are used to project future scaling depths. On the basis of this analysis, it was decided to increase the depth and wellbore diameter in future wells. Later comparisons (see Figure 12.14) show that Kjaran's model provided a reasonable projection of future behavior.

The model of the reservoir as a liquid-filled box with a specified storativity and transmissivity is almost ideal for predictive purposes. It is simple and easy to understand physically, easy to calculate, and matches the reservoir behavior with a minimum of parameters. It is interesting to note that it does not describe all aspects of the reservoir. Kjaran and colleagues use the storativity of 1.5×10^{-6} m/Pa with good results, but the storativity is difficult to interpret. It implies either an unconfined aquifer of 1.2% porosity or a confined one of porosity-thickness 1 km, both difficult to accept. The surface discharge of steam, chemical evidence of boiling, and two-phase fluid encountered in one well suggests the presence of two-phase fluid in the reservoir above the 240°C water. Small pockets of two-phase fluid would create localized patches where the aquifer is unconfined. These conceptual problems do not influence the predictions made.

With time, the presence of two-phase fluid became more apparent. In 1984, a shallow well suddenly switched from liquid to dry steam production. During drilling in 1983, an overpressure was observed at 200 m that had not been present in 1971 or 1972, and there was increasing steam discharge to surface. Björnsson and Steingrimsson (1992) mapped the extent of the two-phase zone in the reservoir, showing that it was localized to a relatively small area and that it extended to ground surface in one place. Tracer tests made in 1982 and 1984 showed rapid fluid returns (Gudmundsson, 1983; Gudmundsson & Hauksson, 1985). The field continued to operate with surface disposal of cooled process fluids.

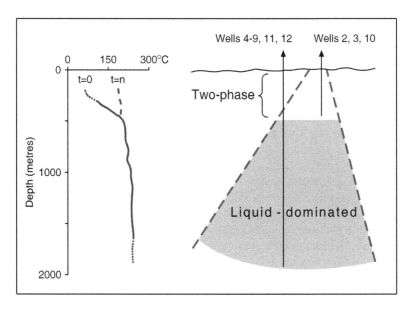

FIGURE 12.11 Conceptual model of Gudmundsson and Hauksson. *Source: Gudmundsson & Hauksson, 2005. Copyright © Geothermal Resources Council.*

Thus, in terms of the conceptual model of the reservoir, the important change at this time is the recognition of a steam zone developing as the pressure drew down. There would have been a small two-phase zone in the natural state, which grew with drawdown and boiling at the reservoir top. The next reservoir models explicitly incorporate such a boiling zone.

Gudmundsson and colleagues (1985) made simple lumped-parameter models based on standard petroleum reservoir models of the production history. They determined the storage coefficient for the reservoir and found that it was too large for a reservoir of confined liquid and corresponded to an unconfined reservoir of 3.5 km^2 (with assumed porosity of 10%). They also found that the long-term pressure response was not linear with cumulative production, and they concluded that there was some lateral recharge. The implicit model behind this is a volume of liquid overlain by steam, with drainage as the water level falls, and lateral recharge in the liquid. They also note that injection was being considered to reduce pressure drawdown.

Gudmundsson and Hauksson (1985) present this revised conceptual model, shown in Figure 12.11, which incorporates a two-phase zone at the top of the reservoir. Note that the two-phase zone is shown as having heated up in response to exploitation. This reflects the observed shallow pressure rise and expansion of surface steam discharge.

Bodvarsson (1988) developed a simulation model that had no impermeable lateral boundaries. Björnsson (1999) developed a stylized radially symmetric

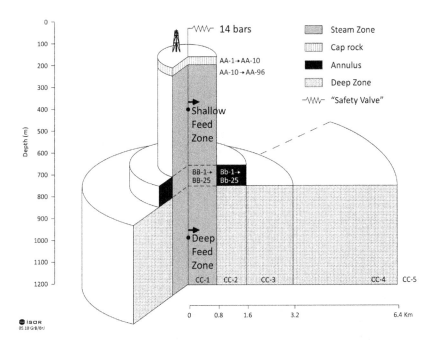

FIGURE 12.12   Model of Björnsson. *Source: Björnsson, 1999.*

model of a steam zone overlying a liquid zone, with lateral recharge in the liquid. This model reflects this new conceptual model of Gudmundsson and Hauksson, as elaborated in Figure 12.12. There is a liquid aquifer at depth, extending to distance radially, while within the reservoir there is a vertical column with a steam zone at the top. The reservoir area increases with depth; the "annulus" shown on Figure 12.12 was needed to match the history. There is production from both the steam zone and the deep liquid.

Ketilsson and colleagues (2008) presented a review of the modeling work on Svartsengi. In contrast to Wairakei, where the main simulation model has been progressively elaborated over the years, with some other independent models from time to time, at Svartsengi different successive models were constructed. There is now 30 years' production history to calibrate a model. Ketilsson and colleagues used a finite-element model, shown in cross section in Figure 12.13. The model is roughly radially symmetric but has 3D fine detail in the well field.

Initial guesses were assigned to the permeability and other parameters, and iTOUGH2 was used to find the best fit. An excellent fit was found to five observed data series:

a) The pressure history
b,c) The reservoir temperature at two elevations
d) The surrounding temperature
e) Production enthalpy

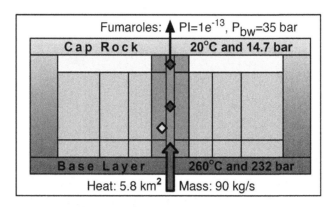

FIGURE 12.13 Model of Ketilsson and colleagues. *Source: Ketilsson et al., 2008.*

They provide a comparison of predictions from the various models for a scenario of constant production of 200 kg/s, shown in Figure 12.14.

The differences between the long-term behavior of the models are largely due to variations in the assumptions about lateral boundaries—that is, the strength of lateral recharge. The least drawdown is given by Bodvarsson's model, which has no impermeable boundaries. Note that Kjaran's model is reasonably accurate in projecting the long-term drawdown, but of course it does not predict the development of the steam zone. They concluded the following: "The results demonstrate the importance of a good conceptual model and the need for caution when assumptions about the system are made."

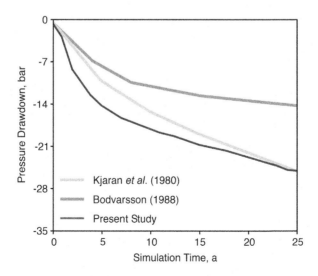

FIGURE 12.14 Pressure match of different models. *Source: Ketilsson et al., 2008.*

12.5. BALCOVA-NARLIDERE

Balcova-Narlidere is a low-temperature field near Izmir in Turkey. Figure 12.15 shows the conceptual model. The well field lies to the north of the Agamemnon fault. Hot water rises on this fault and spreads laterally northward. The well field consists of the earlier wells drilled up to 125 m deep, and later wells drilled 500−1100 m (Satman et al., 2005). Maximum temperature is about 140°C. The field is used for a district heating scheme.

There is production from both shallow and deep wells. There are extensive well tests, showing generally good permeability with kh's in the range 3−121 dm, and a complex system of communicating and noncommunicating fractures/faults (Onur et al., 2005). Reinjection was initially into three shallow wells, but this caused severe cooling (Aksoy & Serpen, 2005). Reinjection was transferred to deep peripheral wells in 2002. There are significant thermal effects from reinjection in shallow wells. When tracer was injected into the shallow well B9, there was observable cooling in well B4 after 12 days and in well B10 after 15 days. Tracer tests between shallow wells showed significant returns and were modeled as flow along a fracture using the method of Bodvarsson (1972).

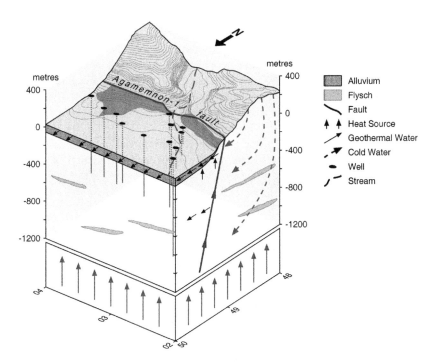

FIGURE 12.15 Simplified conceptual model for Balcova-Narlidere geothermal system. *Source: Aksoy & Serpen, 2005.*

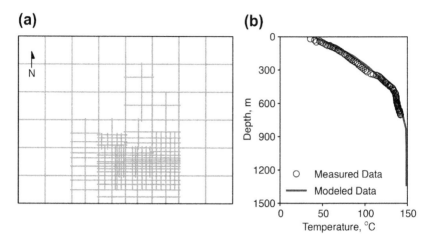

FIGURE 12.16 (a) Model grid. (b) Natural state temperature match. *Source: Gok et al., 2005.*

Lumped-parameter modeling of the field is described by Sarak and colleagues (2005) and Türeyen and colleagues (2007), and simulation is described by Gok and colleagues (2005). A variety of lumped-parameter models were developed, using 1-, 2-, and 3-tank models, open and closed. Different models produced acceptable fits, and it was concluded that a longer production history was needed and the model type should be supported and confirmed by additional geological, geophysical, and hydrological data. The lumped-parameter models fit only pressure and gave no information on temperature changes. Türeyen and colleagues (2007) concluded that a single-tank open model gave the best fit, with $S_M = 8.4 \pm 1.2 \times 10^7$ kg/bar, and recharge coefficient $\alpha = 44 \pm 2$ kg/bar.s.

The reservoir simulation, using a single-porosity model, provided a satisfactory match to the natural state temperatures and production history.

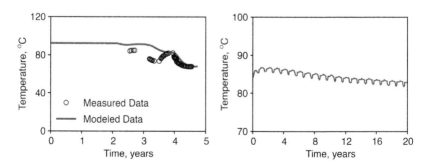

FIGURE 12.17 Temperature history match in well B-4, and forecast temperature. *Source: Gok et al., 2005.*

Figure 12.16 shows the model grid, with finer detail near wells, and one natural state temperature match. Figure 12.17 shows a temperature history match and predicted future temperature. The simulation addressed the field management issues, indicating that injection should be into deep wells to prevent temperature decline, and two additional deep injection wells were recommended. Forecast scenarios indicated production could be maintained for 20 years. Tracer tests were carried out after reinjection was shifted to deep wells and showed multiple paths and fractured media. Deeper wells showed more paths involved in each return (Akin et al., 2010).

12.6. PALINPINON

Palinpinon is located on the south of the island of Negros in the Philippines in mountainous terrain. The field has been developed in sectors, with Palinpinon-1 plant of 112.5 MW at Puhagan, and three plants of Palinpinon-2 totaling 80 MW in Nasuji-Sogongon by 2000. Figure 12.18 shows a map of the field. The reservoir is severely underpressured, with initial water levels more than 400 m below surface. This is because the reservoir lies under high ground, but reservoir pressure in the natural state is controlled by surface discharge in the valley below. The mountainous terrain constrains siting of wells, pipelines, and power stations. Much of the rock above the reservoir water level is cold. Many

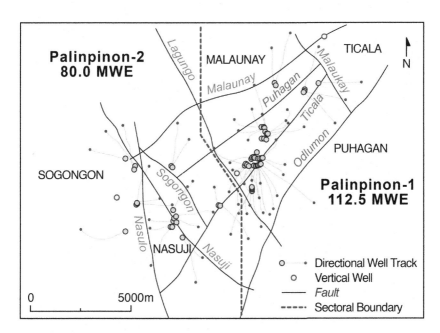

FIGURE 12.18 Palinpinon field, showing division into Palinpinon-1 and Palinpinon-2. *Source: Orizonte et al., 2003. Copyright © Geothermal Resources Council.*

wells were initially difficult to start flowing and required stimulation with steam or two-phase injection.

12.6.1. Early History

The early production history of the field is dominated by the effects of injection returns. The initial development of Palinpinon-1 used infield injectors at Puhagan. From 1983 to 1989, there was drawdown of 25 bar and rapid reinjection returns via the Ticala fault and Puhagan fault splays, causing the loss of production wells. The major flow paths were identified by tracer testing (Urbino et al., 1986). By the end of 1989, most of the injection had been moved further away from production, which resulted in the temporary recovery of most production wells, although some never recovered. By 1993, steam availability had again fallen due to injection returns, and injection locations were again moved further from production. Injection wells identified from tracer tests as communicating with production were preferentially shut. In summary, the dominant reservoir processes in the reservoir have been (Ramos-Candelaria et al., 1997):

1. Injection breakthrough via the Ticala Fault
2. Pressure drawdown enhancing boiling forming a steam zone at 220–240°C
3. Entry of shallow acid fluids in some wells

Production and injection strategy adopted has been as follows:

1. Maximize the use of high-enthalpy wells to reduce injection load
2. Minimize or eliminate use of injection wells with rapid returns
3. Transfer the bulk of injection further outfield
4. Plug the top zone of some injectors to prevent cooling the steam zone
5. Suppress acid inflows by pressure support with heated injection returns
6. Consider injection in high-temperature acid wells

By mid-1997, injection in Puhagan was totally eliminated. This caused excessive pressure drawdown, and although discharge enthalpy increased, steam flow still declined. Some injectors were identified as suitable for pressure support when used at limited flow rates. (Malate & Aqui, 2010).

12.6.2. The Mature Field

Palinpinon-1 field went onto full production in 1991 after the interconnection of Negros and Panay Islands added demand, and drawdown was 50 bar by 1994. With the drawdown, the two-phase zone expanded considerably, and this zone supplies additional steam while reducing the injection load. The two-phase zone improves well performance, since wells will discharge spontaneously, and the higher enthalpy helps to maintain WHP with declining reservoir pressure.

An additional 20 MW plant was considered at Palinpinon-2 after 2000. Orizonte and colleagues (2003, 2005) describe the field assessment using volumetric assessment, correlations of well performance with reservoir pressure, and a lumped-parameter model of the pressure response. Individual well performance and field performance was correlated with reservoir pressure at −1000 m, as shown in Figure 12.19. This figure also shows how the deviation from trend when flow suddenly falls rapidly, indicates deposition. The relation between reservoir pressure and flow rate was fitted to a simple exponential decline model with a relaxation time of four years (i.e., a simple lumped-parameter model), and on this basis it was calculated that an additional 20 MW plant would produce an acceptable additional pressure drop. Figure 12.20 shows the pressure history of Palinpinon-2 field to 2003 and the projected future trend after adding the additional plant.

Shallow steam production has been expanded due to the expanding two-phase zone, with an additional steam flow from this zone of the reservoir of 32 MW since 2004 (Malate & Aqui, 2010). Some wells that had been decommissioned were recommissioned after being shut 10 years, some Puhagan infield injection wells were converted to producers, and steam was produced from a formerly acidic well. These additional steam flows make up for losses due to reservoir pressure drawdown and injection returns.

Tracer testing in Palinpinon-2 again indicates the need to locate injection further from production (Maturgo et al., 2010). With the long history of reinjection returns and chloride measurements, Horne and Szucs (2007) showed that it is possible to correlate well chlorides with different injection wells and hence infer the connections between different producers and injectors. The results are similar to tracer testing but require only the routine chloride monitoring data. Villacorte and colleagues (2010) describe the successful

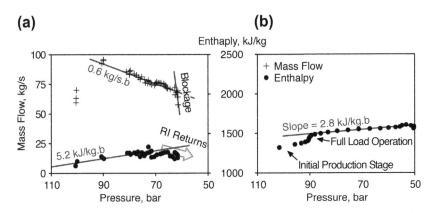

FIGURE 12.19 Correlation of well performance with reservoir pressure. (a) Well NJ-4D. (b) Palinpinon-1 field. *Source: Orizonte et al., 2003. Copyright © Geothermal Resources Council.*

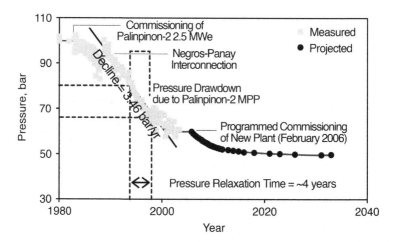

FIGURE 12.20 Pressure history in Palinpinon-2 and projected pressure with additional plant. *Source: Orizonte et al., 2003. Copyright © Geothermal Resources Council.*

application of the method to sectors of the Tongonan field. The method is not applicable in a field with little production history.

Malate and Aqui (2010) summarize that resource management has been a balance between the two major reservoir processes: pressure drawdown and injection returns. Management has been assisted by the expansion of the two-phase region at the top of the reservoir.

12.7. AWIBENGKOK (SALAK)

The initial assessment and development of Awibengkok in Indonesia is described by Murray and colleagues (1995) and later development and management by Acuña and colleagues (2008). Stimac and colleagues (2008) and Ganefianto and colleagues (2010) give an overview of the field. The field is shown in Figure 12.21 and covers an area of about 18 km². In its initial state, the field was liquid dominated with possibly a thin steam cap. Figure 12.22 shows initial conditions, with a well-defined (low permeability) capping formation, shown by conductive temperature gradients. Reservoir temperatures range from 235 to 310°C.

The initial assessment used extensive interference testing, which showed good connectivity in the reservoir with a *kh* of around 150 dm. There also appeared to be some lower-permeability barriers dividing parts of the reservoir. A numerical model was constructed, shown in Figure 12.23. There are two versions: a confined model that represents the volume proven by drilling and an expanded model. It was impossible to match the interference data without expanding the model into undrilled regions. The model in its initial state is entirely liquid. The model simulates only the proven reservoir, not including surrounding or overlying aquifers.

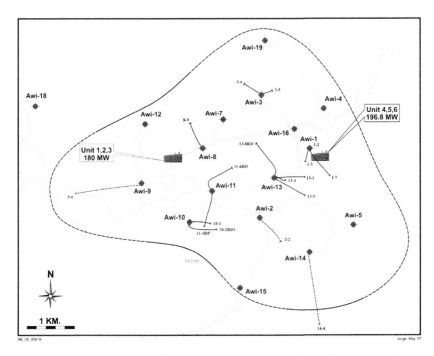

FIGURE 12.21 Map of Awibengkok field. *Source: Acuña et al., 2008. Copyright © Elsevier.*

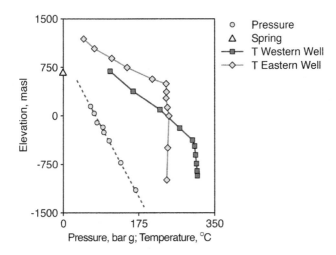

FIGURE 12.22 Initial pressure and temperature. *Source: Acuña et al., 2008. Copyright © Elsevier.*

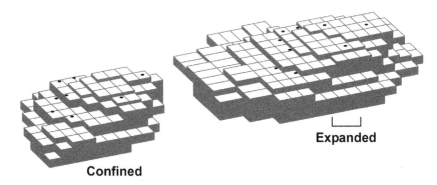

Expanded

Confined

FIGURE 12.23 Models of Awibengkok. *Source: Murray et al., 1995.*

Alternate versions of the model were constructed to deal with uncertainties. Acuña and colleagues (2008) describe the systematic development of these uncertainties. One such uncertainty was whether the additional fluid in the expanded model was hot, warm, or cold. All models showed a similar pattern: that with production and drawdown, a steam zone develops in the upper part of the reservoir. The model was used to simulate production at 330 MW. If the additional fluid is warm or hot, flow is sustained for 30 years, but for only 20 if it is cold. The model also indicated it was necessary to place injection deep and at the reservoir edge. Figure 12.24 shows the simulated and actual response to

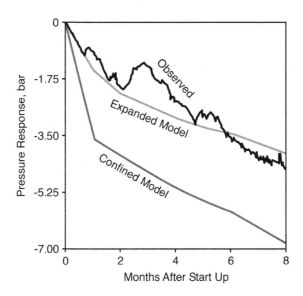

FIGURE 12.24 Response to 110 MWe. *Source: Murray et al., 1995.*

the start of production. The first 110 MW was built in 1994, and a further 220 MW in 1997, with a further increase to 377 MW in 2002.

Tracer tests have shown injection returns within a few days. Chloride concentrations in some wells have shown substantial fractions (up to 50%) of injectate in production fluid but without marked adverse effect on production. High returns were considered undesirable, and injection strategy has been changed periodically. In 1998, injection was moved away from the Awi 10 and Awi 2 pads and, in 1999, from Awi 4 pad. New injection sites were established on the periphery of the field. Discharge enthalpy of shallow-feeding wells has risen with the formation of the steam cap, reducing the injection flow, which has made it easier to manage the injection system. With surplus capacity, it was possible to relocate injection to the least deleterious injectors. Some of the wells that had been used for injection were restored to production. At the present time, it is planned to phase out injection at Awi 9 pad when a viable alternative can be found. Since 2006, the field management strategy has been to move injection locations further away from production, with exploration to the west, north, and south-east of the main reservoir (Ganefianto et al., 2010).

The reservoir simulation provided a good match to many parameters except gravity (Nordquist et al., 2010). Using the simulation to compute gravity changes, comparison with detailed measurements showed an increasing divergence. Some improvements were made by switching to an MINC model with finer vertical and horizontal discretization; however, a mismatch remained in one sector of the field. It was proposed that the mismatch was due to groundwater influx resaturating a shallow zone above the reservoir, not contained within the simulated volume. Simulation results show that field life can be extended significantly by moving some of the injection to outfield locations and some to peripheral locations. The outfield sites have poor permeability, and extensive well stimulation was carried out (Pasikki et al., 2010; see also Chapter 14).

Because of the high permeability within the reservoir, it is more economic to drill wide-diameter production wells, with tie-backs up to 16 inches, and shallow high-angle wells have been drilled to maximize shallow steam production. Well models are used to predict well performance (Acuña, 2003; Libert & Pasikki, 2010). The detailed well models provide more reliable production forecasts than simple trend extrapolation.

12.8. PATUHA AND OTHER HYBRID FIELDS

There are a number of fields that, in their natural states, contained both liquid-dominated and vapor-dominated regions. In some cases the field is basically liquid-dominated, driven by an upflow of liquid from depth, and the vapor-dominated region is relatively small and "parasitic" on the liquid-dominated region. Such fields have a liquid upflow and a relatively minor vapor-dominated zone forms at high elevations. Steam rises from a two-phase upflow and charges the vapor zone. An example is Los Azufres (see Figure 10.7), and another is Olkaria.

There are other fields in which the upflow at depth is steam, as in a vapor-dominated field, but a central steam "jet" pushes through a liquid layer. Figure 12.25 shows a conceptual cross section of Patuha in Indonesia (Layman & Soemarinda, 2003). There is an upflow of steam from depth. This steam rises and spreads laterally in the permeable vapor-dominated reservoir. Beneath the vapor zone there is a liquid-dominated zone. Wells that are sufficiently deep have a water level that is a real water level created by a genuine liquid zone at depth. That there is a genuine liquid zone is shown by the water levels being similar among the three wells that have them and by downhole measurements during discharge. The water level responded to pressure changes, so water flowed into the well when steam was discharged from the upper productive zone. This shows that there was free liquid in the reservoir and that the water was not simply a "dead leg" at the bottom of the well. However, this liquid is not boiling and so does not supply a steam upflow; the steam supply comes from depth through the liquid. Like a true vapor-dominated reservoir, Patuha is driven by an upflow from depth of steam, not liquid. The liquid appears to be primarily condensate. The steam upflow at Patuha is magmatic steam, containing corrosive gases.

Similar fields with a vapor core are Wayang Windu, Karaha-Bodas, and Alto Peak (Reyes et al., 1993; Allis et al., 2000; Moore et al., 2002; Bogie et al., 2008). In all these cases, the central steam upflow is magmatic steam, containing corrosive gases, making the core of the system difficult to exploit.

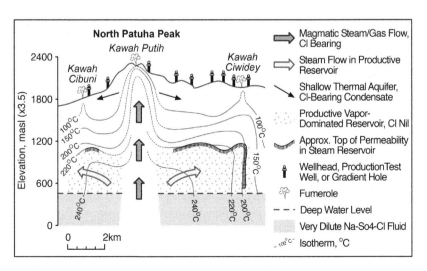

FIGURE 12.25 Conceptual cross section of Patuha geothermal field. *Source: Layman & Soemarinda, 2003.*

Away from the core, the acid gases are neutralized by their passage through rock. It has been conjectured that some of these fields are partway through the evolution from an original liquid-dominated state to a fully vapor-dominated reservoir.

12.9. MAK-BAN

The Mak-Ban (Makiling-Banahaw) geothermal field, also known as Bulalo, is located in the Philippines on the island of Luzon. Its development and history are described by Benavidez and colleagues (1988), Clemente and Villadolid-Abrigo (1993), Sta. Maria and colleagues (1995), and Capuno and colleagues (2010). The discovery well, Bul-1, was drilled in 1974. Three 110 MWe units were installed in 1979, 1980, and 1984. Bottoming units were added in 1994, and 4 × 20 MWe additional steam turbine units were added in 1995−1996.

Figure 12.26 shows a map of the field, and Figure 12.27a shows a cross section. The reservoir in its initial state was liquid-dominated with two-phase conditions in the upper part of the reservoir. Figure 12.28 shows the initial pressure and temperature distribution. There was a central liquid-dominated pressure gradient, with slightly higher pressures in low-permeability marginal wells. Maximum temperatures were at saturation for the pressure. The vertical pressure gradient was 10% above hydrostatic. Under exploitation, the two-phase zone expanded considerably, and enthalpy increased. Wells often contained a two-phase column when shut, due to crossflows from deeper zones to shallower zones (Belen et al., 1999; Menzies et al., 2007). Similar profiles in well BR2 are discussed in Chapter 9.

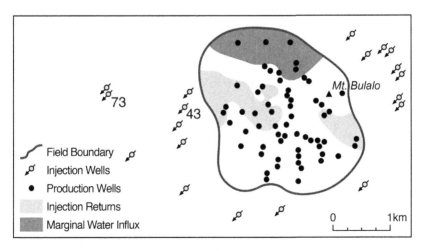

FIGURE 12.26 Map of Bulalo. *Source: Clemente & Villadolid-Abrigo, 1993. Copyright © Elsevier.*

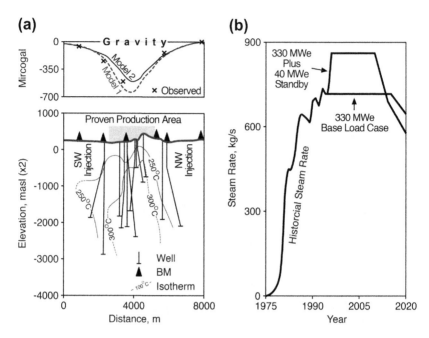

FIGURE 12.27 (a) East-west section showing isotherms and model match to gravity changes. (b) Simulated expansion. *Sources: (a) Protacio, 2000; (b) Sta. Maria et al., 1995.*

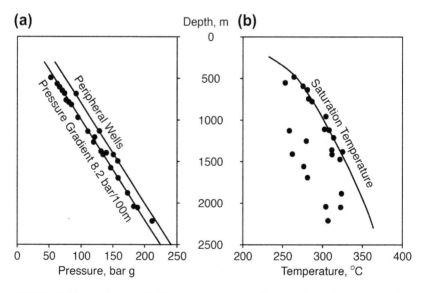

FIGURE 12.28 (a) Pressure (b) Temperature. *Source: Clemente & Villadolid-Abrigo, 1993. Copyright © Elsevier.*

There have been returns of injection water and some inflows of marginal water and groundwater. Figure 12.26 shows inflows in the early 1990s. Some injection was relocated from the western group around Bul-43, further west to the area of Bul-73. Enthalpy has generally increased with time, with an average flash fraction of 50% in 2001, although by 2008 there had been a small decline, with the average flash fraction falling to 43%, due to an increased effect of injection, marginal, and groundwaters. Geochemical monitoring indicated marginal recharge was 30−40% of production in 1999, compared to injectate at 5−10% and negligible amounts of meteoric water (Abrigo et al., 2004).

The first simulation model (Atkinson & Pedersen, 1988) and a later model (Strobel, 1993) were calibrated against initial pressure and temperature distribution, production enthalpy history, and gravity changes. Matching the vertical pressure gradient required low vertical permeabilities of 0.5−15 md. Matching production enthalpy history required a dual-porosity model. In contrast to Awibengkok, the model provided good matches to both the production history and the gravity changes (Nordquist et al., 2004, 2010). There was a qualitative match to tracer tests (Villadolid, 1991), which showed widespread and rapid mixing of injectate in the reservoir. Because of the extensive two-phase region in the reservoir, enthalpy and gravity are sensitive to saturation, and this provides significant constraints for model calibration. The model of Strobel (1993) showed that the reservoir could support additional generation. Figure 12.27b shows the simulated production, with the addition of 80 MW, 40 MW of which was operated as standby generation. This plant was added in 1995−1996.

The field has been an extremely good performer, producing at a power density of 57 MWe/km^2 over the well field area of 7 km^2. The detrimental effects of inflows of injection and marginal water have been manageable, and production has been sustained.

Field Management

13.1. INTRODUCTION

Once a field is developed and production has commenced, management of the production and injection operations to maintain the generation plant at full load is given the highest priority. This is a multidisciplinary task involving production monitoring, geochemistry, reservoir engineering, and simulation to bring together the production and injection data measured at the surface with the subsurface well information. In order to fully understand the processes happening in the reservoir, other specialist geosciences tasks are also called on from time to time, such as microgravity and ground deformation surveys. There will also be well condition monitoring and looking at subsurface corrosion and casing damage. All this information must be brought together into a coherent model that can be used to predict future production and injection capacity and to identify any areas where remedial action must be planned in order to maintain the production plant at full capacity. These actions range from scheduling simple well cleanouts to remove calcite scaling to moving the injection well field to avoid returns of cool fluids between injection and production wells.

Some data will be available frequently: wellhead pressure (WHP), valve settings, separated steam and brine flows, total flow (mass and enthalpy) from separator stations, injection well flows and temperatures, and so on, together

Geothermal Reservoir Engineering. ISBN: 978-0-12-383880-3 DOI: 10.1016/B978-0-12-383880-3.10013-7

with reservoir monitoring data that mainly comprise pressures in observation wells. Other data may be available only intermittently, such as mass flow rate, enthalpy, and chemistry of individual wells.

In production wells, if the production WHP and enthalpy remain constant, the mass flow changes in response to changes in reservoir pressure. Usually this takes the form of a decline, initially rapid and then flattening out as time goes on. Simple curve-fitting of past history often provides the best short-term prediction of well performance and provides a prediction against which to compare actual performance—for example, at Palinpinon (see Chapter 12). A change in trend then indicates some change in the well that may require further investigation. A detailed well model usually provides a better predictor of performance and an explicit ability to model the effects of particular changes, such as deposition or cooling of one feed zone.

Changes in enthalpy indicate a change in the fluid feeding the well. There may be an enthalpy rise due to expansion of two-phase conditions or an enthalpy decline due to cooling from reinjection returns or influx of cooler water in part of the reservoir. These changes will be accompanied by changes in the chemistry of the produced fluids; in fact, changes in the fluid chemistry will usually be a precursor to changes in the physical fluid properties. Occasionally a change of enthalpy, together with a change in mass flow, indicates the blockage of one zone in a well.

13.2. DECLINE AND LUMPED PARAMETER MODELS

This section discusses model or trend-fitting of well performance in order to establish a simple model to project future performance. The methods have underlying physical models but are best regarded as "black box" statistical fits to trend. Such trend analyses provide the best projections for short-term extrapolations, provided that conditions do not change. They are not reliable as models of long-term reservoir performance or capacity because large-scale reservoir processes are not reflected in the fitting process.

13.2.1. Exponential Decline

The commonest assumed form of decline is exponential decline, where the flow decreases by a constant fraction or percentage per year. Exponential decline arises from a simple model, a well producing from a fixed resource. Consider a box with storage coefficient S_M and a well producing at constant WHP so that flow is proportional to the difference between reservoir pressure and operating pressure. Conservation of mass and well flow are:

$$S_M \frac{dP}{dt} = -W \tag{13.1}$$

$$W = \beta(P - WHP) \tag{13.2}$$

which gives:

$$W = W_o \exp(-at) \tag{13.3}$$

where $a = S_M/\beta$ is the decline rate. If this decline continues for a long time, total cumulative production is W_o/a. Plotting well flow against time, with flow rate on a logarithmic scale, produces a straight line.

If there are a group of n identical wells, the flow of each of which follows Eq. (13.2), then:

$$S_M \frac{dP}{dt} = -nW \tag{13.4}$$

and for each well:

$$W = W_o \exp(-nat) \tag{13.5}$$

That is, the decline rate increases proportionately to the number of wells. Total cumulative production of all the wells is unchanged at W_o/a.

If there is a single well or a group of wells operating under constant conditions, the well or group show an exponential decline in flow rate. Note that this is only the case if the number of wells and the operating conditions remain unchanged. If the number of wells increases (i.e., the leak from the reservoir increases), the decline rate increases. This is shown in Figure 13.1, where the decline rate of a well at The Geysers increased rapidly during the period of rapid development in the 1980s. Figure 12.9b shows the use of exponential decline to establish the production trend in order to determine the change in production due to injection.

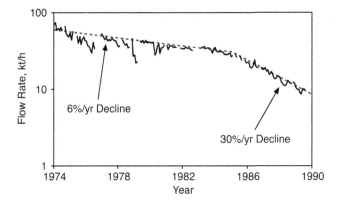

FIGURE 13.1 Increasing decline in Geysers well LF6. *Source: Barker et al., 1991. Copyright © Geothermal Resources Council.*

13.2.2. Other Forms of Decline

There are a range of decline curves used in petroleum reservoir engineering, of which harmonic decline is sometimes used in geothermal (see, for example, Enedy, 1991):

$$W = W_o/(1 + at) \qquad (13.6)$$

Plotting W^{-1} against time produces a linear plot. Similarly for a well flowing at constant pressure in an infinite aquifer, plotting W^{-1} against time with time on a logarithmic scale produces a linear plot.

13.2.3. Lumped Parameter Models

Lumped parameter models often provide a good method of fitting pressure history. They are routinely used in Iceland for modeling pressure in small, low-temperature systems (Axelsson, 1989, 1991; Axelsson et al., 2005a, b). The modeling of Hamar, a small low-temperature system, shows both the strengths and weaknesses of such modeling. There are two lumped parameter models used—an open and a closed model. Figure 13.2a shows the match. Data were matched up to 1993 and then projected ahead and compared with actual pressure. Figure 13.2b shows the projection for 200 years, where the upper line is the open model fit and the lower line the closed model fit.

Either model provides a good match and projection for 10 years or so, but there is a marked divergence at longer time periods. The reservoir storage coefficient S_V is 7×10^3 m^3/bar, and this controls the short-term response. The long-time behavior is controlled by the recharge assumptions, and the divergence is due to an uncertainty that is not resolved by the lumped-parameter model with a relatively short history for calibration—six years in this case. A longer record is required to estimate this parameter. A similar observation was made of the early lumped-parameter modeling of Wairakei, where all models gave excellent fits with markedly different underlying concepts (see Chapter 12 and Grant et al., 1982a).

Lumped-parameter modeling of this form is excellent for forecasting the data on which it is fitted—that is, the pressure-flow history. It has been so applied in a number of high-temperature fields (Vallejos-Ruiz, 2005; see Chapter 12), and it will usually provide the best short-term pressure forecasts. The limitation is that it cannot do anything more complex, such as predict temperature changes, and it cannot allow for reservoir processes that control long-term behavior. That requires a simulation that can represent all the physical processes present. Lumped-parameter modeling has also been used to identify the proportion of reinjection fluid in production flows (Itoi et al., 2003), and Horne and Szucs (2007) use nonparametric regression to achieve a similar result.

13.3. DEVIATIONS FROM TREND

Having established a trend, or pattern, of well performance, ongoing performance is then checked against this trend to observe any deviation. A continued

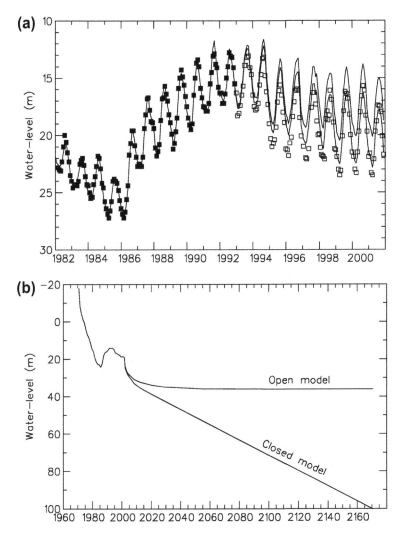

FIGURE 13.2 (a) Hamar pressure history match 1982−1993; prediction 1994−2001. (b) Hamar prediction for 200 years. *Source: Axelsson et al., 2005b.*

deviation normally indicates that some new process is affecting the well, or reservoir, that may require some intervention. Possible significant changes for production wells can be the following:

1. An accelerated decline in mass flow
2. An accelerated decline in reservoir pressure
3. A fall or rise in discharging pressure
4. A fall or rise in enthalpy
5. Unstable performance

For injection wells, possible significant changes are the following:

1. Declining flow rate (at constant pressure)
2. Rising static pressure

In all cases, the first action is usually to take chemical samples of production/ injection fluid and to run downhole surveys to check the wellbore and casing condition and then the reservoir conditions: permeability, pressure, and temperature. The survey sequence should be something like sinker bar, PTS shut, PTS flowing and possibly calliper log, and downhole solid sample or downhole camera.

13.3.1. Deposition

Well performance is frequently affected by mineral deposition, either inside the wellbore or in the formation near the well. The deposition is usually calcite (with a fraction of silica), although silica is sometimes formed in high-enthalpy wells (where the water fraction is $<5\%$), and other more exotic minerals such as sulphides can be formed in wells producing acidic fluids. Deposition has a characteristic effect on well flow. The first solid material is deposited around the inside of the casing or liner, and then additional material is deposited over the initial layer and the wellbore diameter progressively decreases until production stops. The decline rate accelerates with time. Thus, it is important to note early signs of rundown due to deposition. When a well (production or injection) shows unexpected decline in flow rate and there is no other obvious reason such as significant change in reservoir pressure or fluid quality of a production well, deposition is usually the first cause to investigate.

Injection performance is readily monitored using the WHP and flow rate, provided that the downhole temperature is effectively constant and the wellbore is full of water to the wellhead. If there are substantial variations in WHP due to changes in control, the data need to be adjusted to a standard pressure or standard flow rate, or simply plotted up as flow against WHP to see if the characteristic injection curve is systematically shifting with time. A rise in reservoir pressure causes the same rise in WHP at all flow rates, whereas deposition causes an increase in WHP that is greater at higher flow rates. If a change in well performance is suspected, some measurements can be made without interrupting the use of the well. If the injection flow rate can be adjusted by diverting flow to other wells, injectivity and injection performance at standard conditions can be measured to compare with past data. If necessary, downhole surveys can be made to check for scale buildup or to measure changes in injectivity at the feed zone.

If deposition is identified as the cause of declining well performance, a workover will be required. If there is deposition in the wellbore, it is possible that there is also deposition in the formation. Deposition in a production well will normally occur where the fluid first boils. In high-permeability wells, boiling usually starts inside the wellbore and deposition is confined to the perforated liner and casing. In lower-permeability wells with large drawdown, the reservoir fluid may boil in the formation and deposition can form in the fractures at the feed zone. Deposition can also occur due to the mixing of two fluids of different chemistry. Deposition in an injection well is due to the injectate being supersaturated with respect to a mineral, usually calcite or silica. Deposition will then normally occur both in the wellbore and the formation. A mechanical cleanout can effectively remove scale from inside the casing and liner, but the deposition may not be removed from the liner perforations. Where the scale has been deposited in the formation, a mechanical cleanout will not fully recover the well productivity and acid treatment is essential to dissolve scale in the feeding fractures. Pressure transients can be used to measure changes in near well permeability (skin effect) and identify if acid treatment is likely to improve production.

In a production well, unless the flow is single-phase, with current technology there are usually no frequent measurements of flow rate (mass and enthalpy) and WHP, since these will be available only intermittently from TFT measurements (assuming two-phase production from several wells to a single separator). Between these spot measurements, only trends in WHP and perhaps valve setting are available. Deposition will be seen as a gradual decline in WHP or as a progressive opening of the flow control valve needed to maintain the required production flow. Abnormal changes are cause for concern, and the first step is to make a flow measurement by TFT (Tracer Flow Test 8.5.5) or by diverting the well for physical measurement by separator or lip pressure method. If a decline in flow is confirmed, without change in enthalpy, the most likely cause is a fall in reservoir pressure or mineral deposition. Measurements taken in other wells throughout the field should be adequate to define if declining reservoir pressure is a likely cause of declining flow rate. The next step is to shut the well and make downhole measurements to check for blockages and the pressure-temperature conditions at the feed zone, although in real life this decision is usually made too late and the wellbore is blocked to PT surveys by the time the decision to investigate the cause of flow decline has been made.

For a liquid-fed well, deposition will extend upward for some distance above the flashpoint in the wellbore and can be removed mechanically. To sustain production without scaling and associated flow decline, installation of tubing to allow injection of scale inhibiting chemicals below the flashpoint is usually a solution. If the flashing is in the formation, deposition may also be in the formation. Pressure transients measured close to the feed zones while adjusting flow rate can be used to determine the skin factor and compare this with previous measurements. If a significant increase in the skin factor is found, acid treatment to remove scale from production zones near the wellbore is likely to restore well performance.

13.3.2. Changes in Enthalpy

Changing enthalpy will cause changes in the performance of producing wells; in general, an increase in enthalpy causes the total mass flowrate to decrease and the maximum discharging pressure to rise. For constant wellhead pressure conditions, the separated steam flow often shows little change as the overall enthalpy increases. BR2, discussed in Chapter 7, illustrates these changes.

A rise in enthalpy normally indicates boiling in the formation because either a two-phase zone has formed in the reservoir or there is large drawdown around the well and a flashing zone of significant size around the well. Once there is a two-phase zone in the reservoir, wells producing from it typically vary in production enthalpy with time, and the enthalpy history should be monitored for abnormal changes in trend. A fall in enthalpy in a liquid-fed reservoir indicates a fall in temperature of the feed water at one of the well's feed zones. This has an immediate effect on well performance as the fluid column in the wellbore becomes more dense, and the total mass flow and separated steam flows both decline. This could be due to injection returns or incursion of cooler peripheral or shallower fluids.

13.4. TRACER TESTING

Tracer tests are frequently used to test for returns from injection to production wells. Qualitatively, interpretation is straightforward: tracer Returns that are large and rapid imply probable thermal effects. Drawing more specific conclusions has proven to be more difficult. However, for tests between injection and production wells, the source of strong returns is immediately identified and remedial action can be taken by shutting that well.

13.4.1. Normalizing the Data

Tracer concentration $c(t)$ is measured in observation wells. In order to compare between different tests with different flow rates and amounts of tracer injected, the data can be normalized by converting the concentration to an age distribution or distribution of return times:

$$E(t) = c(t)W/M$$

where M is the amount of tracer injected and W is the flow of the observation well so that $c(t)W$ is the amount of tracer recovered per unit time. $E(t)$ has units of 1/time and is simply the fraction of the tracer recovered at the observation well per unit time interval. The integral of $E(t)$ over the entire return—that is, the area under the curve of $E(t)$—is the fraction of the tracer recovered. The following discussion is in terms of concentration, $c(t)$, but if different wells or different tracers were being compared, then they should be normalised to $E(t)$.

13.4.2. Travel Times and Percent Recovery

The most basic interpretation is to catalogue the concentration of tracer returned to each monitoring well and the travel time. The following discussion assumes that there is no significant recycling of the tracer—that is, that the only tracer injected is a single dose at time zero. Let the amount of tracer injected be M. The total tracer recovered at the ith monitor well by cumulative time t is simply the integral over time of tracer recovered:

$$m_i(t) = \int_0^t c_i(t')W_i(t')dt' \tag{13.7}$$

The total recovery of tracer is the recovery after long time, $m_i(\infty)$. This normally involves extrapolating the tail of the recovery curve, since usually at the end of observation the concentration is not zero and there is clearly some more recovery to come. It is convenient to fit the end of the recovery curve to exponential decline, from which the final recovery can be calculated. The fraction of tracer recovered from each well is then $m_i(\infty)/M$. This is also the fraction of fluid injected at the source well that is recovered from the monitor. From this it is possible to calculate the proportion of the production that is returned injectate. If the flow into the injection well is W_I, the amount of this flow that is recovered at the monitor is $m_i(\infty)/M \times W_I$. Then, if the flow of the monitor well is W_i, the fraction of its flow that is returned injectate is $m_i(\infty)/M \times W_I/W_i$. The travel time is usually considered to be either the time of first arrival or the time of peak concentration. The mean residence time, or "first temporal moment" (Shook, 2005), is:

$$\tau_i = \int_0^\infty tc_i(t)dt \bigg/ \int_0^\infty c_i(t)dt \tag{13.8}$$

Figure 13.3 illustrates a tracer test at Wairakei. Figure 13.3a shows the concentrations (in total flow). The values have been normalized using the method of Bixley and colleagues (1995). Figure 13.3b shows $E(t)$. The relative contributions of the different wells change between the two methods of displaying the data because of the different flow rates of the wells. Figure 13.3a

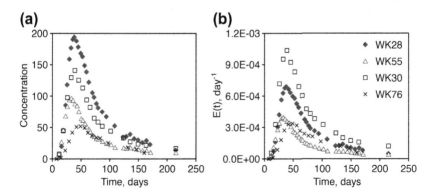

FIGURE 13.3 (a) Scaled concentration, (b) Age distribution. *Source: Contact Energy, Personal Communication.*

compares the tracer concentrations in the different wells, while Figure 13.3b compares the tracer flow recovered from each well. Integrating the area under the curve in Figure 13.3b gives for WK28:

$$\int_0^\infty E(t)dt = 5.2 \times 10^{-2}$$

$$\int_0^\infty tE(t)dt = 5.2 \text{ days}$$

An exponential tail beyond the last value was assumed, and the area under this extrapolated decline added; this makes a significant difference to the second integral. Then the fraction of tracer recovered at WK28 is 5.2×10^{-2}, or 5%, and the mean residence time is $5.2/5.2 \times 10^{-2} = 100$ days. This compares with the first arrival time of less than 12 days and a peak at 39 days.

From the mean residence time, it is possible to calculate the pore volume V_p that has been swept by the tracer (Shook, 2005):

$$V_p = \frac{m_i(\infty)}{M} \times W_I \times \tau_i \qquad (13.9)$$

If some of the tracer is recycled by being injected after being produced, then it is necessary to deconvolve the observed tracer concentrations in order to calculate the tracer returns that would have been observed without this recycling. Assume that a mass M of tracer is injected at the injection well. After some time, t_1 tracer returns to the production well(s), and this is in turn injected again at the injection well. Let the concentration in the injection well be c_I, and let flow rate be W_I. If the observed tracer concentration in a monitor well is c'_i, and c_i is the concentration that would be observed in response to M injected at time zero with no recycling, then:

$$c'_i(t) = C_I(t) + \frac{1}{M}\int_{t_1}^t c_i(t-t')c_I(t')W_I(t')dt' \qquad (13.10)$$

which can be rearranged as:

$$c_i(t) = c'_i(t) - \frac{1}{M}\int_{t_1}^t c_i(t-t')c_I(t')W_i(t')dt' \qquad (13.11)$$

Because the function c_I is zero until time t_1, the integral can be explicitly calculated at each time step, since the integral includes only values of c_i from time $t - t_1$ and earlier. This explicit formulation is subject to increasing errors at longer times, and it is necessary to use a different algorithm to find the best-fitting function $c_i(t)$. The problem is the same as the deconvolution necessary in a pressure transient with several preceding flow rate changes. For an example, see Onur (2010) and Onur and colleagues (2008). Figure 13.4 shows an example from the Habanero tests. The effect of recycling is very important

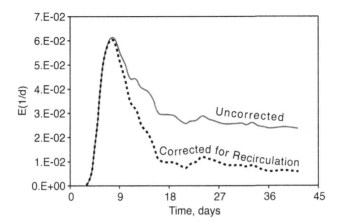

FIGURE 13.4   Tracer return: original and deconvolved data. *Source: Yanigasawa et al., 2009. Copyright © Geothermal Resources Council.*

because a total of 78% of the tracer was returned. The deconvolved data show a first breakthrough in 4 days and a peak at 9 days. Mean residence time was 23.7 days, and the tracer-swept volume was 18,500 m^3 (Yanigasawa et al., 2009).

In a geothermal field, tracer returns decay to below measurement error. If the test is run in a small, confined reservoir, such as an EGS, and all produced tracer is returned, then the tracer may eventually mix uniformly into the reservoir, reaching a finite concentration, $c(\infty)$. Then the reservoir volume is given by $V = M/c(\infty)$ (Rose et al., 2004). For example, in Figure 13.4, it can be seen that the uncorrected concentration is approaching a constant value of around 2.5×10^{-2} d^{-1}. This means that the reservoir volume is equal to the volume of fluid circulated in $1/2.5 \times 10^{-2} = 40$ days, because the concentration is such that the tracer has been equally spread throughout this volume.

13.4.3. Fracture Models

Axelsson and colleagues (2005) describe the transport of tracer along a one-dimensional fracture, with an assumed uniform fluid velocity v along the fracture, and dispersivity K_D. If a mass M of tracer is injected at time $t = 0$, the tracer concentration along the fracture is given by:

$$c(t) = \frac{vM}{w} \frac{1}{2\sqrt{\pi Dt}} e^{-(x-vt)^2/4K_Dt} \qquad (13.12)$$

W is the production rate of the monitoring well, and c is the concentration of tracer (kg/kg). This equation provides a model for the concentration as a function of time that can be fitted to the observed returns. Dwikorianto and colleagues (2005) describe the fitting of this formula to vapor tracer returns at Kamojang. With assumptions about the thickness of the fracture zone, it is possible to calculate thermal changes. Sambrano and colleagues (2010) describe in detail the fitting of tracer tests and calculation of thermal returns in Mindanao. The predicted time of thermal breakthrough varies greatly with the assumed porosity.

13.5. INCORPORATION IN SIMULATION

The limitation of the previous sections is the lack of connection between the tracer observations and predictions of temperature change. Even with an excellent fit to the pattern of tracer return based on flow in a fracture or pipe, the surface area through which heat exchange occurs is completely unknown. This is not calibrated until some temperature change is observed. The best way to approach this deficiency is to fit the tracer returns within a simulation model of the reservoir. The fit to the tracer history may be less perfect, but the advantage of conforming to what is known of the reservoir structure far outweighs this.

Nakao and colleagues (2007) present a simulation of the two-phase Uenotai geothermal field with a match to observed production enthalpy and tracer return. They found it was necessary to make fracture porosity (product of fracture-zone porosity and fracture-zone volume fraction) very small—less than 1% to achieve a good match between measured and modeled tracer returns.

13.5.1. Ribeira Grande

The modeling of this field is described by Ponte and colleagues (2009a, b, 2010) and Pham and colleagues (2010), and the field history is described by Kaplan and colleagues (2007). The field is located on the island of São Miguel in the Azores. There are two separate projects: Pico Vermelho in the north (PV wells) and Ribeira Grande in the south (CL wells). The reservoir is relatively narrow, being an outflow from an upflow further south, as shown in Figure 13.5, and in the north the outflow is only a few hundred meters thick. Figure 13.6a shows a map of the field and simulation grid. The first development was a 3 MW noncondensing turbine at Pico Vermelho, starting generation in 1981. Production was limited by well damage and scaling. The 5 MW Phase A Ribeira Grande plant was installed in 1993—1994, and Phase B 9 MW plant in 1997. In 2006, a 10 MW Pico Vermelho plant replaced the noncondensing turbine. With the last development, a simulation was commissioned to manage reservoir depletion.

The model was constructed using the dual porosity, dual permeability option of TETRAD. The initial state temperature was matched with an error typically less than $10°C$. Production enthalpy history was matched with

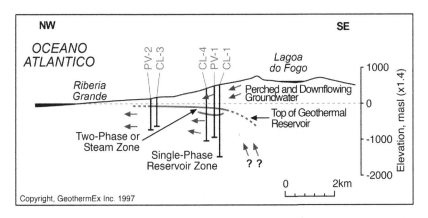

FIGURE 13.5   Conceptual model of Ribeira Grande. *Source: Granados et al., 2000.*

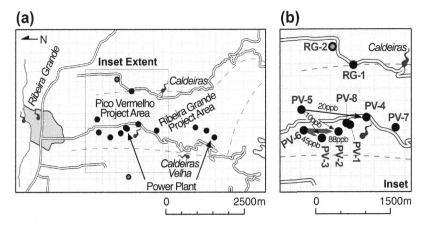

FIGURE 13.6   (a) Field map of Ribeira Grande and simulation grid. (b) Pattern of returns from injection wells PV-5 and PV-6. *Source: Ponte et al., 2009a. Copyright © Geothermal Resources Council.*

a typical error less than 50 kJ/kg; well CL-5 had significant excess enthalpy, and a good match was made. A tracer test was carried out, and there were significant returns from wells PV-5 and PV-6, as shown in Figure 13.6b. A good match was made to the returns, one of which is shown in Figure 13.7. Simulating operation of the present plant and with the Pico Vermelho 10 MW expansion, the model predicted significant cooling due to injection returns from PV-5 and PV-6, and it was recommended that injection be relocated to the east. If the injection-related cooling could be adequately addressed, the expansion was supported. Cooling was limited to the northern part of the field (Pico Vermelho), with no significant cooling in the south (Ribeira Grande).

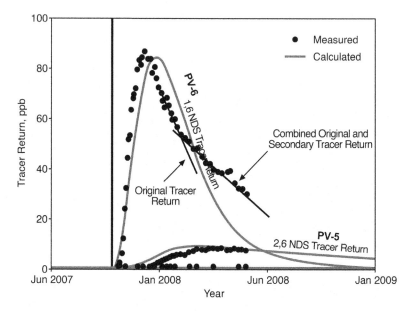

FIGURE 13.7 Match to returns in PV-3 from wells PV-5 and PV-6. There were zero observed and simulated returns to injection in CL-4. *Source: Ponte et al., 2009a. Copyright © Geothermal Resources Council.*

13.6. SURFACE EFFECTS

Surface activity is divided according to whether the fluid supporting the features derives from reservoir liquid or steam/gas. Liquid-fed features include springs, geysers, and hot pools. The discharge from these surface features contains water from the geothermal reservoir, usually recognized by chloride content, and are often called chloride springs. Steam-heated features include fumaroles, solfataras, mud pools, kaipohan (cold gas vents), and acid springs. Fumaroles, solfataras, and kaipohan are discharges of steam and gas from the reservoir, which in the case of kaipohan has been conductively cooled to leave only the gas content. Mud pools and acid springs are formed when steam and gas from the reservoir condense into surface waters, forming acid water due to the dissolved CO_2, or the oxidation of dissolved H_2S.

The different types of features respond differently to pressure changes in the reservoir. High-chloride, water-fed features are the simplest to explain, since they are connected to the reservoir by hydrostatic pressure. If the deep reservoir pressure changes, flow of these features will respond accordingly. If the deep pressure declines, the spring flow also declines. The Ohaaki ngawha (pool), the major spring in Ohaaki field, shows a response to production from nearby geothermal wells (Glover et al., 1996, 2000; Hunt & Bromley, 2000). Figure 13.8 shows the history of this feature during the initial discharge from

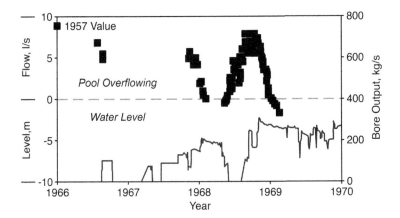

FIGURE 13.8   Changes in Ohaaki Pool. Graph shows flow rate when spring flowing and water level when not flowing. *Source: Glover et al., 1996.*

the deep reservoir at Ohaaki. The flow of the pool declines when deep wells discharge, and it recovers when discharge stops. With further discharge, the water level in the ngawha falls below the overflow, and thereafter the water level in the pool responds to well discharge. Steam-heated features in a vapor-dominated reservoir similarly respond to reservoir pressure changes. Figure 13.9 shows the simulated surface discharge history of Larderello. As the reservoir pressure declines, so does the surface discharge.

Steam-heated features occur in liquid-dominated fields, where there is some degree of separation of the upflow. Rising steam forms a vapor-dominated zone, and steam from this zone supplies steam-fed features. The flow of steam to the surface from a vapor-dominated zone at depth changes as reservoir pressure changes. If the steam flows to the surface along a simple path that can be considered as a constant resistance, falling steam pressure results in

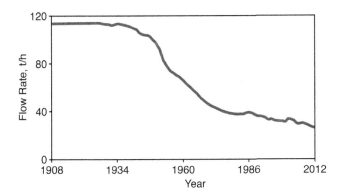

FIGURE 13.9   Simulated history of surface discharge at Larderello. *Source: Barelli et al., 2010b.*

a proportional reduction of steam flow to the surface. There are other effects also present. When the steam upflow mixes into, or passes through, surface groundwater aquifers, falling steam pressure can result in the channels being flooded, stopping the steam upflow.

In a liquid-dominated reservoir, when pressure declines due to production, there is a characteristic pattern of change. Water-fed features dry up in response to the falling deep reservoir pressure. This is a simple response to the pressure change as just discussed. However, increased boiling can result in an increased steam flux toward the surface. In addition, drainage of water from shallow layers can open more passages for steam flow. This drainage of water is the most important factor modifying the steam flux. The upward fissures in the natural state are almost entirely occupied by water, as shown in Figure 2.3a. Drainage of water makes the fissures much more permeable to steam, as shown in Figure 2.3b, and steam upflow can increase severalfold, and steam discharge can appear where there was initially none. The result is that as the deeper reservoir pressure declines steam-heated features can increase in flow and expand in area and liquid-fed springs may convert to steam-heated features. Chapter 12 describes some of the changes in surface activity at Wairakei-Tauhara that have followed this pattern. A similar pattern of decline in water-fed features and expansion of steam discharge has been observed in some exploited liquid-dominated fields in the western United States (Sorey, 2000). Figure 11.5 shows the history of heat discharge at Karapiti, an area of steaming ground, at Wairakei. The observed increased heat flux to surface imposes significant constraints on the shallow structure of the simulation model.

Injection also can have an effect of surface activity. Injection in shallow aquifers has been observed to revive springs at Tongonan (Bolaños & Parrilla, 2000). In New Zealand, shallow injection has created a new spring and an area of heated ground at Rotokawa, and at Mokai produced chloride water flow and a hydrothermal eruption at a steam-heated feature.

13.7. SUBSIDENCE

Geothermal exploitation has caused surface subsidence in most fields. The underground changes in pressure and temperature cause contraction (or expansion) of the rock, and this in turn causes a change in surface elevation.

A decline in fluid pressure causes contraction of the reservoir rock, since the pore pressure helps to support the rock, and removal of this support causes contraction of the rock. Similarly a decline in temperature causes thermal contraction of the rock. If the elastic properties of the rock are reasonably similar throughout and above the reservoir, the contraction is roughly proportional to the pressure (or temperature) change. Thus, with a general widespread drawdown in reservoir pressure, there is a similar widespread pattern of subsidence. Figure 13.10 shows subsidence over a 20 year period at Mak-Ban.

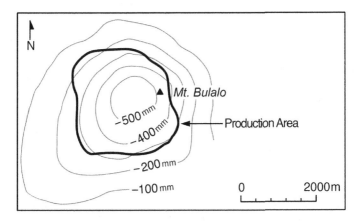

FIGURE 13.10 Elevation changes 1979–1999 at Mak-Ban. *Source: Protacio et al., 2000.*

There is a wide pattern of subsidence across the field, roughly similar to the pattern of pressure drop at depth. Where injection causes a (local) rise in pressure, there may be inflation around the area of pressure increase.

In most fields, subsidence follows such a pattern. This means that it is not an environmental issue unless there is some structure that depends on maintaining existing levels, such as an irrigation canal. Figure 13.11 shows the relation between maximum rate of subsidence and average pressure drop in a number of fields. There are two outliers, Wairakei and Ohaaki, that have a broad pattern of subsidence similar to other fields, but superimposed upon this is are localized areas of greater subsidence. Test drilling in these areas of localized increased

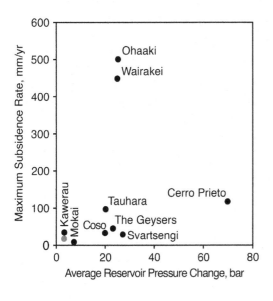

FIGURE 13.11 Relation between average reservoir pressure drawdown and maximum rate of subsidence. *Source: Clotworthy et al., 2010.*

subsidence has shown the presence of relatively shallow layers of material with anomalous compressibility (Bromley et al., 2010).

13.8. INJECTION MANAGEMENT

The general principles of injection planning are that injection wells should be "as far as possible" from production wells and if within the reservoir, as deep as possible. In a few cases, in fields that are liquid with relatively low temperature, injectors have been sited close in to provide pressure support, accepting the resulting thermal degradation. This was done at East Mesa, a sedimentary-hosted reservoir where the primary permeability is relatively isotropic compared with fields hosted in fractured volcanics (Bodvarsson & Stefansson, 1988). Similarly, limited infield injection at Palinpinon has been used for pressure maintenance (see Chapter 12.6). In high-temperature fields, pressure support is usually not needed to maintain well performance, so the dominant issue is to prevent thermal degradation of the production area. Summaries of injection experience are given by Stefánsson (1997) and Sanyal and colleagues (1995), and both observe that there are common difficulties with designing reinjection-production layouts, but these constraints usually do not cripple a development proposal.

13.8.1. Injection Well Siting

The following are some guidelines for siting injection wells and the expectations:

1. At the edge of the field—low permeability
2. Deeper than production—expensive wells
3. Spread out—expensive pipelines
4. Off the structural axis in the field—minimize return of injected fluids
5. Outfield (outside the geothermal reservoir)—low permeability

In most projects where injection has been implemented from the start of production, the injection wells have been relocated from positions relatively close to production areas, or along structural trends, to areas within the reservoir but further away from production or outfield and totally isolated from production areas. Management of reinjection fluids was a major issue in Palinpinon and is described in Chapter 12.6.

For high-temperature two-phase resources, or those that have become substantially two-phase as a result of exploitation, injection has little value as pressure support. Pressure support is not needed because with sufficient enthalpy, the pressure drop in the flowing well, from feedpoint to surface, is relatively low, and good well performance is obtained despite the reservoir being significantly underpressured. Where there is substantial boiling in the reservoir with an attendant increase in average production enthalpy, injection of cool fluids into the two-phase zone or suppression of the two-phase zone by pressure increase or thermal returns can decrease the ability to extract extra heat from the rock by the boiling process. At Salak, for example, the simulation showed a benefit if some reinjection could be relocated outfield (see Chapter 12.7).

13.8.2. Augmentation Injection

In a few fields, additional water has been sourced in order to support production or mitigate environmental impacts by supporting pressure. This is now accepted as essential in heavily exploited vapor-dominated fields where the fluid required to transfer heat from the rock to surface has been severely depleted. At The Geysers, production rates have been supported by injecting additional water (Chapter 12.3). At Larderello, similar injection into super-heated regions was tested in 1979, with condensate being injected from 1984 and supplemental injection started in 1994 (Capetti et al., 1995). At Dixie Valley, a liquid-dominated reservoir at about 230°C, some additional water has been injected to maintain reservoir pressure (Benoit et al., 2000). At Ngawha, a small additional water flow has been injected to maintain reservoir pressure.

Chapter 14

Well Stimulation and Engineered Geothermal Systems

14.1. INTRODUCTION: FRACTURING ROCK

The term Engineered Geothermal System (EGS) is used to refer to a range of experimental projects that involve modifying rock permeability in order to create a geothermal reservoir or to enhance an existing one. The first experiments along this line were done at Los Alamos, under the name Hot Dry Rock (HDR). Those experiments were unsuccessful and later abandoned. The ideas have continued. In most cases, the concept is to create fractures around a well in order to enhance local permeability or to create it in rock of little permeability. This chapter is only a brief introduction to a very extensive global research project. Before further discussing these experiments, the basic concepts of fracturing are discussed.

In the simplest form of the method, to drill a well into rock that has little permeability. Fluid is pumped into the well at sufficient pressure, and the rock fractures. More fluid is pumped, and the fracture or fracture network extends. Hydrofracking is a standard technique in petroleum reservoirs where it is used to stimulate a well—that is, to improve the flow of a well drilled into a formation where the existing permeability is too poor to support sufficient flow. Rather than pumping only water, specialized fluids are used. These usually have high viscosity to provide more back pressure and thus make it easier to reach an elevated pressure at the wellface. Packers can be used to isolate a zone of interest,

Geothermal Reservoir Engineering. ISBN: 978-0-12-383880-3 DOI: 10.1016/B978-0-12-383880-3.10014-9

although the high temperature conditions in geothermal wells limit the possible methods (e.g., packers that can handle high temperatures are not available), and the simplest approach is usually to pump water.

How the fracture develops depends on the rock stress. Pressure in a column of water increases with depth on a hydrostatic gradient. The weight of rock defines a lithostatic gradient of pressure equal to the weight of overlying rock. If the fluid pressure in the rock pores is increased to lithostatic, then the fluid will lift the overlying rock. However, fractures normally occur at a lower pressure: the "fracture gradient." They happen at a lower pressure because rock stress in a horizontal direction is usually less than the lithostatic gradient. The fluid will fracture the rock in the easiest direction; it pushes in all directions, but the rock will part most readily in the direction of the least stress, and the fracture develops normal to this direction.

Figure 14.1 shows the pressures in a hypothetical well before and during fracturing. It is drilled to 2000 m and cased to 1000 m. The solid line shows hydrostatic pressure with depth, and the long dashes show the lithostatic pressure. The short dashes show the fracture gradient—the pressure at which the rock will fracture. The dotted line shows the pressure in the well when it has been pressurized sufficiently to equal the fracture gradient at the casing shoe. If the pressure is further increased, the formation will fracture. Comparing the pressure in the well with the fracture gradient, it can be seen that the fracture

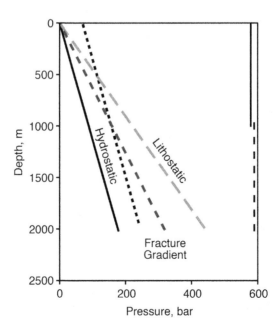

FIGURE 14.1 Fracture gradient with depth.

will preferentially occur near the top of the open interval exposed to the pressure. However, the precise location will depend on local geology and local weakness in the rock or on preexisting fractures. If more pressure is applied, the fracture will grow. It will tend to grow upward rather than downward. Because of the difference between the pressure gradient in the fluid and the fracture gradient, there is more pressure excess, relative to the fracture gradient, at the top of the fracture than at the bottom.

Summarizing, a fracture will form when downhole pressure exceeds the fracture gradient. It will tend to form near the top of the open interval, and it will be oriented normally to the direction of the least principal stress.

In most parts of the world, the least principal stress will be in a horizontal direction, so the fracture will be vertical or nearly so. Traditional papers on fracturing petroleum reservoirs typically show a vertical fracture spreading out from the well. In Central Australia, where the continent is under compression, the least stress in the deep basement rocks can be in the vertical direction, and in this case the fracture will be horizontal. The direction of the fractures is important because an EGS will normally want multiple fractures, so the well must be drilled at an angle to the direction of fracturing. If the fractures are vertical, it is necessary that the well be deviated.

The fracture gradient is often determined during drilling. After setting casing, it is common to perform a "leak-off test." After drilling a short distance beyond the casing water is pumped slowly into the well until a pressure is reached at which it "leaks off." Figure 14.2a shows the fracture gradient at Bulalo as determined from leak-off tests. The fracture pressure at 1000 m is 135

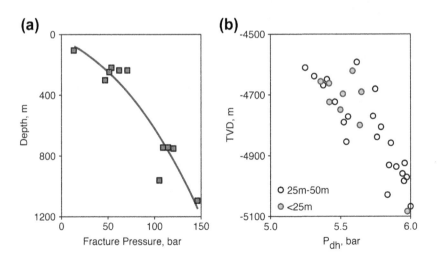

FIGURE 14.2 (a) Fracture pressure versus depth, based on leak off tests at Bulalo. (b) Borehole pressure at time and depth of near-borehole seismic events at Soultz-sous-forêts. *Sources: (a) Menzies et al., 2007. Copyright © Geothermal Resources Council. (b) Mégel et al., 2005.*

bar, or a gradient of 0.135 b/m. Measurements in GPK3 at Soultz-sous-forêts determined a failure pressure of 520 bar at 4700 m, 540−570 bar at 4750−4780 m, and 600 bar at 5100 m (Mégel et al., 2005). Figure 14.2b shows the pressure at which seismic events were generated as a function of depth. The fracture pressure at 5000 m is 600 bar, or 0.12 bar/m. Yoshioka and Stimac (2010) report a fracture gradient of 0.12 bar/m at Awibengkok, and also describe a detailed geomechanical model of the fracture process.

If there is any permeability at the bottom of the casing, the test will not measure the fracture gradient because fluid will be lost into that permeability. If there isn't any permeability, the test will measure the pressure at which the rock fractures. Measurement of the fracture gradient is important, since wells should normally be operated so that downhole pressure never exceeds the fracture gradient at the casing shoe, unless, of course, it is intended to create a fracture. This limitation can impose a constraint on injection pressures.

The start of fracturing or the closing of a fracture can be recognized by a change in the pressure trend. Figure 14.3 shows injectivity measurements at a low-permeability peripheral injection well under stimulation at Awibengkok (Yoshioka & Stimac, 2010; Pasikki et al., 2010). The initial relation between pressure and injection flow shows two linear segments, with higher incremental injectivity at higher flow rates. The change in incremental injectivity (arrowed) is interpreted as the opening of a fracture. Subsequent measurements, after differing periods of stimulation, show further improvement in well performance.

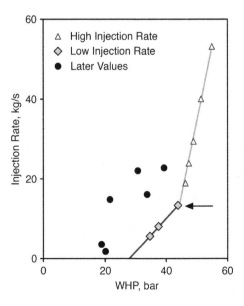

FIGURE 14.3 Injectivity change with pressure and time. *Source: Pasikki et al., 2010.*

14.2. THERMAL STIMULATION

The simplest form of stimulation in high-temperature formations is achieved by pumping cold water into a well, usually at pressures too low to cause hydraulic fracturing. This usually results in a pattern of increasing injectivity with time. Such stimulation is often shown by injection wells, which typically improve their performance with continued injection, provided that deposition is not occurring. The mechanism is a result of thermal contraction of the rock, causing fractures to open (Stefansson & Steingrimsson, 1980; Benson et al., 1987). A theoretical study is given by Nygren and colleagues (2005), who also observe that fracture aperture change in an EGS environment is largely controlled by thermoelastic effects. The effect is reversible, at least in part, since injectivity decreases if the well is allowed to warm up and then increases again under new injection, as illustrated in Grant and colleagues (1982b). The reverse effect has also been demonstrated as injectivity decreases with time when hot water is injected into cooler rock. Because the effect is reversed if the well warms up, it is typically applied to injection wells. Figure 14.4 shows the history of injection well 4R1 at Tongonan. Injectivity generally increases with time, until deposition becomes a problem in 1981. Cold water stimulation has the advantage of being cheap, usually making it the most economic form of stimulation, even if poorly understood (Flores et al., 2005; Axelsson & Thórhallsson, 2009). It is also unclear how much of the permeability increase in hydrofrac experiments is actually due to thermal stimulation of the new fracture; once the fracture is opened and cold water enters, some thermal stimulation occurs. Yoshioka and Stimac (2010) described a geomechanical simulator, including the effects of both fluid pressure and temperature, concluding that created fracture volume is proportional to the fluid volume injected and injection efficiency increases with injection pressure and temperature difference between injectate and reservoir.

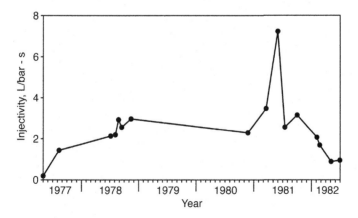

FIGURE 14.4 Injectivity history of well 4R1, Tongonan. *Source: Sarmiento, 1986. Copyright © Elsevier.*

There is the possibility that thermal stress from cooling causes some cracking or that spalling of the fracture face creates proppants so that fractures are held open when the rock heats up again, and there is a permanent increase in permeability. For these reasons there is the possibility of improving production well performance. An experiment on a production well at Bouillante (Tulinius et al., 2000) showed an improvement in production performance. The fluid injected was seawater with an inhibitor to prevent deposition of anhydrite. Injectivity improved from 0.9 to 1.4 kg/b.s, and production that was previously 80 t/h and unstable increased to 140 t/h and was stable. Similarly Kitao and colleagues (1990) reported improvement in productivity of two out of three production wells stimulated by cold water injection, with proppants, at Sumikawa, and Zúñiga (2010) reported the improved injectivity and production flow of a well at Borinquen. With proppants, the idea is to expand the fracture by thermal stimulation and place proppants in the fracture to prevent its closing when the rock heats up again.

Extended cold water injection has also been used at low wellhead pressure when the stress conditions in the rock meant that fracturing occurred. In this case it was also thought that the water dissolved silica from the formation over an extended period of flow (Rose et al., 2005).

In a well with very poor injectivity, it may be necessary to pump at high pressure initially—not to fracture but to get flow into the fractures to cool them. Once this is achieved and some cooling occurs, the fracture opens and further stimulation is easier. An example from Los Humeros (Flores-Armenta & Tovar-Aguado, 2008) documents an injection well improved from an initial acceptance of 1.4 to 30 kg/s. Thermal stimulation as described here is a feature of fractured rock. However, even in sandstone, permeability is significantly greater at lower temperatures (Weinbrandt et al., 1972).

14.3. ACID STIMULATION

Stimulation of a geothermal well by acid is very similar to acid stimulation of any other well. Acid is injected into the formation and dissolves some rock or material filling the fractures. The acid is injected through tubing to the depth of the permeable zone to be treated to avoid damage to the casing. Inhibitors are used to minimize damage to the liner, but inevitably some metal loss occurs. Sometimes mechanical washing or flushing of the well is also done.

The acid will have the most effect at and near the well face, so acid treatment is best suited to a well where the permeability is impaired near the wellbore. Typical examples are a well that has been drilled with mud or an injection well with deposition. Acid stimulation is also used on wells with poor permeability without such impairment. A resistance near the well face should appear on transient testing as a positive skin. Identification of such well damage and monitoring the changes made by stimulation are the most important uses of detailed pressure transient analyses in geothermal reservoir engineering.

The testing of a well should include pressure transients before and after stimulation. It is important to do a thorough transient analysis to obtain a good estimate of skin, and the best way to do this is using a computer transient analysis package. If there are a number of wells, the transient testing can be used to select the best candidate well; ideally the candidate well has high positive skin and reasonable transmissivity so a good flow can be expected if the skin is removed. The possible improved flow can be calculated by assuming that skin is reduced to zero. Acid cannot create permeability in an impermeable rock; it is used to remove an obstruction preventing access to the permeability that does already exist.

In practice, the results of acid stimulation are somewhat variable, and often the transmissivity changes as well as skin. If a zone is completely blocked and it is opened by stimulation, this will result in increased transmissivity. If there are drilling losses at a depth that does not show as a feed zone in the completed well, this suggests blockage of the zone, which makes it a target for stimulation with acid or cold water.

As the simplest estimate, injectivity index is given by:

$$\frac{1}{II} = \frac{\mu}{2\pi\rho kh}\left[2.303 \log_{10}\left(\frac{r_o}{r_w}\right) + s\right] \tag{14.1}$$

where r_o is an outer radius far from the well where pressures are undisturbed—typically assumed to be 10^3 or 10^4 times the well radius. This can give an estimate of the change in injectivity index that can be expected from a change in skin.

Barrios and colleagues (2007) and Epperson (1983) describe typical stimulation of injection wells in Berlin and Beowawe fields. The acid stimulation program was as follows:

1. A preflush of HCl to dissolve calcium and carbonate minerals
2. A mainflush of HCl/HF to dissolve carbonates and silicate minerals and drilling mud
3. A postflush of clean water to displace acids and reaction products away from the wellbore to prevent damage

Injection wells were treated with mechanical and/or chemical cleaning. The wells' performances had decreased with time due to deposition and mud damage during drilling was also suspected as a contributing factor. In all cases the result was an improved injectivity index, kh values improved, ΔP_{skin} decreased and injectivity index increased. Similar results have been reported from the Philippines and Mexico, where acid stimulation with or without mechanical cleaning is routinely used (see, e.g., Yglopaz et al., 2000; Fajardo & Malate, 2005; Flores-Armenta et al., 2006, 2008). Flores-Armenta (2010) provides an evaluation of 17 acid treatments in Mexican fields, finding that the gain in capacity was equivalent to 13 new wells and a return of the investment was achieved in less than a year.

Pasikki and Gilmore (2006) described the acid stimulation of a production well at Awibengkok. The well had poor production and had possibly been

TABLE 14.1 Injectivity of Well Zones, Before and After Test

Entry Depth, m MD	*II* Preacidizing, kg/b.s	*II* Postacidizing, kg/b.s	*ΔII*, kg/b.s
1340	1.50	5.20	3.70
1610	1.10	3.17	2.07
1770	0.48	0.64	0.16
1905	0.97	1.54	0.57
1925	0.62	1.45	0.83

Source: Pasikki & Gilmore, 2006.

TABLE 14.2 Permeability of AWI 8-7, Before and After Stimulation

	kh*(dm)	s	II (kg/s.b)
Pretest	76	2.2	4.7
Posttest	123	−1.2	12

*Using properties of injected water
Source: Pasikki & Gilmore, 2006.

mud-damaged. Modeling a PTS run gave injectivity indices at five feed zones. Acid stimulation was carried out, with an improvement in permeability. A PTS run showed improved injectivity at all zones, as shown in Tables 14.1 and 14.2. Well production increased from 4.2 MWe to 9.8 MWe.

14.4. STIMULATING EXISTING RESERVOIRS: DEEP SEDIMENTARY AQUIFERS

This is the simplest type of the unconventional or enhanced geothermal reservoirs. Sandstone or other aquifers, relatively deep in regions of high geothermal gradient, can be hot enough to economically supply water for heating or power generation. There is no convective geothermal system present, but rather an aquifer at depth heated by the normal conductive geothermal heat flux. Such aquifers have supplied district heating schemes in Europe and China for some time. Their use for power generation or power generation in conjunction with heating is more recent. Aided by high tariffs for green power, such projects are proliferating, particularly in Europe, and this industry is growing rapidly.

The reservoir engineering of such a project is very similar to that for a petroleum reservoir or a conventional groundwater aquifer, with the addition of thermal changes in the reservoir due to injection. It may be necessary to stimulate the wells to provide sufficient flow; the reservoir permeability is sufficient for the flow through the bulk of the reservoir, but there is too much pressure gradient near the wells. In contrast to the "true EGS," where fracturing is used to create the reservoir, in this case the reservoir already exists, and it is necessary only to improve the well performance. Figure 10.2 shows a sketch of the project at Landau, Germany. One of the wells was stimulated by hydraulic fracturing.

Well performance is crucial to such a project. With the comparatively low reservoir temperature, power generated per unit of water flow is relatively small. The system requires pumps to maintain the flow. There may be a downhole pump in the production well, a surface pump for the injection, or both. Power to run the pumps is often a significant fraction of the generation and depends on the injectivity and productivity of the wells. Economic operation translates into a required minimum well performance.

Reservoir engineering issues revolve around permeability and well performance. Pressure transients from tests conducted during and after drilling can be interpreted using the full range of models from petroleum engineering, since the aquifer is a homogeneous medium. If the well performance is insufficient for economical operation this can be improved by stimulation.

Optimal pump choice is determined by a tradeoff between the increased flow due to the additional drawdown that the pump creates and the parasitic power needed to run the pump. Often it is economic to have pumps as large as possible (for examples, see Sanyal et al., 2007). Sanyal (2009) shows that technological improvements in pumps would improve EGS economics.

14.5. EGS: CREATING A RESERVOIR

The full EGS is when both permeability and reservoir are artificially created. The concept was first called HDR with the first pilot at Fenton Hill (Robertson-Tait et al., 2000; Brown, 2009). A summary of past EGS experiments is provided by MIT (2007). The principle is that production and injection wells are drilled and fractures are created in the rock around them, providing a path between injection and production wells. The fracture system is intended to pervade a sufficient volume that a reservoir is created, not just a path between the two wells. Water is circulated down the injection well through the rock and back up the production well. Pumps are needed in one or both wells to drive the flow.

The fractures created may be either new fractures or the enlargement of existing joints or fissures in the rock. EGS trials have found that there are always some existing fractures and some formation fluid, so the rock is never "dry" or totally solid. Growth of the fractures during stimulation is typically monitored by mapping the locations of microseismic events, the spatial distribution of which provides a map of the fractured region. Fracture intersections with the wellbore are found from geophysical logs and detailed PTS logs during injection and production.

For a reservoir engineer there are two parameters that summarize the performance of an EGS reservoir: the resistance to flow between the two wells and the rate of temperature drawdown of the produced water. The flow resistance determines the pumping power required and should be as low as possible.

More difficult to predict is the temperature drawdown with time. In short-term tests there will be little if any change in production temperature, but these data alone are insufficient to project decline over plant lifetime. A target of $10°C$ drawdown at abandonment was proposed by MIT (2007). The objectives of lowering resistance between the wells and minimizing temperature drawdown are in conflict, since a direct path between wells would give least resistance but most rapid cooling. Ideally what is required is the creation of a volume of fractured rock, the reservoir, with fractures sufficiently pervasive that a large surface area of rock is exposed to the passage of water but without any major preferential path between producer and injector.

Much effort has gone into the well stimulation method and geological and geophysical means of measuring the results. The reservoir engineer's concern is to predict the pressure and temperature changes. The resistance between the wells is easily measured, although it will vary with pressure, with temperature and time due to thermal stimulation, and changes with deposition. To date, predicting the temperature changes can only be done with the aid of a detailed model of the fracture network, together with data to calibrate this model. An example of detailed fracture mapping is given by Sausse and colleagues (2008).

Figure 14.5 shows the Hijiori experiment. There are three deep wells plus a shallow well, with two reservoirs that have been created by fracturing. HDR-2a and HDR-3 are production wells. HDR-1 and SKG-2 are injection wells. A long-term circulation test was carried out using these wells. SKG-2 was shut for the first part of the test. Figure 14.6 shows a model match to production temperature at well HDR-2a in Hijiori, Japan. The simulation model uses a stacked set of fractures, with individual parameters for each fracture. Permeability of the main fracture increased fivefold during the test. The temperature drawdown is rapid, so this experiment did not create a viable reservoir, but the wells are very close together. The project was closed due to the failure to create a reservoir (Matsunaga et al., 2005b).

In contrast, at the Cooper Basin project in Australia, modeling based on tests between the Habanero wells projects a temperature decline of $20-40°C$ in 20 years, at flows of $15-25$ l/s (Chen & Wyborn, 2009). Figure 14.7 shows a section through the wells where they intersect the fractured granite resource. The reservoir rock has existing fractures with geopressured reservoir fluid at

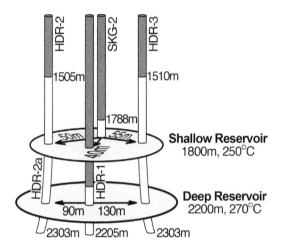

FIGURE 14.5 Hijiori Hot Dry Rock system. *Source: Tenma et al., 2001. Copyright © Geothermal Resources Council.*

around 250°C containing 20,000 ppm dissolved solids. At a depth of 4 km, the vertical (principal minimum) stress is 900 bar, and the minor and major horizontal stresses 1100 and 1400 bar. Pore pressure is 750 bar. Well Habanero #2 had an initial productivity of 0.1 kg/b.s, which was expected to improve with stimulation (Wyborn et al., 2005). The "main fracture" has been created by stimulation. Reservoir volume increased with more stimulation. A tracer test produced results similar to tests at Hijioiri (Yanagisawa et al., 2009).

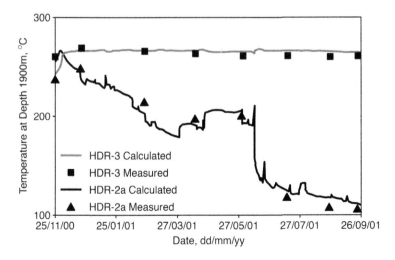

FIGURE 14.6 Calculated and measured production temperature in well HDR-2a, Hijiori, during long-term circulation test. *Source: Tenma et al., 2002. Copyright © Geothermal Resources Council.*

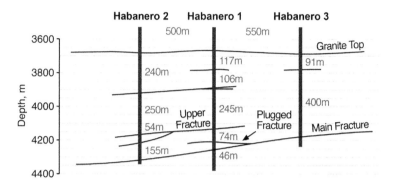

FIGURE 14.7 Habanero well field. *Source: Chen & Wyborn, 2009. Copyright © Geothermal Resources Council.*

Seventy-eight percent of the tracer was returned, with a mean residence time of 23.7 days and a tracer-swept pore volume of 18,500 m³, similar to the 16,000 m³ volume of the Soultz reservoir (Rose et al., 2006). The Soultz tracer-swept volume is much less than the volume indicated by acoustic emissions (MIT, 2007), showing that flow does not uniformly permeate the fractured volume. A three-dimensional model of the Habanero reservoir was developed assuming a horizontal homogeneous main fracture layer, with parameters calibrated on flow testing and the tracer test. The fracture was modeled with transmissivity of 2 dm and porosity-thickness 2.7−3.5 cm and longitudinal dispersivity 20 m. The results are shown in Figure 14.8 for the flow between Habanero #1 and #3. There is appreciable surface temperature drawdown after 25 years at a flow rate of 25 l/s and less drawdown at 12.5 l/s. Options for improving performance are to increase the distance between wells and stimulate multiple-layered fracture zones. The first option was investigated by simulation. With wells separated by 2.8 km, the maximum separation available within the stimulation area, and transmissivity of 20 dm, a flow of 70 l/s is predicted to produce 5°C drawdown after 20 years.

The modeling used for EGS systems revolves entirely around flow in one or a number of fractures and conductive heat transport into the fracture. It uses different formulation and coding to the simulation of conventional reservoirs. At the European project at Soultz-sous-forêts, there has been very detailed mapping of the fracture system and modeling of the fluid flow and mechanical response of the rock to the fluid injection (Kohl & Mégel, 2005, 2007). This modeling was only possible because of the excellent data quality at the project.

Extensive analysis of pressure testing and microseismic observations at Basel showed "an irreversible permeability improvement by two orders of magnitude," but "the reservoir has evolved along a distinct fracture zone confined to a relatively narrow plane of a few tens of metres" (Häring et al., 2008; Ladner & Häring, 2009). An economic EGS reservoir must have

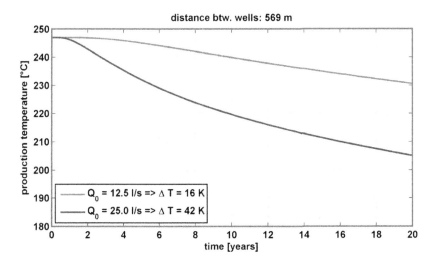

FIGURE 14.8 Modeled temperature drawdown. *Source: Wyborn, Personal Communication.*

substantial thickness, and achieving fracturing over a wide interval remains a challenge. Stored heat estimates have been used to calculate the potential size of the EGS resource (e.g., MIT, 2007). Such estimates are subject to an unknown recovery factor, since there is no operational experience on which it can be calibrated.

Pressure Transient Analysis

A1.1. INTRODUCTION

In this appendix the theory of pressure transient analysis and its application to geothermal wells are reviewed. The theory has been developed in great detail in the groundwater literature, starting with the work of Theis (1935). Groundwater techniques were applied sporadically in the 1950s in various geothermal fields, with their first systematic use in the early 1960s, to analyze the results of a field discharge at Pauzhetsk, Kamchatka (Sugrobov, 1970). The following discussion usually assumes that geothermal "aquifers" are in volcanic rock, with the permeability controlled by faults and fractures. There are some geothermal fields in sedimentary rock, with matrix permeability, such as fields in the Imperial and Mexicali valleys, and low-temperature aquifers in Europe. For these fields these reservations about the use of pressure transient theory do not apply.

Pressure transient theory is also developed in the petroleum literature. While groundwater methods are restricted to single-phase flow of a fluid of constant properties in a single aquifer, the petroleum formulation includes multiphase flow and varying fluid properties. In this appendix the development of transient analysis as used in the petroleum industry is followed, in common with nearly all geothermal literature. A comprehensive exposition of petroleum techniques is given by Matthews and Russell (1967) and Earlougher (1977).

The techniques are described only briefly, and instead the text concentrates on examples of their application in geothermal wells and on the many practical problems that can confuse or obscure pressure measurements. The application of a theory that assumes homogeneous permeability to geothermal reservoirs with fracture permeability, three-dimensional structure, and fluid entry at a few limited points might seem a major assumption. In practice the theory often works well with transients conforming to theory over several log cycles of time.

A1.2. BASIC SOLUTION

The basic aquifer model for pressure transients is of a well that fully penetrates an aquifer of uniform and homogeneous permeability. The fluid is uniform and of constant compressibility. Flow to the well is radial and horizontal. Pressures at the middepth of the aquifer represent the depth-averaged behavior. At any other depth, pressures differ by a constant amount, so pressures at all depths change by equal amounts with time. The equation for pressure change, ignoring matrix compressibility, is derived in Chapter 3.

If the reservoir rock has significant compressibility compared to the fluid, the expression φc must be replaced throughout by $c_m + \varphi c$. This expression is the total compressibility of the rock-fluid combination. Rock compressibility is usually ignored but may be significant if the reservoir fluid is compressed liquid. The compressibility c is often replaced by c_t, the total compressibility of the aquifer resulting from all mechanisms.

A1.2.1. Line Source Solution

If an aquifer is initially at rest, and at time $t = 0$, a well begins discharge at rate q (l/s) or $W = \rho q$ (kg/s), the line source solution is given by Eqs. (3.26)–(3.29). If the well radius is known, it is then possible in principle to identify two parameters: kh/μ (the transmissivity) and φch (the storativity). Note that it is only these parameter combinations that are identified, not the separate parameters k, h, φ, c, μ. Here arises one of the distinctive problems of geothermal well tests. A groundwater or petroleum aquifer is clearly defined geologically: thickness is known, porosity and permeability may be measured by tests on cores, and viscosity and fluid compressibility are similarly available from lab tests or tabulated values dependent on fluid composition, pressure, and temperature. For a geothermal "aquifer" it is seldom clear what the thickness of a geothermal aquifer is, and it is frequently the case that the permeability structure is independent of geological boundaries. The wellbore usually intersects a fracture over a narrow interval, a fraction of a meter to a few tens of meters, but through this, the well presumably draws on a much greater thickness of fractured rock. The depth over which production has been found in a reservoir may be known, but permeable fractures may extend to considerably greater depth. Thus, the thickness is not known a priori but must be determined. Similarly the porosity is vague. It could be total porosity or just the pore space in and near the fractures. Fluid properties can at times be unclear if there is doubt as to whether single-phase or two-phase fluid is present at the well or nearby. Because so much is unknown, it is important to be clear and to report which parameter groups are measured by a particular test. Usually the transmissivity is of greatest interest, since it controls the ability of the reservoir to deliver fluid. Note that the permeability-thickness kh is sometimes referred to as transmissivity.

Before the advent of personal computers, transient analysis was done manually using graphical techniques based on semilog and log-log plots. Only the simplest of these manual techniques are now used, mostly semilog analyses. Any more thorough analysis is done with a computer fitting package that will have the option to try all of the standard variants or configurations of permeability near the well. The results are normally presented in the same semilog or log-log format of manual analyses.

A1.2.2. Semilog Analyses

The exponential integral E_1 has, for small x (long time), the asymptotic form

$$E_1(x) \approx -\ln(x) - \gamma = 2.303 \log_{10}(x) - \gamma \qquad (A1.1)$$

where $\gamma = 0.5772$ is Euler's constant. Then

$$-\Delta P = Po - P = \frac{q\mu}{4\pi kh}\left[2.303 \log_{10}\left(\frac{4kt}{\varphi\mu cr^2}\right) - 0.5772\right] \qquad (A1.2)$$

$$= m\left[\log_{10}(t) + \log_{10}\left(\frac{4kt}{\varphi\mu cr^2}\right) - 0.251\right] \qquad (A1.3)$$

where

$$m = \frac{2.303 q\mu}{4\pi kh} \qquad (A1.4)$$

is measured as the pressure change per log cycle, and this unit is written as b/~; and r is the radius of observation.

Thus, when pressure is plotted against time since the flow change on a semilogarithmic scale, an asymptotic straight line should be obtained. This line is characterized by two numbers: its slope, m, and the value at some particular time, t. When the slope is identified, the permeability-thickness can be found:

$$kh = \frac{2.303 \, q\mu}{4\pi m} \qquad (A1.5)$$

assuming that a volume flow is specified. In geothermal wells it is usually more convenient to specify the mass flow W. Then Wv is substituted for $q\mu$, and Eq. (A1.5) is replaced by

$$kh = \frac{2.303 Wv}{4\pi m} \qquad (A1.6)$$

Using the value of the drawdown ΔP at some time t, Eq. (A1.3) gives

$$\frac{\Delta P}{m} = -\log_{10}\left[\left(\frac{4kh}{\mu}\right)\left(\frac{1}{\varphi ch}\right)\frac{t}{r^2}\right] + 0.251 \qquad (A1.7)$$

or

$$\varphi ch = 2.25 \left(\frac{kh}{\mu}\right)\left(\frac{t}{r^2}\right)10^{\Delta P/m} \qquad (A1.8)$$

A1.2.3. Example: Interference BR19-BR23

Figure A1.1 shows an example from an interference test at Ohaaki of such an analysis (well locations are shown in Figure 9.1). When tested in 1980, wells

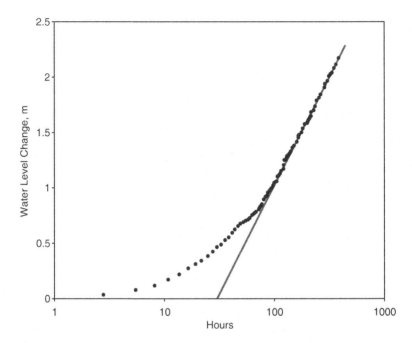

FIGURE A1.1 Interference response at BR23. *Source: Contact Energy, Personal Communication.*

BR19 and BR23 communicated through an aquifer containing liquid water at 270–280°C. At this temperature $\mu_w = 99$ μPa.s. BR19 was opened and produced a flow of 64 kg/s = 0.084 m³/s. (The volume flow rate is computed at reservoir conditions, not at the wellhead, where it is a steam-water mixture.) The pressure response to opening BR19 was measured in BR23 was measured as a water level change. The slope of semilog straight line is 1.98 m/~. The water at the top of the well is cold, so that 1.98 m ≈ 0.195 bar. Then

$$\frac{kh}{\mu} = \frac{(2.303)(0.094)}{4\pi(0.195 \times 10^5)} = 7.9 \times 10^{-7} \text{ m}^3/\text{Pa.s}$$

$$kh = 7.8 \times 10^{-11} \text{ m}^3 = 80 \text{ dm}$$

Alternatively, the calculation could have been made using the mass flow of 64 kg/s:

$$\frac{kh}{v} = \frac{(2.303)(64)}{4\pi(0.195 \times 10^5)} = 4 \times 10^{-4} \text{ m.s}$$

Storativity can be evaluated using any point on the straight line. It is conventional in petroleum analyses to use $t = 1$ hour. In this example the

straight line intersects $\Delta P = 0$ at $t = 11$ h $= 39{,}600$ s. Evaluating here, with interwell distance $r = 350$ m
or

$$\varphi ch = 2.25(7.9 \times 10^{-7})\left(\frac{39600}{350^2}\right) = 5.7 \times 10^{-7} \text{ m/Pa}$$

Using the compressibility of liquid water, $c_w = 1.9 \times 10^{-9}$ Pa^{-1} gives $\varphi h = 300$ m, which may or may not be physically realistic.

As an example an interference test was used to avoid problems with skin. The radius r in Eqs. (A1.2), (A1.3), (A1.7), and (A1.8) is the radial distance of the observation point from the origin $r = 0$. For an interference test this is the interwell distance.

A1.2.4. Superposition

Since the pressure transient equation is linear, solutions can be superimposed. The most useful case is when a well is shut after producing for a period at a constant rate. The solution is a sum of the pressure changes caused by the flow increase at the discharge start and the decrease at shut-in:

$$\Delta P = -\frac{q\mu}{4\pi kh} E_1 \frac{\varphi\mu cr^2}{4k(t + \Delta t)} + \frac{q\mu}{4\pi kh} E_1 \left(\frac{\varphi\mu cr^2}{4k(\Delta t)}\right) \quad \text{(A1.9)}$$

where t is the time flowing and Δt is the time shut. If the asymptotic form (Eq. (A1.1)) is valid for both exponential integrals,

$$\Delta P = \frac{2.303 \, q\mu}{4\pi kh} \log_{10}\left(\frac{t + \Delta t}{\Delta t}\right) \quad \text{(A1.10)}$$

A plot of ΔP against $(t + \Delta t)/\Delta t$ on a semilog scale is known as a "Horner plot" (Earlougher, 1977), and $\Theta = (t + \Delta t)/\Delta t$ is sometimes called the "Horner time." When there are more flow rate changes, more complicated expressions for the "superposition time" are defined.

The drawdown caused by several different wells at different locations can also be superimposed. In addition, some types of barriers or other reservoir discontinuities can be represented by image wells, whose drawdown is superimposed upon that of real wells.

Figure A1.2 shows a well and an adjacent plane boundary. Three types of boundaries can arise: an impermeable boundary, a constant-pressure boundary, and a free surface. An impermeable boundary is one across which there is no flow. The effect of the boundary is equivalent to that of an image well with exactly the same flow as the real well, in a medium of infinite extent. At the boundary, the two wells balance each other so there is no flow, as is required at an impermeable boundary. A constant-pressure boundary could be produced by a fault or some other feature of much greater permeability than the reservoir. This is equivalent to an image well of opposite sign. On the boundary, the

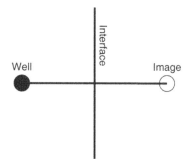

FIGURE A1.2 Image well created by plane interface in reservoir.

drawdown of the well and image well exactly cancel, so there is no drawdown and pressure is maintained constant. Finally, if the source well withdraws fluid from beneath an aquifer with a free surface, this is equivalent to two images: one at the image point and a double image of reverse sign that moves away with constant velocity (Zais & Bodvarsson, 1980).

A1.2.5. Dimensionless Variables

The pressure change in a test depends on the particular flow rates, permeability, and other parameters. These parameters can be absorbed into the definition of time to define dimensionless variables:

$$P_D = \frac{2\pi kh}{q\mu} \Delta P \tag{A1.11}$$

$$t_D = \frac{kt}{\varphi\mu cr^2} \tag{A1.12}$$

Then the drawdown equation is:

$$P_D = P_D(t_D) = \tfrac{1}{2}E_1(4\,t_D) \tag{A1.13}$$

The function P_D is defined independently of the flow rate, transmissivity, or storativity. It does depend on the reservoir geometry. The dimensionless variables P_D, t_D are principally of use in more complex situations where correspondingly more complex forms of drawdown can be represented in a standard form.

The asymptotic (long-time) form of the dimensionless pressure for a well in an infinite homogeneous aquifer is Eq. (A1.3):

$$P_D = -1.151 \log_{10}(t_D) + 0.351 \tag{A1.14}$$

The log slope is 1.151 per cycle in dimensionless form. The dimensionless time is defined on the basis of the well radius r_w. Other dimensionless times that are dependent on other relevant lengths can be used, but they are denoted by some other appropriately varied notation.

A1.2.6. Type-Curve Matching

Given the single function P_D (t_D), the drawdown in an actual situation is related to it by the scaling (Eq. (A1.11)), (Eq. (A1.12)) of the pressure, and time variables. Thus, if P_D is plotted against t_D on log-log scale, and the actual data ΔP, t are plotted on log-log scale, the two graphs should be related to each other by a constant displacement. Prior to the advent of computers, these graphs were matched by plotting the measured data on transparent paper and overlaying this on pre-printed type-curves to obtain the best fit. It is now used as a way of representing the results of computer matching. Standardized log-log type curves of $P_D(t_D)$ and the pressure derivative $t_D \frac{dP_D}{dt_D}$ are used as a means of representing the difference between type curves for different situations. There are distinctive "diagnostic" features of semilog and log-log plots, both of pressure and the derivative, corresponding to different structural models of the aquifer which are useful in choosing models for transient fitting.

Figure A1.3 shows a type-curve match to data of Figure A1.1. Note that the log-log plot shows additional information not apparent on the semilog plot—the data for less than 20 hours deviates from the later data. Transients and fits to them should be plotted in both semilog and log-log format as the different plots provide different emphases to the data.

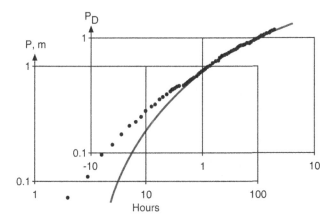

FIGURE A1.3   Type-curve match to BR19-BR13 interference. *Source: Contact Energy, Personal Communication.*

A1.3. WELLBORE STORAGE AND SKIN

Two effects that can occur in or near a well may affect pressure changes measured in the well. These are wellbore storage and skin. *Wellbore storage* is the wellbore's capacity to store fluid. With an increase in pressure, more fluid is stored. Thus, if a well is shut at wellhead, some *afterflow* continues into the wellbore. The flow from the reservoir does not stop instantaneously, but tapers off.

Skin refers to the possibility that immediately adjacent to the well there is a region of different permeability, most often caused by the side effects of drilling. This is idealized as a resistance, possibly negative, concentrated at the wellface.

A1.3.1. Wellbore Storage

Wellbore storage is defined in terms of a coefficient:

$$C = \frac{\Delta V}{\Delta P} \tag{A1.15}$$

where ΔV is the change in fluid volume, at wellbore conditions, for pressure change ΔP. A dimensionless coefficient C_D can be defined, which is the actual storage coefficient divided by the storage capacity of the well volume if occupied by reservoir $C_D = C/(\pi r^2 \varphi c_t L)$, $L =$ length of wellbore. In most groundwater and petroleum wells filled with liquid, if the wellbore volume is V and the compressibility of the fluid in the wellbore is c, $C = Vc$. However, in geothermal wells with flashing in the wellbore, a different effect usually controls wellbore storage. After shut-in, fluid continues to enter the well. For steam wells the wellbore is cooler during discharge because the downhole pressure is lower than in the surrounding formation. As the pressure recovers, the wellbore and adjacent rock reheats, and steam needed to supply this heat condenses in the well. This results in much more storage than would be achieved by simple compression (Barelli et al., 1976). Typical values of wellbore storage coefficient in such wells are $C_D \approx 10^4$. The wellbore storage effect cannot be interpreted literally to give a wellbore volume. Wellbore storage is basically a nuisance effect, affecting the form of pressure transients, which must be recognized in order to make an accurate analysis. For example, the correct straight line cannot appear on a semilog plot until $1\frac{1}{2}$ log cycles after the end of the period affected by wellbore storage, which is the period of unit slope on log-log plot. Often a semilog plot will seem to show a straight line before this time, but this is misleading.

A1.3.2. Skin

Skin is defined as an additional pressure drop ΔP_{skin} at the wellface:

$$\Delta P_{skin} = \frac{q\mu}{2\pi kh}.s \tag{A1.16}$$

where s is dimensionless. This is added to the pressure drop in the homogeneous medium away from the wellbore, so that the drawdown equation becomes:

$$\Delta P = -\frac{q\mu}{4\pi kh}\left[E_1\left(\frac{\varphi\mu cr^2}{4(t)}\right) + 2s\right] \tag{A1.17}$$

The presence of skin does not alter the evaluation of transmissivity in semilog analysis. It does affect storativity. Equation (A1.8) must now be written

$$\varphi c h e^{-2s} = 2.25\left(\frac{kh}{\mu}\right)\left(\frac{t}{r^2}\right)10^{\Delta P/m} \qquad (A1.18)$$

$$s = 1.151\left[\frac{\Delta P}{m} - \log_{10}\left[\left(\frac{4kh}{\mu}\right)\left(\frac{1}{\varphi c h}\right)\frac{1}{r^2}\right] + 0.251\right] \qquad (A1.19)$$

In these equations ΔP is the difference between the pressure at time t and the pressure immediately before the flow rate change—that is, if the transient is for drawdown it is $P_o - P$; if for buildup $\Delta P = P - P_{wf}$, where P_{wf} is the flowing pressure before shut-in.

In petroleum and groundwater, φ, c, h may be known from geologic structure, coring, and fluid samples. In this case the skin can be evaluated unambiguously from Eq. (A1.19). In geothermal, where none of these parameters are well constrained, the uncertainty means that skin can only be identified unambiguously from log-log type curves or better from computer matching. Figure A1.4 shows an example match.

Figure A1.4a shows a log-log plot of the data and a match made using Saphir™. The data are plotted as solid circles, and the solid line is the match. The crosses are the derivative, and the dashed line is the match to the derivative. Figure A1.4b is a semilog plot and the match. The semilog plot apparently shows a straight line, but the unit slope on the log-log plot does not end until 100 seconds, so a semilog straight line will not develop until after 3000 seconds, which is after the end of the data. Although the semilog plot apparently shows a straight line period, this is deceptive, as can be seen from the extrapolated match. The match gives $s = -6$. Geothermal wells normally have negative skin, which is usually presumed to be a consequence of rock breaking away around the fracture zone during drilling. The wellbore is often physically enlarged at

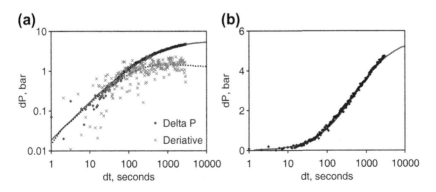

FIGURE A1.4   Transient amatch. *Source: Mighty River Power, Personal Communication.*

a fracture zone, but the size of the negative skin (e.g., −6 in the preceding example) indicates stimulation must extend further into the formation.

Transient matching gives two parameters of interest: the permeability kh and skin s. If a positive skin is obtained, flow of the well is obstructed, perhaps by drilling mud or cuttings. In a production well, mineral deposition at or near the wellface can change the near-well permeability and increase the skin value. Flow should be increased if this skin could be removed by acid treatment or other stimulation. The potential for well damage can be reduced by using different drilling practice, such as using only water for circulation or by maintaining balanced or underbalanced pressure conditions such that formation fluids are produced and remove drilling waste from the permeable fractures.

At high flow rates, skin can increase due to turbulent flow. In addition to the normal pressure drop proportional to the flow, there is an additional pressure drop proportional to the square of the flow rate. This can be detected if pressure transients at different flow rates show a skin that increases with flow rate.

A1.3.3. Productivity

Following the line source solution, Eqs. (A1.2)−(A1.4), pressure changes increasingly slowly with time as time increases, so that a quasi-steady flow is reached. In this quasi-steady state the drawdown is related to the flow rate by the productivity PI:

$$W = PI \times \Delta P \qquad (A1.20)$$

Comparing with Eq. (A1.2):

$$\frac{1}{PI} = \frac{v}{4\pi kh}\left[2.303 \log_{10}\left(\frac{4kt}{\varphi \mu c r^2}\right) - 0.5772 + 2s\right] \qquad (A1.21)$$

An alternative expression is obtained if there is a distant radius r_o at which pressure is held constant. Then flow does ultimately stabilize, with:

$$\frac{1}{PI} = \frac{v}{2\pi kh}\left[2.303 \log_{10}\left(\frac{r_o}{r_w}\right) + s\right] \qquad (A1.22)$$

For large values of t or r_o/r_w, the expression for PI is not very sensitive to r_o. Then Eq. (A1.21) or (A1.22) can be used to obtain a rough estimate for permeability-thickness. A similar expression is used for injection, defining injectivity II.

A1.4. INJECTION

So far, the analysis has been for discharge where reservoir fluid is flowing into the wellbore. Since injection is, in principle, the reverse process, the equations

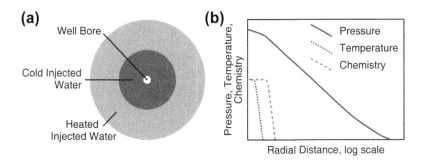

FIGURE A1.5 Distribution of fluid, pressure, and temperature around an injection well where relatively cool fluid is being injected into hotter surrounding formation.

are altered only by changing the signs of ΔP and W. Injection is a simple inverse of production if the fluid injected is the same as that produced. Usually this is not the case, especially during well testing where cold water is injected into a hotter reservoir. Then the injected water has different viscosity and compressibility to the reservoir fluid.

For short time scales (such as govern pressure tests), Figure A1.5 shows the fluid distribution in the reservoir for injection of cold water into an aquifer of homogeneous fluid. Near the well is a bank of cold water. Beyond that is a bank of injected water that has been heated by contact with the rock, and beyond that is reservoir fluid. The reservoir "sees" an expanding volume of water at reservoir temperature. The region of disturbed pressure extends far beyond the region so that the bulk of the fluid controlling the pressure transient is close to reservoir conditions. Thus, the appropriate fluid parameters are those of reservoir fluid, not injected fluid. There may be a skin due to the cooler water near the wellbore. The injected fluid acts as a volume source of reservoir fluid, and with this specified volume source injection yields the transmissivity kh/μ and storativity φch of the undisturbed aquifer, not the properties of the small region around the well containing injectate.

Over longer times, other effects can occur if sufficient cold water accumulates to affect pressure gradients or if mixing of the cold injected water into reservoir fluid, when thermal effects can then affect pressure changes. Notable examples are that commonly permeability measured during injection of cold water is higher than measured during production (injectivity is greater than productivity), which is normally ascribed to contraction of the rock and opening of fractures; and that injection of cold water into two-phase conditions can cause a fall in pressure in the region near the well. For further discussion of permeability changes with temperature see Chapters 7 and 14. Note that in Figure A1.5b the pressure gradient as drawn is less in the cold region, reflecting thermal stimulation of the rock. If this stimulation is absent, the gradient would be higher due to the greater viscosity of cold water.

A1.5. TWO-PHASE FLOW

A frequent occurrence in geothermal fields is that two phases (steam and water) flow together. This may occur in a liquid-dominated reservoir or a vapor-dominated reservoir (in which case the liquid phase may be immobile). The pressure transient techniques remain valid provided that the fluid properties are appropriately defined for the mobile fluid. Compressibility is given by:

$$\varphi c_t = \frac{\rho_t C_t}{H_{sw}} \frac{dT_s}{dP} \frac{\rho_w - \rho_s}{\rho_w \rho_s} \tag{A1.23}$$

$$= 50\, P^{-1.66} \tag{A1.24}$$

where pressure is in bar and compressibility in bar^{-1}, and $\rho_t C_t$ is taken as 2.5×10^6 J/m^{3}K.

Fluid viscosity is defined by:

$$\frac{1}{\beta_t} = \frac{k_{rw}}{\beta_w} + \frac{k_{rs}}{\beta_s}, \ \beta = \mu, \nu \tag{A1.25}$$

and density by:

$$\frac{H_{sw}}{\rho_t} = \frac{H_t - H_w}{\rho_s} + \frac{H_s - H_t}{\rho_w} \tag{A1.26}$$

These imply:

$$\frac{k_{rw}}{k_{rs}} = \frac{\nu_w}{\nu_{rs}} \cdot \frac{H_s - H_t}{H_t - H_w} \tag{A1.27}$$

H_t is the enthalpy of the flowing steam-water mixture, and all other thermodynamic variables are evaluated at undisturbed reservoir conditions. Equations (A1.25) and (A1.26) also satisfy $\mu_t = \rho_t \nu_t$.

Computing the viscosities requires knowledge of the relative permeabilities. Unfortunately, these are poorly known for fractured geothermal rock. It is usual to assume the rock matrix has similar properties to sandstones, such as the "Corey" relative permeabilities. In the fractures it is often also assumed that the phases impede each other less, possibly to the extent that $k_{rw} + k_{rs} = 1$.

A common example of a two-phase transient is in a vapor-dominated reservoir, where the flowing fluid is steam. Only the compressibility needs to be adjusted from single-phase. An example of a two-phase transient analysis with both phases flowing is given by Grant (1980b) and Grant and colleagues (1982a).

The principal practical problem in a two-phase analysis is the value of discharge enthalpy used, since this normally varies with both time and flow rate. The standard technique of evaluating at undisturbed reservoir conditions (i.e., conditions distant from the wellface) implies that it should be the flowing enthalpy of the undisturbed reservoir that is used. This would normally be the enthalpy at low flow rates. The validity of two-phase analyses has been checked, primarily by analysis of simulated data, finding reasonable agreement (Moench & Atkinson, 1978; Garg & Pritchett, 1981).

A1.6. PSEUDOPRESSURE

As gas flows to a well, pressure may vary sufficiently that the assumption of nearly constant density may not be valid. This problem can be overcome by the definition of a pseudopressure, a modified pressure function (Al-Hussainy et al., 1966). The equation for isothermal flow of a dry gas is:

$$\varphi c\rho \frac{\partial P}{\partial t} = \nabla\left(\frac{k}{v}\nabla P\right) = k\left[\frac{\partial}{\partial P}\left(\frac{1}{v}\right)\right](\nabla P)^2 + \frac{k}{v}\nabla^2 P \qquad (A1.28)$$

The normal linearization omits the first term on the right-hand side. It can, however, be accommodated by defining the pseudopressure $m(p)$:

$$m(P) = \int \frac{dP}{v} \qquad (A1.29)$$

Then Eq. (A1.28) becomes, exactly:

$$\varphi\mu c \frac{\partial m}{\partial t} = k\nabla^2 m. \qquad (A1.30)$$

There remains nonlinearity in that μ and c are pressure dependent. This is ignored by evaluating these parameters at undisturbed conditions. In the region of large drawdown, the rate of change with time is small so that the left of Eq. (A1.30) is near zero anyway. The pseudopressure is a uniform approximation, valid in regions of both large and small drawdown.

It is commonly assumed that the dynamic viscosity μ varies little. Then:

$$m(p) = \int \frac{\rho}{\mu} dP = \frac{1}{\mu} \int \frac{M_s P}{R(T+274)z} dP = \left(\frac{1}{2}\frac{M_s}{\mu R(T+274)Z}\right) P^2 \qquad (A1.31)$$

M_s is the molecular weight of steam (the gas), and Z is the gas law deviation factor. The density is proportional to pressure, and this makes the pseudopressure proportional to pressure squared. A dimensionless pseudopressure m_D can be defined and liquid solutions P_0 used in place of m_D.

$$m_D = \frac{\pi k h M_s}{W\mu Z R(T+274)} P^2 \qquad (A1.32)$$

The expression $ZR(T+274)/M_s = P/\rho$ is approximately constant for saturated steam at 1.9×10^5 Pa.m^3/kg.

Expressed in terms of P^2, the line source solution becomes:

$$\Delta P^2 = P_o^2 - P^2 = \left(\frac{W\mu}{2\pi kh}\right)\left(\frac{R(T+274)Z}{M_s}\right)E_1\left(\frac{\varphi\mu cr^2}{4kt}\right) \qquad (A1.33)$$

Note that ΔP^2 is the change in pressure-squared, not $(\Delta P)^2$. Then, if there is a slope of M on a slope of pressure-squared against time:

$$kh = 2.303\left(\frac{W\mu}{2\pi M}\right)\left(\frac{R(T+274)Z}{M_s}\right) \qquad (A1.34)$$

$$\varphi che^{-2s} = 2.25\left(\frac{kh}{\mu}\right)\left(\frac{t}{r^2}\right)10^{-\Delta P^2/M} \qquad (A1.35)$$

In geothermal "dry" steam wells, flow is usually not isothermal but follows the saturation line, indicating that the steam is in contact with liquid water. However, Eq. (3.54) for flow of steam with immobile water is identical in form to Eq. (A1.28), using the two-phase compressibility, and the development of the pseudopressure remains valid. The pseudopressure is used only for "dry" steam wells—that is, wells flowing only steam—in a vapor-dominated reservoir.

Comparing Eqs. (A1.22) and (A1.33), a productivity can be defined in terms of pseudopressure, giving:

$$W = PI'\Delta P^2 \qquad (A1.36)$$

where:

$$PI' = PI/(2P_r) \qquad (A1.37)$$

This expression, with flow proportional to ΔP^2 is more accurate at large pressure changes than the normal productivity which is linear in pressure.

A1.7. VARIABLE FLOW RATE

It can often be difficult to maintain a constant mass flow for a geothermal discharge. If there are a series of step changes in flow, the pressure change can be superimposed:

$$\Delta P = \frac{v}{2\pi kh}\sum_{t>t_i}\Delta W_i P_D(t_D - t_{Di}) \qquad (A1.38)$$

and a continuous variation can be represented by an integral:

$$\Delta P = \frac{v}{2\pi kh}\int_0^t P_D(t_D - t_D')dW' \qquad (A1.39)$$

Given actual observations at varying flow rates, the function P_D must be reconstructed to compare it with standard solutions; the pressure history must be deconvolved. This can be done numerically (Barelli & Palama, 1981; Onur et al., 2008). Otherwise, the most common method is to ignore the flow rate variation on the assumption that it is not important. Thus, for example, if the flow is slowly running down, one can attempt to obtain the form of the drawdown for constant flow by plotting $\Delta P/W$. This should remove some of the variation caused by nonconstant flow.

One important case of variable rate is flow at constant pressure. The pressure (at wellface) is held constant and the flow changes. This can arise if a well is in production at constant pressure and the pressure drop from downhole to wellhead does not change much. A common case is a well at low or moderate permeability flowing wide open for a period.

If the pressure is changed suddenly, the flow rate changes in a form that is asymptotically similar to the line source solution:

$$\frac{1}{q} = \left(\frac{\mu}{4\pi kh}\right)\left(\frac{1}{\Delta P}\right)\left[2.303 \log_{10}\left(\frac{4kt}{\varphi\mu cr^2}\right) - 0.5772\right] \quad \text{(A1.40)}$$

Thus, one plots $1/q$ or $1/W$ against time on a semilog graph. If M_W is the slope of the plot of W^{-1}:

$$kh = \left(\frac{v}{4\pi M_W}\right)\left(\frac{2.303}{\Delta P}\right) \quad \text{(A1.41)}$$

Note that Eq. (A1.38) is identical in form to the line source solution. However, the straight line is approached much more slowly.

A1.8. FRACTURED MEDIA

Two types of pressure transient are specifically directed to the study of fractured media. For a medium that is fractured throughout, characteristic changes occur at a time scale dependent on block size and other parameters. Alternatively, if a well penetrates a medium that is homogeneous except for a fracture

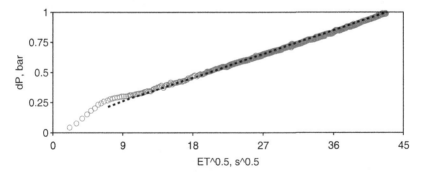

FIGURE A1.6 Pressure buildup showing $t^{0.5}$ dependence. *Source: Ladner & Häring, 2009.* © *Geothermal Resources Council.*

intersecting the wellbore, as may be produced by hydraulic stimulation, there is a distinctive history that is best represented by type curves (Gringarten & Ramey, 1975; Cinco-Ley & Samaniego, 1981).

Figure A1.6 shows an example. The data are from an EGS well at Basel. The fractured well has a characteristic period of one-half slope on a log-log plot. There may also be a unit slope preceding it, reflecting wellbore storage. It will also produce a linear plot when pressure is plotted against the square root of time, as in the figure. The $t^{0.5}$ dependence is one of the distinctive features of a well intersecting a major fracture and reflects the period when flow effects near the plane of the fracture dominate. The time when the correct straight line for semilog analysis starts is at twice the pressure of the end of the half-slope period on the log-log plot—the double ΔP rule (Wattenbarger & Ramey, 1969). The rule refers to liquid flow, so for steam wells it is double the value of ΔP^2. Given that geothermal reservoirs are fractured, it might be expected that such pressure transients would be standard, but in practice they are surprisingly uncommon.

A1.9. WELLBORE THERMAL AND FLOW EFFECTS

In previous sections, the basic pressure transient theory was very briefly reviewed. There are a considerable number of physical processes peculiar to geothermal wells that do not occur in reservoirs (such as groundwater and petroleum) of small vertical extent, or in effect two-dimensional, in which flow is single-phase and isothermal. These effects can render pressure transients useless or misleading, and it is necessary to be able to recognize them in order to obtain good quality and reliable data analyses. When they are strong, it is easily recognized that the analysis of the transient will be nonsense. Even more dangerous is a weak effect that produces a biased interpretation.

The pressure transient analyzed by theory is the pressure history, at well-face, of a homogeneous medium. Thus, the true transient is measured downhole at the depth where the well responds to reservoir pressure: the well's principal feed point. For the ideal well with a single feed zone, the pressure gauge must be placed there. However, this is the exception, and for most wells there are multiple feed points and there must be a compromise in selecting the point at which to measure transient pressure changes. This choice will not always be correct, with the result that the pressure record may be affected by temperature change and interzonal flow to the extent that the data cannot be analyzed.

A1.9.1. Condensation in Steam Wells

Steam wells maintain a column of steam to wellhead, so it is convenient to measure pressure at wellhead. The pressure recovery at wellhead is a good measure of downhole recovery provided that the wellbore remains full of steam down to the principal feed. It can happen that condensation causes a water level to rise above this point, making the wellhead pressure record useless from that time.

A1.9.2. Flashing Column in Wellbore

In a liquid-dominated system wellhead pressures are not generally used for transients (except in observation wells with a stable water column) because the column of water in the well can vary between a continuous water column, a water column with changing temperature, and a low-density two-phase mixture. This variation can also affect transients measured at an inappropriate depth such as bottomhole. Figure A1.7 shows a buildup after discharge. The pressure profile changes so bottomhole pressure only moves in parallel with the feed point when there is a column of water at constant temperature between them. Early time data at bottomhole is incorrect because the density of the fluid column between bottomhole and the feed zone is changing with time, creating a changing pressure difference between bottomhole and feed zone. If the well contains two-phase fluid during discharge and in its stable shut state, bottomhole pressure never moves in parallel with the main feed point.

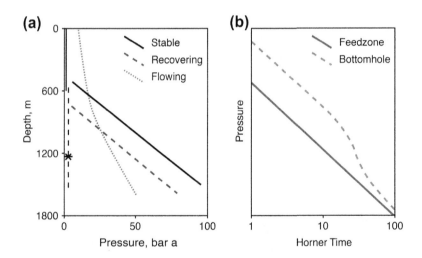

FIGURE A1.7 Pressure recovery in well with flashing fluid during discharge.

A1.9.3. Injection Tests

When pressure transients are measured under injection, the main risk of incorrect analysis arises from interzonal flow. Figure A1.8 shows temperature profiles when injection is stopped in a hypothetical well with interzonal flow and pressure changes. The hydrostatic gradient in the well changes with the changing temperature profile, and a constant profile is not present. The well has a hot inflow at 550 m and an outflow at 1150 m. Profile PT1 is taken while injecting, and the subsequent runs at increasing time after shut. The pressure

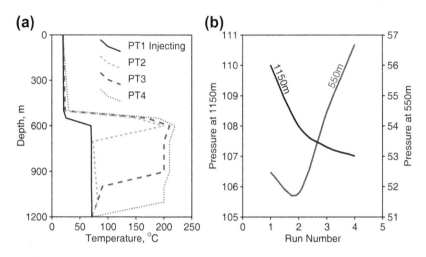

FIGURE A1.8   Temperature profiles in an injection well with interzonal flow during and after injection, and pressure at upper and lower feed zones.

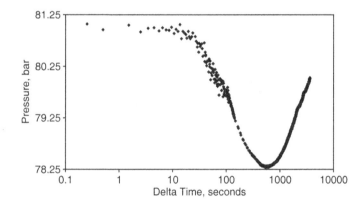

FIGURE A1.9   Rebound in pressure falloff. *Source: Rotokawa Joint Venture, Personal Communication.*

at the lower zone falls in a normal manner, but pressure at the upper zone initially falls and then rises as the water column in the well gets hotter. This situation typically shows up on a transient as rebound or oscillation. Figure A1.9 shows a typical example: a pressure falloff when injection was stopped when the pressure is recorded above the main feed. At first pressure falls, as would be expected. But then the continuing inflow of hot water at the upper zone heats the water column, reducing its density. The pressure difference between the tool and the feed zone decreases, raising the pressures above the feed zone.

A1.9.4. Use of Profiles

In all of the preceding cases, the problems with pressure transient data are caused by fluid of variable density in the wellbore. This situation can be detected by running PTS profiles as part of the transient program (particularly at a later time, when closely spaced pressure data are not essential). Running a spinner is particularly useful in unequivocally identifying flow in the wellbore, as long as at least one survey is done in each direction. This will show the presence of interzonal flow or two-phase fluid or liquid conditions, sometimes found at the bottom of a "steam" well.

A1.9.5. Wellbore Thermal Storage Effects

It was earlier mentioned that heating of the wellbore and condensation of steam produce a spuriously large wellbore storage effect. In essence, through condensation, two-phase compressibility applies to the wellbore fluid. Other effects can also arise (Miller, 1980a, b). The heating or cooling of the wellbore can change fluid density sufficiently to cause a storage effect that lasts much longer than simple compression of liquid. In addition, transmissivity of geothermal wells can be so large that downhole pressures change significantly over the time that a pressure pulse takes to ascend the wellbore, and during transients the fluid in the wellbore may not be in vertical equilibrium. This can produce pressure transients, plotted log-log that have initial slope greater than unity, or oscillate. On standard analysis, the initial log-log slope cannot exceed unity, but in practice if often does.

If flashing fluid is present in the wellbore during flow, it persists for some time after shutting, producing transient changes in the vertical pressure distribution and corresponding changes in the pressure difference between the feed zone and the wellhead or bottomhole. The long period of time before a liquid column forms is caused solely by the transient energy and mass transfer within the wellbore itself.

A1.10. BAROMETRIC, TIDAL, AND OTHER EFFECTS

Changes in barometric pressure and the strain of the Earth's crust on Earth tides apply signals to aquifers, including geothermal reservoirs. They produce a response in the reservoir and any well penetrating it. The pressure changes are relatively small, but can be significant in interference testing. It is necessary to recognize and remove them.

A1.10.1. Tidal Responses

Geothermal reservoirs can exhibit a response to the changes in strain in the rock caused by Earth tides. The theory of tidal response in an aquifer is described by

Bödvarsson (1970). The magnitude of the tidal response is less than 10 mb. Hanson (1979) describes observed responses in Raft River and Salton Sea fields. The tidal strain is a superposition of sinusoidal variations with different periods. The main tides have periods of $\frac{1}{3}$, $\frac{1}{2}$, 1, and 16 days. If such a periodic signal appears to be present, a Fourier analysis of the data will show it, and this signal can then be removed.

A1.10.2. Barometric Response

Barometric pressure changes are usually long-period pressure changes applied at the surface of the Earth, although at midlatitudes pressure changes of 0.03 bar can happen in a few hours. This is transmitted through both the well itself and the rock, due to the increasing weight on the ground surface, and a corresponding pressure change occurs in any aquifer containing pore fluid. The change in reservoir pressure is related to the barometric pressure change ΔP_{atm} by the barometric efficiency BE:

$$\Delta P = (1 - BE)\Delta P_{atm} \qquad (A1.42)$$

The barometric efficiency is given by:

$$BE = \varphi c_t / (c_m + \varphi c_t) \qquad (A1.43)$$

In a well standing open with a water level, that level is controlled by a balance between atmospheric and reservoir pressure. The water level (measured upwards, i.e., increasing when the level rises) changes by an amount $\Delta \eta$:

$$\Delta \eta = -(BE/\rho_w g)\,\Delta P_{atm} \qquad (A1.44)$$

The barometric efficiency lies between 0 and 1 and approaches 1 for fluids of high compressibility. Thus, barometric effects are greatest in wells feeding from liquid conditions. The pressure changes due to barometric effects can be as large as barometric variations—that is, as much as 0.1 bar. These are certainly enough to affect interference testing. When barometric noise is present, it is simplest to fit the well response to both a modeled interference signal and barometric pressure. More detailed modeling (Kiryukhin & Kopylova, 2009) shows that the barometric efficiency depends not only on the compressibilities, but also on permeability near the well. It can be the wellbore itself which is the primary link for atmospheric pressure variation to affect the reservoir. There can be practical problems in determining the barometric efficiency, so the best way to analyze and interference test with barometric effects present is to fit the response to both the interference signal and barometric pressure. Figure A1.10 shows an example. Interference was monitored using water levels. Water level is increasing downward. Barometric pressure also changed rapidly at the beginning of the flow period, producing a signal similar in shape for several days to a pressure transient. If the barometric noise

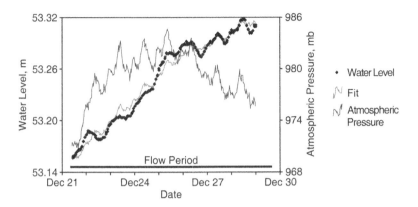

FIGURE A1.10   Interference test with barometric noise. *Source: Rotokawa Joint Venture, Personal Communication.*

were ignored, the interference results would be biased because of this similarity. The data were fitted by regression to the line source solution and the barometric signal, producing good estimates for reservoir transmissivity and storativity, and barometric efficiency.

A1.10.3. Other Effects

Other external signals have been observed to influence pressures in geothermal fields. Reykjavik shows a response to oceanic tides (Thorsteinsson & Eliasson, 1970), which was explained by assuming that the aquifer extended under the sea and responded to the changing load of seawater above. Seferihisar on the coast of Turkey shows a significant tidal response (Palaktuna et al., 2010). Momotombo in Nicaragua shows responses to rainfall, indicating that the reservoir must be open (Dykstra and Adams, 1978). Subsequent field performance has confirmed this open nature, with significant cold recharge (Porras & Bjornsson, 2010). Wells at Matsukawa up to 400 m deep and reaching 240°C similarly showed temperature changes correlated with rainfall (Mori, 1970).

In an operating geothermal field, changes in well flow send a signal into the reservoir. For example, a daily signal can be due to daily control changes in the steam field and consequent flow changes rather than tidal effects. A diurnal signal can also be introduced into pressure records by temperature effects on the surface instrumentation due to freezing or high temperatures where the equipment is directly exposed to sunlight.

In interference testing, it is common for there to be a preexisting trend in pressure due to the nearby production or injection, or the residual effects of previous well testing operations. Ideally a background record before a test, at least as long as the transient record, followed by a similar record after the test is needed to eliminate such drift effects. If pressure is observed at wellhead or by measuring the depth to the water level, there may be a trend due to cooling of the fluid column in the wellbore after a previous flow. It is necessary to start observations before the source wells are flowed in order to establish any previous drift in the pressure record, which must be subtracted from the interference observations.

A1.11. TEMPERATURE TRANSIENTS

During drilling, and in any subsequent cold water injection, the wellbore is cooled by the flow of water or drilling mud. After this ceases, the fluid in the wellbore warms up, quickly or slowly, to a final temperature close to the stable temperature in the surrounding formation. Sometimes it is useful to be able to determine this final temperature without waiting for the complete time of temperature stabilization, which will usually be more than 30 days.

One approach to make this prediction has been to use a Horner plot. The well is cooled for a time t_{pc} by drilling. Thus, t_p is the time that the formation, at the depth under study, has been exposed to circulating fluid. This would usually be the time, since the drill bit passed the particular depth. Then circulation is halted, and the temperature measured at several times Δt afterward. The data are plotted on a Horner plot and extrapolated to $\Delta t = \infty$ —that is, $\Theta = (t_{pc} + \Delta t)/\Delta t = 1$ to obtain an estimate of final temperature.

The validity of the Horner plot is based on the observation the equation for heat conduction is:

$$\rho_t C_t \left(\frac{\partial T}{\partial t} \right) = K \, \nabla^2 T \tag{A1.45}$$

that is, the diffusion equation, which is the same form as the pressure transient equation. This governs the cooling and warming of the well provided that conduction in the radial dimension (i.e., 2D conduction) is the dominant mechanism of heat transfer. It is not valid at any zone of fluid loss or gain or at any permeable zone, or if there is internal circulation of fluid in the wellbore passing the depth of observation, or at the bottom of the hole where the heat flow is three-dimensional. If there is any convection, this will overwhelm conduction as a means of heat transfer.

There is an additional problem. The condition imposed at the wellbore during circulation is, approximately, $T = $ constant rather than heat flux $=$ constant. The problem is analogous to pressure recovery after flow at constant pressure rather than after flow at constant rate. The temperature recovery is analogous to a pressure recovery after flow at constant pressure. The Horner plot will in this case give an underestimate of the final temperature.

An improved method is given by Roux and colleagues (1979): a Horner plot is made and extrapolated to an apparent final temperature T^*_{ws}. Then the final temperature is computed as:

$$T = T^*_{ws} + m \, T_{DB}(t_{PD}) \tag{A1.46}$$

where m is the slope of the Horner straight line and T_{DB} is a dimensionless correction term, dependent on the dimensionless buildup time t_{PD} and the Horner time $\Theta = (t_{pc} + \Delta t)/\Delta t$. The term t_{PD} is defined as:

$$t_{PD} = \left(\frac{K}{\rho_t C_t r_w^2}\right) t_{Pc} \tag{A1.47}$$

Since rock conductivity and heat capacity do not vary greatly, a reasonable average value for a well of radius 0.1 m is:

$$\left(\frac{K}{\rho_t C_t r_w^2}\right) = 0.4 \text{ hr}^{-1}$$

The function T_{DB} is given by (generalizing the results of Roux and colleagues):

$$T_{DB} = 0.03 \, \Theta^{1.678} \, t_{PD}^{-0.373} \tag{A1.48}$$

Figure A1.11 shows an example adapted from Menzies (1981). The extrapolation gives $T^*_{ws.} = 238°C$, and the slope is $m = 194$ K per cycle. The slope is defined over the range $\Theta = 2$-4, giving a (geometric) midpoint of 2.8, and the circulating time was 10 hours, giving $t_{PD} = 4$. Then Eq. (A1.48) gives $T_{DB} = 0.10$, and $T_i = 238 + 194 \times .10 = 258°C$. Later measurement found a downhole temperature of 265°C, but this may have been affected by an internal flow that developed after completion of the well. For another fully worked example, see Chapter 6.

Under most circumstances, extrapolated temperatures are not accurate to more than 5–$10°C$. Temperature extrapolation may be the only way of estimating a reservoir temperature in wells where the warmed-up well later contains an internal flow. If the flow is absent during warm-up, these data can

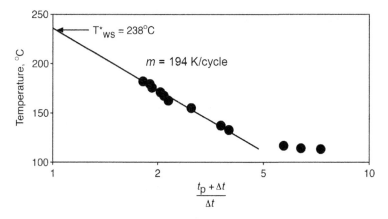

FIGURE A1.11 Temperature recovery in well MG-1, Tongonan. *Source: Menzies, 1981.*

be extrapolated to obtain a good estimate of formation temperature at a time when the wellbore temperatures are not controlled by internal flow.

A1.12. CONVERSION OF GROUNDWATER UNITS

Pressure transient analysis in groundwater has different units but the same mathematical form. There are two unit differences:

1. Pressure is measured in head of water.
2. Viscosity is amalgamated into the permeability.

In the following discussion, the subscript G is used to denote a property in groundwater units.

A1.12.1. Relation Between Pressure and Head

$$\Delta P = \rho g \Delta h$$

With pressure in bar:

$$\Delta P = 10^{-5} \rho g \Delta h$$

When water level is measured in a well with a temperature profile, there is a question about the relevant temperature. When pressure changes, water enters or exits at the feed zone. Thus, for short time (days), the loss or gain is water at reservoir temperature. But over long periods of time, the water column heats or cools and reequilibrates with the surrounding rock. Then the gain or loss is water at the surface—that is, the relevant temperature is at the water level. This applies to wells used for long-term reservoir monitoring over time periods as long as months. At intermediate times, it's better to use tubing and measure pressure directly at the feed zone.

A1.12.2. Permeability

Darcy's Law: $v = \frac{k}{\mu} \nabla P$
In groundwater: $v = K_G \nabla h$
giving $K_G = \frac{k}{\mu} \rho g = \frac{k}{\nu} g$
where ν is the kinematic viscosity.

With k in Darcy $(= 10^{-12} \text{ m}^2)$ $k = 10^{12} \nu K_G / g$ Darcy
And for transmissivity $T_G = K_G h$; $kh = 10^{12} \nu T_G / g$ dm

A1.12.3. Storativity

Groundwater: Storativity S_G = volume change/unit area/unit head
Absolute Storativity $S = \varphi c h$ = volume change/unit area/unit pressure
that is, the difference is the pressure scaling

$$\varphi c h = S_G / \rho g$$

Gas Correction for Flow Measurements

A2.1. EFFECT OF NONCONDENSABLE GAS

Noncondensable gases are present in all geothermal discharges. Where these are present in significant amounts, the effect of the gas on the flow measurements must be allowed for to obtain accurate mass flow and enthalpy data. Methods to determine the concentration of noncondensable gases in steam-gas mixtures and in two-phase pipelines are described by Ellis and Mahon (1977). For separated steam-gas flows, the samples can be taken directly from the pipeline, and for two-phase fluids, a miniseparator is used to provide samples of the steam-gas mixture. The resulting sample is then analyzed for noncondensable components. The gas content in a separated steam-gas flow can also be determined directly by physical measurements by simultaneously measuring the condensed steam and gas flow rates from a representative sample of pipeline flow (Blair & Harrison, 1980). For high-temperature geothermal systems, carbon dioxide usually makes up the bulk of the noncondensables (>90%), and for most engineering purposes it can be assumed that CO_2 is the only gas present when making enthalpy and flow rate corrections.

At pressure-temperature conditions normally encountered in geothermal well testing, all of the noncondensable gases can be assumed to be in the vapor phase for calculation of well flow characteristics. When the flash fraction is small (less than ~2%), it will be necessary to include the gas dissolved in the water phase to obtain an accurate total gas flow rate. The total (measured) pressure in a steam or two-phase pipeline is made up of a steam partial pressure and gas partial pressures, in ratios according to Dalton's Law. When measuring steam-gas flow rates, the effect of noncondensable gases on the specific volume of the mixture must be allowed to obtain an accurate measurement of the total flow rate using the orifice calculation (see Eq. (8.10)).

The following assumptions are made:

1. The temperature of the steam-gas mixture is the saturation temperature at the partial pressure of the steam.

2. The measured pressure is the sum of the partial pressure of steam and the partial pressure of noncondensable gases:

$$P = P_s + P_g \tag{A2.1}$$

3. The steam and noncondensable gas act as ideal gases, and the partial pressure of the gas is proportional to its mole fraction in the mixture:

$$P_g = P \times x, \ P_s = P \times (1 - x) \tag{A2.2}$$

4. The specific volume v_v of the vapor-phase mixture is given by:

$$v_v = v_s \times (1 - f_v) \tag{A2.3}$$

where v_s is the specific volume of steam at the partial pressure of the steam and f_v is the mass fraction of gas to steam plus gas.

5. The enthalpy of the mixture for the purposes of well evaluation is that of the steam and water alone.

6. It is assumed that there has been sufficient steam formed that all gas is in the vapor phase; the residual gas dissolved in water can be ignored.

A2.2. GAS CORRECTION FOR THE SEPARATOR METHOD

Given a noncondensable gas content of x moles gas to moles steam plus gas, at a sampling pressure of P (note that gas analyses are usually in the form *moles gas:moles steam,* and these need to be recalculated to *moles gas:total moles gas + steam* in order to apply the various gas laws):

1. Calculate the average molecular weight of the noncondensable gas mixture, using a recent analysis:

$$M_g = \frac{\sum M_{gi} \times x}{\sum x_i} \tag{A2.4}$$

where the sum is over the different gas species present. Usually the noncondensable gas is more than 90% CO_2, and the molecular weight of the gas can be taken as 44, that of carbon dioxide.

2. Assuming all the gas is CO_2, calculate the average molecular weight of the vapor phase (steam + gas):

$$M_v = M_g \times nx + M_s \times (1 - x) = 44x + 18(1 - x) = 18 + 26x \tag{A2.5}$$

The mass fraction of gas in the vapor phase is then given by:

$$f_v = \frac{M_g \times x}{M_v} = \frac{44 \times x}{18 + 26 \times x} \tag{A2.6}$$

3. Calculate the partial pressure of steam at sampling pressure:

$$P_s = P \times (1 - x) \tag{A2.7}$$

4. Calculate the specific volume of the steam-gas mixture (v_S is taken from the steam tables at pressure P_S from Eq. (A2.7)):

$$v_v = v_s \times (1 - f_s) \qquad \text{(A2.8)}$$

5. Calculate the flowrate of the steam-gas mixture:

$$W_v = W_S + W_g = C\sqrt{\frac{\Delta P}{V_v}} \qquad \text{(A2.9)}$$

6. Calculate the steam flow alone:

$$W_s = W_v \times (1 - f) \qquad \text{(A2.10)}$$

7. Calculate the separated water flow from the separator by the normal method (Eq. (8.5) or (8.10)), then recalculate the enthalpy and mass flows from the separator using the *partial pressure of steam* to obtain the sensible and latent heat values for water (H_w and H_{ws}) from the steam tables (enthalpy and mass flow are assumed to be for the flowing steam plus water, excluding the gas):

$$H = H_w + H_{ws} \times \frac{W_s}{W_s + W_w} \qquad \text{(A2.11)}$$

8. Calculate the noncondensable gas flow in the separated steam:

$$W_g = W_v \times f_v \qquad \text{(A2.12)}$$

Ignoring the gas dissolved in the water, this is the *total gas flow*.

Example: Well EX12 is producing at 12.2 bar g wellhead pressure to a steam-water separator. The following readings are made at orifice plates installed on the separated steam and water lines. The separator pressure is 10.2 bar g, at the steam orifice plate the ΔP is 207 millibar and upstream pressure is 10.1 bar g; at the water orifice the ΔP is 298 millibar and temperature is 184°C. The equation for the steam orifice is $W_s = 10.525 \times \sqrt{\Delta P/v}$ and for the water orifice is $W_w = 5.69\sqrt{\Delta P/v}$, with flow in kg/s, ΔP measured in millibar, and specific volume in cm3/gm. A gas sample is taken from the steam pipeline at pressure 10.2 bar g, and gives an analysis of 5500 millimoles gas to 1 mole steam plus gas, or $x = 5500 \times 10^{-5}$. Atmospheric pressure is 1.0 bar absolute. Procedure:

1. Using equation (A2.6) calculate the mass fraction, f_v, of gas to steam plus gas:

$$f_v = \frac{44}{18 + 26 \times x} \times x = \frac{44}{18 + 26 \times 0.055} \times 0.0550 = 0.125$$

2. Calculate the partial pressure of steam at sampling pressure:

$$P_s = P \times (1 - x) = (10.2 + 1.0) \times (1 - 0.055) = 10.6 \, \text{bar}$$

3. Calculate the specific volume of the steam-gas mixture (v_s is taken from the steam tables at partial pressure of steam P_s):

$$v_v = v_s \times (1 - f_v) = 185 \times (1 - 0.125) = 162 \text{ cm}^3/\text{gm}$$

4. Calculate the flowrate of the steam-gas mixture (assume $\varepsilon = 1$):

$$W_v = 10.525\sqrt{\Delta p/v_v} = 10.525\sqrt{207/162} = 11.9 \text{ kg/s}$$

5. Steam and gas flows:

$$W_s = W_v \times (1 - f_v) = 11.9 \times (1 - 0.125) = 10.4 \text{ kg/s}$$
$$W_g = 11.9 - 10.4 = 1.5 \text{ kg/s}$$

6. Water flow:

$$W_w = 5.69 \times \sqrt{\Delta P/v_w} = 5.69\sqrt{298/1.133} = 92 \text{ kg/s}$$

7. Calculate enthalpy using separation conditions at the partial pressure of steam, from step 2, 10.6 bar:

$$H = X \times H_{ws} + H_w = [10.4/(10.4 + 92) \times 2005 + 773] = 977 \text{ kJ/kg}$$

8. Mass flow (steam and water flows at the same separation pressures):

$$W = W_s + W_w = 10.4 + 92 = 102 \text{ kg/s}$$

The partition function for carbon dioxide at 184°C is 563, using Eqs. (A3.26) and (A3.27). Thus, the concentration of gas in the liquid phase is 0.125/562 = 0.00022, and the flow of gas in the liquid phase is 0.00022 × 102 = 0.022 kg/s, which is negligible compared to the gas in the vapor phase.

A2.3. GAS CORRECTION FOR THE LIP PRESSURE METHOD

Since the lip pressure method has been developed on an empirical basis, there is no simple physical principle providing a correction for the effect of noncondensable gas. Grant and colleagues (1982) have suggested the following method based on the critical flow model of Fauske given in Karamarakar and Cheng (1980). This correction method has been checked by simultaneous separator and lip pressure measurements with gas contents up to 2% by weight of total mass flow.

Applying the method requires some preliminary calculations, since the gas content is usually determined at a pressure other than the lip pressure. As with the separator method, the correction can be simplified by assuming that all the gas is CO_2. When using the lip pressure method, the noncondensable gas content is determined by sampling the two-phase flow using a miniseparator and before calculating the enthalpy correction, the gas content in the separated steam at the sampling pressure must be recalculated to lip pressure using the preliminary enthalpy value. The enthalpy correction can then be applied and mass flow recalculated.

A preliminary estimate of the flowing enthalpy (H_c) and mass flow rate (W_C) are first calculated assuming no gas (Eq. (8.15)). Then after calculating the mass fraction of gas in the vapor phase at the lip pressure, the enthalpy correction factor is calculated with Eq. (A2.13), and the preliminary enthalpy corrected as in Eq. (A2.14):

$$\Delta H_c = f_{lip} \times \frac{H_c \times (2675 - H_c)}{3070 - 0.11 \times H_c} \qquad (A2.13)$$

$$H = H_c - \Delta H_c \qquad (A2.14)$$

Procedure:

1. Calculate well enthalpy (H_c) and flow rate, assuming no gas.
2. Convert the gas analyses to the standard format: x, mole fraction of gas to steam plus gas at sampling pressure, P (bar absolute).
3. Calculate the dryness at lip pressure (X_{lip}) and dryness at sampling pressure (X_{sep}). Use enthalpy (H_c) obtained from step 1 and steam table values for both sample and lip pressure at measured pressure. As the ratio of X_{lip}/X_{sep} is used in Eq. (A2.16), the small errors arising from not using the true enthalpy and allowing for partial pressures cancel out, with an accumulated error of less than 5% in the recalculated f_{lip} value.
4. Calculate mass fraction of gas:steam+gas at sampling pressure as outlined in Eqs. (A2.4)–(A2.6) for the separator gas correction. Note that this equation can be simplified by assuming the great majority of the gas (>95%) is CO_2:

$$f_{sep} = \frac{44 \times x}{18 + 26 \times x} \qquad (A2.15)$$

5. Calculate mass fraction of gas:steam+gas at lip pressure:

$$f_{lip} = f_{sep} \times \frac{X_{sep}}{X_{lip}} \qquad (A2.16)$$

6. Calculate the gas fraction in total well flow:

$$f_t = f_{sep} \times X_{sep} \qquad (A2.17)$$

7. Calculate the enthalpy correction (ΔH_c) using Eq. (A2.13).
8. Calculate the corrected flowing enthalpy using Eq. (A2.14).
9. Correct mass flow using corrected enthalpy H, using Eq. (8.16), where atmospheric pressure is 1 bar abs.
10. Calculate gas flowrate:

$$W_g = W \times f_t \qquad (A2.18)$$

Example: Well EX12 is being flow tested through a 153 mm diameter lip pressure pipe. At wellhead pressure 20.9 bar gauge the lip pressure is 4.50 bar absolute and water flow measured over a sharp-edged V-notch weir is 48 kg/s. A gas sample is taken at sampling pressure 10.2 bar gauge, and gives a mole fraction of $x_v = 0.055$ moles gas:moles steam+gas. Assume all the noncondensable gas is CO_2.

Procedure:

1. Using Eq. (8.16), calculate Y value:

$$Y = \frac{W'_w}{AP^{0.96}_{lip}} = \frac{48}{\pi(153/20)^2 \times 4.5^{0.96}} = 0.0617$$

2. Calculate H_c using Eq. (8.17):

$$H_c = \frac{2675 + 3329 \times Y}{1 + 28.3 \times Y} = \frac{2675 + 3329 \times .0617}{1 + 29.3 \times .0617} = 1049 \text{ kJ/kg}$$

3. Calculate a first estimate W_c of the mass flow using Eq. (8.18*):

$$W_c = \frac{W'_w \times 2258}{2675 - H_c} = \frac{48 \times 2258}{2675 - 1049} = 67 \text{ kg/s}$$

4. Calculate the dryness at lip pressure 4.5 bar (X_{lip}) and dryness at sampling pressure 11.2 bar (X_{sep}), using the initial enthalpy estimate (H_c) obtained from step 1:

$$X_{lip} = \frac{H_c - H_{wlip}}{H_{wslip}} = \frac{1049 - 623}{2120} = 0.201$$

$$X_{sep} = \frac{H_c - H_{wsep}}{H_{wssep}} = \frac{1049 - 785}{1997} = 0.132$$

5. Calculate mass fraction of gas: total vapor = steam + gas at sampling pressure:

$$f_{sep} = \frac{44 \times x}{19 + 26 \times x} = \frac{44 \times 0.055}{19 + 26 \times 0.055} = 0.124$$

6. Calculate mass fraction of gas:vapor at lip pressure:

$$f_{lip} = f_{sep} \times \frac{X_{sep}}{X_{lip}} = 0.124 \times \frac{0.132}{0.201} = 0.0821$$

7. Calculate the gas fraction in total well flow:

$$f_t = f_{sep} \times X_{sep} = 0.124 \times 0.132 = 0.0165$$

8. Calculate the enthalpy correction (ΔH_c):

$$\Delta H_c = f_{lip} \times \frac{H_c \times (2675 - H_c)}{3070 - 0.11 \times H_c} = 0.0821 \times \frac{1049 \times (2675 - 1049)}{3070 - 0.11 \times 1049}$$

$$= 47 \text{ kJ/kg}$$

9. Calculate the corrected flowing enthalpy:

$$H = H_c - \Delta H_c = 1049 - 47 = 1002 \text{ kJ/kg}$$

10. Correct mass flow using corrected enthalpy H:

$$W = \frac{W'_w \times 2258}{2675 - H_c} = \frac{48 \times 2258}{2675 - 1002} = 65 \text{ kg/s}$$

11. Calculate gas flowrate:

$$W_g = W \times f_t = 65 \times 0.0165 = 1.07 \text{ kg/s}$$

Equations of Motion and State

A3.1. INTRODUCTION

The equations of flow through porous media have been developed, often almost independently, in the petroleum, groundwater, and soil science literature. Their application to geothermal reservoirs requires that in addition to fluid transport, heat and possibly chemical species or gas transport must also be described. A full development of the flow equations can be found in many places. Here the derivation of McNabb is followed.

A3.2. CONSERVATION EQUATIONS

For any quantity X that is conserved, in a medium without sources or sinks, the conservation law takes the form:

$$\text{Rate of gain (at one point)} + \text{Net outflow} = 0 \qquad \text{(A3.1)}$$

If ρ_x is the density of the quantity X and $\boldsymbol{u}_x$ is the flux density as a result of flow through the porous medium, the conservation law is:

$$\frac{\partial}{\partial t}(\rho_x) + \nabla \cdot \boldsymbol{u}_x = 0 \qquad \text{(A3.2)}$$

This equation may be applied to mass, energy, or chemical species, which are the three relevant quantities conserved in geothermal systems. The conservation laws apply also to quantities such as heat and chemical species, which may be transferred between phases; such transfer or reaction affects the distribution of the quantity between fluid and solid phases but does not alter its conservation within each unit volume. Since the fluid may exist as liquid or vapor alone, or as a two-phase mixture, the equations are expanded for both single- and two-phase conditions.

A3.2.1. Conservation of Mass, Single-Phase

In this case the single fluid saturates the pore space. The density of fluid mass per unit reservoir volume (that is, per unit volume of rock and fluid) is $\varphi \rho$, where φ is the porosity of the rock matrix and ρ is the density of the fluid. Then if u is the fluid mass flux density (the mass flowing through unit cross-sectional area in unit time), conservation of mass is:

$$\frac{\partial \rho}{\partial t} + \nabla \cdot u = 0 \tag{A3.3}$$

In groundwater hydrology the volume flux density v is more commonly used. It is related to the mass flux density by:

$$u = \rho v \tag{A3.4}$$

A3.2.2. Conservation of Energy, Single-Phase Flow

Energy, such as heat, is contained in the rock, as well as in the fluid. The total energy contained in unit reservoir volume is the sum of two components, $(1 - \varphi)\rho_m U_m$ for the rock and $\varphi \rho U$ for the fluid. U_m and U are the internal energies of the rock and fluid, respectively. Energy is transported through the reservoir in two ways: by the motion of the fluid and by conduction through the rock and fluid. The energy flux density carried by the fluid is uH, where H is the enthalpy of the fluid. The conductive flux density is K ∇T, where K is the conductivity of the wetted rock. Then conservation of energy is:

$$\frac{\partial}{\partial t}[(1 - \varphi)\rho_m U_m + \varphi \rho U] + \nabla \cdot [uH + K \nabla T] = 0 \tag{A3.5}$$

Note that the internal energy U has been used in the storage terms of Equation (A3.5), but enthalpy H in the flux terms. This formulation is exact, since the energy flux due to fluid flow is the flux of internal energy plus the work done by the pressure gradient, which is equal to uH. For many purposes it is convenient to remove one thermodynamic function, U or H, and formulate the equations in terms of only one of these variables. Some additional terms are thereby generated.

A3.2.3. Conservation of Mass, Two-Phase Flow

In rock saturated with a mixture of water and steam, a fraction of the pore space is filled with each phase. The fraction of the pore space occupied by water is S_w, and the remainder $S_s = 1 - S_w$, is occupied by steam. These fractions, S_w and S_s, are termed the water and steam saturations. The water saturation is often referred to as "the saturation" and is simply denoted by S. The unit reservoir volume then contains in its pore space a mass $\varphi \rho_w S_w$ of water and a mass $\varphi \rho_w S_w$, giving a total fluid mass per unit volume of $\varphi(\rho_w S_w + \rho_s S_s)$. Both liquid and vapor phases can move independently through the medium. There are separate mass flux densities u_s, u_w for the steam and water. Then conservation of mass takes the more complex form:

$$\varphi \frac{\partial}{\partial t}(\rho_w S_w + \rho_s S_s) + \nabla \cdot [u_s + u_w] = 0 \qquad (A3.6)$$

A3.2.4. Conservation of Energy, Two-Phase Flow

There is a natural analogue of Eq. (A3.5) by comparison with Eqs. (A3.3) and (A3.6). The fluid energy in situ is that of the liquid plus that of the steam, and the convective heat transfer is the sum of the transfer by liquid and vapor phases. The equation has the form:

$$\frac{\partial}{\partial t}[(1 - \varphi)\rho_m U_m + \varphi S_w \rho_w U_w + \varphi S_s \rho_s U_s] + \nabla \cdot [u_s H_s + u_w H_w + K \nabla T] = 0$$
$$(A3.7)$$

A3.2.5. Conservation of a Chemical Species

If f is the concentration (mass/mass) of a chemical species in the fluid, and there is no reaction with the reservoir matrix, the single and two-phase conservation equations are respectively:

$$\frac{\partial}{\partial t}(\rho f) + \nabla \cdot (uf) = 0 \qquad (A3.8)$$

$$\varphi \frac{\partial}{\partial t}[S_w \rho_w f_w + S_s \rho_{sfs}] + \nabla \cdot [u_s f_s + u_w f_w] = 0 \qquad (A3.9)$$

The conservation equation for each quantity (mass, energy, chemical species) is a balance between the rate of change of the density of the quantity and the rate at which it is transported. To this balance must be added Darcy's Law, which gives the flux density as a function of the pressure field and the constitutive relations for the fluid(s) occupying the pore space.

A3.3. DARCY'S LAW

For most flow-through homogeneous porous media, the simple linear flow equation first established by Darcy is now used in nearly all petroleum, groundwater, and geothermal studies.

A3.3.1. Single-Phase Flow

For single-phase flow, Darcy's law takes the form:

$$v = -\mathbf{k} \cdot \frac{1}{\mu}(\nabla P - \rho \mathbf{g}) \qquad (A3.10)$$

or

$$u = -\mathbf{k} \cdot \frac{1}{\nu}(\nabla P - \rho \mathbf{g}) \qquad (A3.11)$$

The permeability k is, for generality, represented here as a tensor. In the future, we shall assume that this tensor has its principal axes oriented vertically and horizontally. In the vertical direction, for single phase flow in the vertical direction:

$$u_z = \frac{k_z}{\nu}\left(\frac{\partial P}{\partial z} - \rho g\right) \qquad (A3.12)$$

And in the horizontal direction, for the x-axis:

$$u_x = \frac{k_x}{\nu}\frac{\partial P}{\partial x} \qquad (A3.13)$$

with a similar expression in the y axis. The assumption of horizontal anisotropy may be invalid if faulting supplies a preferential direction of flow or barriers to flow.

A3.3.2. Two-Phase Flow and Relative Permeabilities

If two phases are present in a porous medium, each phase occupies a part of the pore volume and part of the medium's permeable passages. In consequence, each reduces the flow of the other phase below what it would be if it fully saturated the medium. For two-phase, two-component flow, an empirical variation of Darcy's Law has been established for some decades. The mass flux density of each phase is given by:

$$u_w = -\mathbf{k} \cdot \frac{k_{rw}(S_w)}{\nu_w}(\nabla P - \rho_w \mathbf{g}) \qquad (A3.14)$$

$$u_s = -\mathbf{k} \cdot \frac{k_{rs}(S_w)}{\nu_s}(\nabla P - \rho_s \mathbf{g}) \qquad (A3.15)$$

The difference between these equations and the single-phase Darcy's Law lies in the introduction of the relative permeabilities k_{rw} and k_{rs}, shown here as a function of liquid saturation. These functions are experimentally determined for homogeneous porous media such as sandstone. As each phase obstructs the flow of the other, the relative permeabilities are normally less than unity if both phases are present, and they approach unity as the saturation of the other phase approaches zero. It is normally the case that $k_{rw} + k_{rs} < 1$, if both phases are present and mobile.

Each relative permeability may also approach zero even though the saturation for that phase is not zero. The fluid component is then immobile, and the saturation of the other component at which this occurs is termed the *residual saturation*. The other component remains mobile. The obvious example is a vapor-dominated reservoir with immobile water in the matrix pore space.

The commonly assumed relative permeabilities for homogeneous media are the Corey expressions (Corey, 1972):

$$
\begin{aligned}
k_{rw} &= S^{*4} \\
krs &= (1 - S^*)^2 (1 - S^{*2})(S^* < 1)
\end{aligned}
\tag{A3.16}
$$

$$
S^* = (S - S_{wr})/(S_{sr} - S_{wr})
$$

where S_{sr} and S_{wr} are the residual saturations for steam and water, respectively. There is considerable doubt that these expressions apply to fractured media, since two fluids in a fracture obstruct each other less effectively than pockets trapped in pore space. It has been common to assume that flow in fractures follows the "X-curves" (Horne et al., 2000):

$$
\begin{aligned}
k_{rw} &= S^* \\
k_{rs} &= 1 - S^*
\end{aligned}
\tag{A3.17}
$$

Figure 3.4 shows the two sets of curves, with a residual liquid permeability of 30% for the Corey curves and no residual permeability for the X-curves.

A3.4. CONSTITUTIVE RELATIONS

The reservoir contains at any time a distribution of pressure and temperature (for single-phase), or pressure-temperature and saturation (for two-phase). To these must be added any chemical species variable (concentration or partial pressure). All other thermodynamic variables must be expressed as a function of the basic two, plus any chemical species. The properties of water are tabulated in steam tables, usually available as convenient computer programs or plug-ins.

A3.4.1. Single-Phase Liquid

Viscosity, density, enthalpy, and internal energy must be specified as functions of pressure and temperature. The enthalpy and internal energy are needed, primarily as functions of temperature, to determine the energy content of the reservoir and the fluid produced. The variation of density with pressure is important; although it changes little with pressure, this small compressibility is important in pressure transients. It is measured as the fractional rate of change of density ρ_w (or specific volume v_w) with pressure:

$$
c_w = \frac{1}{\rho_w}\left(\frac{d\rho_w}{dP}\right)_T = \frac{1}{v_w}\left(\frac{dv_w}{dP}\right)_T
\tag{A3.18}
$$

A3.4.2. Single-Phase Vapor

This is similar to liquid except that the density of steam varies more strongly with pressure. The PVT relation of steam is approximately described by the perfect gas law. It can be more accurately represented by the imperfect gas law:

$$PV = \rho ZR(T + 274)/M \qquad (A3.19)$$

where Z is the gas law deviation factor. Then the compressibility is given by:

$$c_s = \frac{1}{P} - \frac{1}{z}\left(\frac{dZ}{dP}\right)_T \qquad (A3.20)$$

The first term is usually the only significant one.

A3.4.3. Two-Phase Fluid

The existence of steam and water in contact means that the pressure and temperature are related by the saturation (Clausius-Clapeyron) curve:

$$P = P_{sat}(T) \qquad (A3.21)$$

or

$$T = T_{sat}(P) \qquad (A3.22)$$

Note that "saturation" is used here in a different sense than in Appendix 2, where it refers to the liquid fraction in the pore space. Both uses are standard terminology. The coupling of pressure and temperature through the saturation relation means that the mass and energy conservation equations are strongly coupled. The thermodynamic properties (density, viscosity, enthalpy, and internal energy) of each phase are now functions of a single variable: pressure or temperature.

A3.4.4. Noncondensable Gases

A vapor-soluble species such as a noncondensable gas can create greater complexity in mathematical analysis. The pressure of the vapor phase is now a sum of steam partial pressure and gas partial pressure:

$$P = P_s + P_g \qquad (A3.23)$$

For both single-phase vapor (dry steam) and two-phase reservoirs, cases occur where the gas partial pressure is significant compared to the steam partial pressure. Assuming that the gas and steam act as perfect gases, the molar ratio x_s of gas in vapor to H_2O in vapor is proportional to the ratio of the partial pressures (assuming no other gases present):

$$\frac{x_v}{P_g} = \frac{1 - x_v}{P_s} \qquad (A3.24)$$

The mass concentration of gas is derived from the molar concentration:

$$f = \frac{M_g x}{M_g x + 18(1 - x)} \quad (A3.25)$$

where M_g is the molecular weight of the gas ($M_s = 18$). The PVT relation of each gas phase is evaluated at the temperature of the vapor phase and partial pressure of that gas.

The molar ratio of CO_2 to H_2O in vapor phase is equal to the ratio in liquid phase, multiplied by the distribution or partition function B:

$$\frac{x_v}{1 - x_v} = B \times \frac{x_L}{1 - x_L} \quad (A3.26)$$

This partition relation is equivalent to Henry's Law. For small concentrations, Eq. (A3.26) gives the concentration of gas in liquid as:

$$x_L = x_v/B = P_g/(B \times P_s) \quad (A3.27)$$

Giggenbach (1980) gives fits of the form:

$$\log_{10}(B) = a + bT \quad (A3.28)$$

Note that the temperature in Eq. (A3.28) is degrees Centigrade, not Kelvin. For carbon dioxide:

$$\log_{10}(B) = a4.7593 - 0.0192 \times T \quad (A3.29)$$

As an example, consider water at 227°C containing 1.2% by weight of carbon dioxide. What are the saturation pressure of this liquid and the gas concentration of the vapor?

At 227°C, $P_s = 26.46$ bar. The molar concentration of gas in liquid phase is

$$x_L = \frac{0.012/44}{0.012/44 + 0.988/18} \quad 0.0049$$

From A3.28, $B = 191$. Then using Eq. (A3.26)

$$\frac{x_v}{1 - x_v} = B \times \frac{x_L}{1 - x_L} = 191 \times \frac{.0049}{.9951} = 0.948$$

The partial pressure of gas is, from Eq. (A3.24),

$$P_g = P_s \times \frac{x_v}{1 - x_v} = 24.46 \times 0.948 = 25.08 \text{ bar}$$

The mass fraction of gas in the vapor phase is

$$f_v = \frac{44 \times x_v}{44 \times x_v + 18 \times (1 - x_v)} = 0.698$$

The total saturation pressure of the liquid is $26.46 + 25.08 = 51.5$ bar. The vapor phase is 70% by weight of gas.

A3.5. BOILING-POINT FOR DEPTH MODEL

Following the assumptions of the upflow model in Chapter 2, the fluid distribution in an undisturbed geothermal reservoir is determined by the natural upflow of fluid. The pressure distribution is given by Darcy's Law. Assuming one-dimensional vertical flow and if u is the mass flux density of the natural flow:

$$u_w = u = -\frac{k_v}{v_w}\left(\frac{dP}{dz} - \rho w g\right)$$ (A3.30)

if the upflow is wholly liquid. The pressure increases slightly more than hydrostatically (for that temperature). The extra gradient is needed to drive the upflow through the reservoir.

In the boiling zone, the steam/water ratio in the upflow can be calculated from conservation of energy. If the effects of dilution and heat conduction are ignored, conservation of mass and energy gives:

$$u_w + u_s = u$$ (A3.31)

and

$$u_w H_w + u_s H_s = u H$$ (A3.32)

where u_w and u_s are the water and steam mass flux densities and H_w and H_s are the water and steam enthalpies at the temperature at that depth. H is the enthalpy of the deep upflow. The pressure and saturation distribution are then defined by applying Darcy's Law to each phase:

$$u_w = -k_v \cdot \frac{k_{rw}(S_w)}{v_w}\left(\frac{dP}{dz} - \rho w g\right)$$ (A3.33)

$$u_s = -k_v \cdot \frac{k_{rs}(S_w)}{v_s}\left(\frac{dP}{dz} - \rho_s g\right)$$ (A3.34)

And temperature and pressure are related by the saturation relation $T = T_{sat}(P)$.

For many purposes, a simplification of the fluid and pressure distribution is possible. It is frequently the case that the pressure distribution in an undisturbed liquid-dominated reservoir is near hydrostatic. Then the driving head for the steam phase, $\frac{dP}{dz} - \rho_s g$, is relatively large, so that the k_{rs} must be small. This means that the reservoir is close to being liquid-saturated. The pressure profile in the reservoir matrix is approximated by a hydrostatic profile:

$$\frac{dP}{dz} = \rho_w g$$ (A3.35)

and

$$T = T_{sat}(P) \qquad (A3.36)$$

This pressure profile is that of a column of liquid water, which is at boiling point for the pressure at all depths, for pure water. It is modified by the presence of dissolved solids and gases. It is often useful to plot this BPD profile on downhole pressure-temperature graphs as a reference, giving roughly the temperature that may be expected at a given depth. Measured pressures and temperatures may be more or less than BPD. For example, a well containing an upflow of boiling liquid will show a boiling-point profile relative to the actual pressure at which fluid enters the well.

Figure A3.1 shows a boiling-point profile in a well at Ohaaki, New Zealand. The well is on bleed. The isothermal profile below 700 m identifies water at 261°C entering the well. The temperature curve shows that the water starts to boil at 67 bar. This is 20 bar above the boiling point for pure water and is due to the high gas content of the upflowing fluid. Above the boiling point is the typical smoothly rounded profile characteristic of a BPD profile, here modified by the gas content.

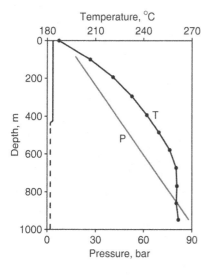

FIGURE A3.1 Boiling point profile in gassy well. *Source: Contact Energy, Personal Communication.*

Geothermal Fields

Ahuachapan, El Salvador
Alto Peak, Leyte, Philippines
Awibengkok (Salak), Java, Indonesia
Bacon-Manito (Bac-Man), Luzon, Philippines
Balcova-Narlidere, Izmir, Turkey
Basel, Switzerland
Beowawe, Nevada, United States
Borinquen, Costa Rica
Broadlands (Ohaaki), New Zealand
Bulalo (Mak-Ban), Luzon, Philippines
Cerro Prieto, Baja California, Mexico
Cooper Basin, Australia
Darajat, Java, Indonesia
Dixie Valley, Nevada, United States
East Mesa, Imperial Valley, California, United States
El Tatio, Chile
Hamar, Iceland
Hatchobaru, Kyushu, Japan
Hijiori, Honshu, Japan
Kamojang, Java, Indonesia
Karaha-Bodas, Java, Indonesia
Kawerau, New Zealand
Kizildere, Kirikkale, Turkey
Landau, Rhineland-Palatinate, Germany
Larderello, Tuscany, Italy
Lihir, New Ireland, Papua New Guinea
Los Azufres, Michoacan, Mexico
Mahanagdong, Leyte, Philippines
Mak-Ban (Makiling-Banahaw, Bulalo), Luzon, Philippines
Mammoth, California, United States
Matsukawa, Honshu, Japan
Miravalles, Costa Rica
Mokai, New Zealand

Mt. Amiata, Tuscany, Italy
Nesjavellir, Iceland
Ngatamariki, New Zealand
Ngawha, New Zealand
Oguni, Kyushu, Japan
Ohaaki (Broadlands), New Zealand
Olkaria, Rift Valley, Kenya
Palinpinon, Negros, Philippines
Patuha, Java, Indonesia
Puna, Hawaii, United States
Ribeira Grande, Azores, Portugal
Rotokawa, New Zealand
Salak (Awibengkok), Java, Indonesia
Salton Sea, California, United States
Soultz-sous-forêts, Alsace, France
Steamboat Springs, Nevada, United States
Sumikawa, Honshu, Japan
Svartsengi, Iceland
Tauhara, New Zealand
The Geysers, California, United States
Tiwi, Luzon, Philippines
Tongonan, Leyte, Philippines
Travale, Tuscany, Italy
Uenotai, Kyushu, Japan
Waiotapu, New Zealand
Wairakei, New Zealand
Wayang Windu, Java, Indonesia
Yangbajan, Tibet, China
Yellowstone, Wyoming, United States

List of Symbols

SYMBOLS

Note: Symbols that are used in only one or two sections are defined there. Symbols in bold are vectors—for example, u is vector volume flux density.

c	compressibility
C	specific heat, J/kg.K
d,D	diameter, m
E_1	exponential integral function (Equation 3.27)
f	gas content, mass/mass; frequency
f_M	friction factor
g	acceleration due to gravity, m^2/s
h	thickness, m
H	enthalpy, J/kg
II	injectivity index, kg/b.s
K	conductivity, W/mK
k	permeability, m^2; d $= 10^{-12}$m^2
L	length, m
m	slope of semilog plot, Pa/cycle
M	molecular weight
P	pressure, Pa; bar $= 10^5$ Pa
PI	productivity index, kg/b.s
q	volume flow, m^3/s
Q	heat flow, J/s $=$ W; amount of heat, J
r	radial distance, m
S	saturation
s	skin
T	temperature, K $=$ °C
t	time, s
u	volume flux density, m^3/m^2.s $=$ m/s
v	specific volume, m^3/kg; mass flux density, kg/m^2.s
W	mass flow, kg/s
x	gas content, mol/mol
X	flash fraction
Y	lip pressure variable
z	depth, m
α	recharge coefficient, kg/Pa.s
β	diameter ratio d/D
ϵ	expansivity
ε	roughness
κ	diffusivity, m^2/s

μ dynamic viscosity, Pa.s
ν kinematic viscosity, m^2/s
φ porosity
ρ density, kg/m^3

SUPERSCRIPTS

'   at atmospheric pressure

SUBSCRIPTS

b block
D dimensionless
f fissure or fracture; flowing
g gas
L liquid
lip at lip pressure
m matrix, mixture
o initial value
r reservoir, relative (in relative permeability)
s steam, at saturation
sep at separator or sample pressure
T tracer
t total flow
v vapor (steam + gas)
WH wellhead
w water, well
ws well shut, latent heat ($H_{ws} = H_s - H_w$)

References

Abrigo, M.F.V., Molling, P.A., Acuña, J.A., 2004. Determination of recharge and cooling rates using geochemical constraints at the Mak-Ban (Bulalo) geothermal reservoir, Philippines. Geothermics v33 (1/2), 11–36.

Abramowitz, M., Stegun, I.A., 1965. Handbook of mathematical functions. Dover, New York.

Acuña, J., Pasaribu, F., 2010. Improved method for decline analysis of dry steam wells. World Geothermal Congress paper 2275.

Acuña, J.A., 2003. Integrating wellbore modeling and production history to understand well behavior. Proceedings, 29th Workshop on Geothermal Reservoir Engineering, Stanford University, pp. 16–20.

Acuña, J.A., 2008. A new understanding of deliverability of dry steam wells. Transactions, Geothermal Resources Council v32, 431–434.

Acuña, J.A., Parini, M., Urmeneta, N., 2002. Using a large reservoir model in the probabilistic assessment of field management strategies. Proceedings, 27th Workshop on Geothermal Reservoir Engineering, Stanford University, pp. 8–13.

Acuña, J.A., Stimac, J., Sirad-Azwar, L., Pasikki, R.G., 2008. Reservoir management at Awibengkok geothermal field, West Java, Indonesia. Geothermics 37, 332–346.

Acuña, J.A., Acerdera, B.A., 2005. Two-phase flow behaviour and spinner analysis in geothermal wells. Proceedings, 30th Workshop on Geothermal Reservoir Engineering, Stanford University, pp. 245–252.

Adams, M.C., Moore, J.N., Bjornstad, S., Norman, D.I., 2000. Geologic history of the Coso geothermal system. World Geothermal Congress paper 0105.

AGEG, 2008. Geothermal lexicon for resources and reserves definition and reporting.

Akin, S., Parlaktuna, M., Sayik, T., Sexer, H., Karahan, C., Bakraç, S. Interpretation of the tracer test of Balçova geothermal field. World Geothermal Congress paper 2309.

Aksoy, N., Serpen, U., 2005. Reinjection management in Balcova geothermal field. Proceedings, World Geothermal Congress paper 1206.

Al-Hussainy, R., Ramey Jr., J.J., Crawford, P.B., 1966. The flow of real gases through porous media. Soc. Pet. Eng. J. v6, 624–636.

Allis, R.G., James, C.R., 1980. A natural convection promoter for geothermal wells. Transactions, Geothermal Resources Council v4, 409–412.

Allis, R., Moore, J.N., McCulloch, J., Petty, S., DeRocher, T., 2000. Karaha-Telaga-Bodas, Indonesia: a partially vapor-dominated geothermal system. Transactions, Geothermal Resources Council v24, 217–222.

Allis, R.G., James, R., 1979. A natural convection promoter for geothermal wells. Geo-Heat Center, Klamath Falls Oregon.

Allis, R.G., 2000. Insights on the formation of vapor-dominated geothermal systems. Proceedings, World Geothermal Congress, pp. 2489–2496 [0610].

Amistoso, A.E., Orizonte, R.G., 1997. Reservoir response to full load operation Palinpinon production field, Valencia. Negros Oriental, Philippines. Proc, 17th PNCO-EDC conference, pp. 1–14.

ANSI B40. 1M. Gauges-pressure, Indicating dial type-Elastic element (metric).

Aqui, A.R., Aragones, J.A., Amistoso, A.E., 2005. Optimisation of Palinpinon-1 production field based on exergy analysis — the Southern Negros geothermal field, Philippines. World Geothermal Congress paper 1312.

Aragón, A., Izquierdo, G., Arellano, V., 2009. Analysis of well productivity as related to changes in well damage. Transactions, Geothermal Resource Council v33, 789–793.

Arellano, V.M., Torres, M.A., Barragan, R.M., 2005. Response to exploitation of the Los Azufres (Mexico) geothermal reservoir. Proceedings, World Geothermal Congress paper 1115.

ASME PTC 19.11 Part II, Water and steam in the power cycle (purity and quality, leak detection, and measurement) Instruments and apparatus. Supplement to Performance and Test Codes. The American Society of Mechanical Engineers.

Asturias, F., 2005. Reservoir assessment of Zunil I and II geothermal fields, Guatemala. Proceedings, World Geothermal Congress paper 1112.

Atkinson, P.G., Perdersen, J.R., 1988. Using precision gravity data in geothermal reservoir engineering modelling studies. Proceedings, 13th Workshop on geothermal reservoir engineering, Stanford University, pp. 35–40.

Aunzo, Z.P., Björnsson, G., Bödvarsson, G.S., 1991. Wellbore models GWELL, GWNACL, and HOLA. Lawrence Berkeley Laboratory report LBL-31428.

Axelsson, G., Thórhallsson, S., 2009. Review of well stimulation operations in Iceland. Transactions, Geothermal Resources Council v33, 794–800.

Axelsson, G., 1989. Simulation of pressure response data from geothermal reservoirs by lumped parameter models. Proceedings, 14th Workshop on Geothermal Reservoir Engineering, Stanford University, pp. 257–263.

Axelsson, G., 1991. Reservoir engineering of small low-temperature systems in Iceland. Proceedings, 16th Workshop on Geothermal Reservoir Engineering, Stanford University, pp. 143–149.

Axelsson, G., Björnsson, G., Montalvo, F., 2005a. Quantitative interpretation of tracer test data. Proceedings, World Geothermal Congress paper 1211.

Axelsson, G., Björnsson, G., Quijano, J.E., 2005b. Reliability of lumped parameter modelling of pressure changes in geothermal reservoirs. Proceedings, World Geothermal Congress paper 1179.

Axelsson, G., Stefansson, V., Björnsson, G., Liu, J., 2005c. Sustainable management of geothermal resources and utilisation for 100–300 years. Proceedings, World Geothermal Congress paper 0507.

Bangma, P., 1961. The development and performance of a steam-water separator for use on geothermal bores. Proc UN Conference on New Sources of Energy. v3.

Banwell, C.J., 1957. Borehole measurements. Bull NZ Dept. Sci. Ind. Res. 123, 39–72.

Barelli, A., Palama, A., 1981. A new method for evaluating formation equilibrium temperature in holes during drilling. Geothermics v10 (2), 95–102.

Barelli, A., Ceccarelli, A., Dini, I., Fiordelisi, A., Giorgi, N., Lovari, F., et al., 2010a. A review of the Mt. Amiata geothermal system (Italy). World Geothermal Congress paper 0613.

Barelli, A., Cei, M., Lovari, F., Romangoli, P., 2010b. Numerical modelling for the Larderello-Travale geothermal system (Italy). World Geothermal Congress paper 2225.

Barelli, A., Celati, R., Manetti, G., 1977. Gas-water interface rise during early exploitation test in Alfina geothermal field (Northern Latium, Italy). Geothermics v6, 199–208.

Barenblatt, G.E., Zheltov, I.P., Kochina, I.N., 1960. Basic concepts in the theory of seepage of homogeneous liquids in fissured rocks. J. Appl. Math. Mech. v24, 1286–1303.

Barker, B.J., Gulati, M.S., Bryan, M.A., Riedel, K.L., 1991. Geysers reservoir performance. Transactions, Geothermal Resources Council Special Report no 17, 167—177.

Barrios, L.A., Quijano, J., Guerra, E., Mayorga, H., Rodriguez, A., Romero, R., 2007. Injection improvements in low permeability and negative skin wells using mechanical cleanout and chemical stimulation, Berlin geothermal field, El Salvador. Transactions, Geothermal Resources Council v31, 141—146.

Beall, J.J., Adams, M.C., Smith, J.L.B., 2001. Geysers reservoir dry out and partial resaturation evidencedby twenty-five years of tracer tests. Transactions, Geothermal Resources Council v25, 725—729.

Bear, J., 1972. The dynamics of fluids in porous media. Am. Elsevier, New York.

Behrens, H., Ghergut, I., Sauter, M., 2010. Tracer properties, and tracer test results — Part 3: Modification to Shook's flow-storage method. Proceedings, 35th Workshop on Geothermal Reservoir Engineering, Stanford University.

Belen, R. Jr., Aunzo, Z., Strobel, C., Mogen, P., 1999. Reservoir pressure estimation using PTS data. Proceedings, Workshop on Geothermal Reservoir Engineering, 24th University 24.

Benavidez, P.J., Mosby, M.D., Leong, J.K., Navarro, V.C., 1988. Development and performance of the Bulalo geothermal field. Proceedings, 10th New Zealand geothermal workshop, Auckland University, pp. 55—60.

Benoit, D., 1978. The use of shallow and deep temperature gradients in geothermal exploration in northwest Nevada using the Desert Peak thermal anomaly as a model. Transactions, Geothermal Resources Council v2, 45—46.

Benoit, D., 1992. A case history of injection through 1991 at Dixie Valley, Nevada. Transactions, Geothermal Resources Council v16, 611—620.

Benoit, D., Johnson, S., Kumataka, M., 2000. Development of an injection augmentation program at the Dixie Valley, Nevada geothermal field. Proceedings, World Geothermal Congress, 819—824.

Benson, S.M., Dagget, J.S., Iglesias, E., Arellano, V., Ortiz-Ramirez, J., 1987. Analyses of thermally induced permeability enhancement in geothermal wells. Proceedings, 12th Workshop on Geothermal Reservoir Engineering, Stanford University, pp. 57—65.

Bertani, R., Bertini, G., Cappetti, G., Fiordelisi, A., Marocco, B.M., 2005. An update of the Larderello/Radicondoli deep geothermal system. Proceedings, World Geothermal Congress paper 0936.

Bignall, G., Milicich, S., Ramirez, E., Rosenberg, M., Kilgour, G., Rae, A., 2010. Geology of the Wairakei-Tauhara geothermal system, New Zealand. Proceedings, World Geothermal Congress.

Bixley P.F., Grant M.A., 1981. Evaluation of pressure-temperature profiles in wells with multiple feed points. New Zealand Geothermal Workshop 3, 1981.

Bixley, P.F., Clotworthy, A.W., Mannington, W.O., 2009. Evolution of the Wairakei geothermal reservoir during 50 years of production. Geothermics v38, 145—154.

Bixley, P.F., Dench, N., Wilson, D., 1998. Development of well testing methods at Wairakei 1950—1980. Porceedings, 20th New Zealand Geothermal Workshop, pp. 7—12.

Bixley, P.F., Glover, R.B., McCabe, W.J., Barry, B.J., Jordan, J.T., 1995. Tracer calibration tests at Wairakei geothermal field. World Geothermal Congress, 1887—1891.

Björnsson, A., 2005. Development of thought on the nature of geothermal fields in Iceland from medieval times to the present. Proceedings, World Geothermal Congress paper 0005.

Björnsson, G., Steingrímsson, B., 1992. Fifteen years of temperature and pressure monitoring in the Svartsengi high-temperature geothermal field in SW-Iceland. Trans Transactions, Geothermal Resources Council v16, 627—632.

Björnsson, G., 1999. Predicting the future performance of a shallow steam zone in the Svartsengi geothermal field, Iceland. Proceedings, 24th Workshop on Geothermal Reservoir Engineering, Stanford University.

Björnsson, G., 2008. Review of generating capacity estimates for the Momotombo geothermal reservoir in Nicaragua. Transactions, Geothermal Resources Council v32, 341–345.

Blackwell, D.D., Golan, B., Benoit, D., 2000. Temperatures in the Dixie Valley, Nevada geothermal system. Transactions, Geothermal Resources Council v24, 223–228.

Blackwell, D.D., Leidig, M., Smith, R.P., Johnson, S.D., Wisian, K.W., 2002. Exploration and development techniques for Basin and Range geothermal systems: examples from Dixie Valley, Nevada. Transactions, Geothermal Resources Council v26, 513–518.

Blair, C.K., Harrison, R.F., 1980. Development of an instrument to measure the non-condensable gases in geothermal discharges. Lawrence Berkeley Laboratory, LBL-11499 (GREMP-13).

Bödvarsson, G., 1951. Report on the Hengill thermal area. J. Eng. Ass. Iceland 36, 1.

Bödvarsson, G., 1964. Physical characteristics of natural heat sources in Iceland. Proc. UN Conf. New Sources of Energy, Sol. Energy, Wind Power Geotherm. Energy 1961, v2, Pap G/6, 82–90.

Bödvarsson, G., 1970. Confined fluids as strain meters. J. Geoph. Res. v75 (14), 2711–2718.

Bödvarsson, G., 1972. Thermal problems in siting of reinjection wells. Geothermics 2.

Bodvarsson, G.S., 1988. Model predictions of the Svartsengi reservoir, Iceland. Water Resources Research v24, 1740–1746.

Bodvarsson, G.S., Stefansson, V., 1988. Reinjection into geothermal reservoirs. in Okandan 1988, pp. 103–120.

Bodvarsson, G.S., Bjornsson, S., Gunnarsson, A., Gunnlaugsson, E., Sigurdsson, O., Stefansson, V. et al., 1988. A summary of modeling studies of the Nesjavellir. Geothermal Field, Iceland, 13th Workshop on Geothermal Reservoir Engineering, Stanford University, pp. 83–91.

Bodvarsson, G.S., Bjornsson, S., Gunnarsson, A., Gunnlaugsson, E., Sigurdsson, O., Stefansson, V., et al., 1990a. The Nesjavellir geothermal field, Iceland, Part 1. Field characteristics and development of a three-dimensional numerical model. Geothermal Science and Technology 2 (3), 189–228.

Bodvarsson, G.S., Bjornsson, S., Gunnarsson, A., Gunnlaugsson, E., Sigurdsson, O., Stefansson, V., et al., 1991. The Nesjavellir geothermal field, Iceland, 2. Evaluation of the generating capacity of the system. Geothermal Science and Technology 2 (4), 229–261.

Bodvarsson, G.S., Gislason, G., Gunnlaugsson, E., Sigurdsson, O., Stefansson, V., Steingrimsson, B., 1993. Accuracy of reservoir predictions for the Nesjavellir geothermal field, Iceland, 18th Workshop on Geothermal Reservoir Engineering. Stanford University, pp. 273–278.

Bodvarsson, G.S., Pruess, K., Haukwa, C., Ojiambo, S.B., 1990b. Evaluation of reservoir model predictions for Olkaria East geothermal gield, Kenya. Geothermics 19 (5), 399–414.

Bodvarsson, G.S., Pruess, K., Stefansson, V., Bjornsson, S., Ojiambo, S.B., 1987a. East Olkaria geothermal field, Kenya, 1 History match with production and pressure decline data. Journal of Geophysical Research 92 (B1), 521–539.

Bodvarsson, G.S., Pruess, K., Stefansson, V., Bjornsson, S., Ojiambo, S.B., 1987b. East Olkaria geothermal field, Kenya, 2. Predictions of well performance and reservoir depletion. Journal of Geophysical Research 92 (B1), 541–554.

Bogie, I., Kusumah, Y.I., Wisnandary, M.C., 2008. Overview of the Wayang Windu geothermal field West Java. Geothermics v37 (3), 347–365.

Bolaños, G.T., Parrilla, E.V., 2000. Response of Bao-Banati thermal area to development of the Tongonan geothermal field, Philippines. Geothermics v29 (4–5), 499–508.

Bolton, R.S., 1970. The behaviour of the Wairakei geothermal field during exploitation. UN symposium on the development and use of geothermal resources, Special Issue. Geothermics v2, 1426−1439.

Boseley, C., Cumming, W., Urzúa-Monsalve, L., Powell, T., Grant, M., 2010. A resource conceptual model for the Ngatamariki geothermal field based on recent exploration well drilling and 3D MT resistivity imaging. World Geothermal Congress.

Boyd, T., 2009. Design of a convection cell for a downhole heat exchanger system in Klamath Falls, Oregon. Proceedings, 34th Workshop on geothermal reservoir engineering, Stanford University.

Brigham, W.E., Morrow, C.B., 1977. P/Z behavior for geothermal steam reservoirs. Soc. Pet. Eng. J. v17 (no5), 407−412.

Bromley, C., Currie, S., Ramsay, G., Rosenberg, M., Pender, M., O'Sullivan, M., et al., 2010. Tauhara stage II geothermal project: Subsidence Report. GNS Science Consultancy Report 2010/151. p. 154.

Bromley, C.J., 2009. Groundwater changes in the Wairakei-Tauhara geothermal system. Geothermics v38 (1), 134−144.

Brown, D.W., 2009. Hot dry rock geothermal energy: important lessons from Fenton Hill. Proceedings, 34th Workshop on Geothermal Reservoir Engineering, Stanford University.

Browne, P.R.L., 1979. Minimum age of the Kawerau geothermal field, North Island, New Zealand. J. Volcanol. Geotherm. Res. 6, 213−215.

BS 1042: Section 1.1: 1981., Methods of measurement of fluid flow in closed conduits.

BS1041: Section 2.1: 1985., Temperature Measurement.

Butler, S.J., Sanyal, S.K., Klein, C.W., Iwata, S., Itoh, M., 2005. Numerical simulation and performance evaluation of the Uenotai geothermal field, Akita prefecture. Proceedings, World Geothermal Congress paper 0939.

Butler, S.J., Sanyal, S.K., Robertson-Tait, A., Lovekin, J.W., Benoit, D., 2001. A case history of numerical modelling of a fault-controlled geothermal system at Beowawe, Nevada. Proceedings, 26th Workshop on Geothermal Reservoir Engineering, Stanford University.

Capetti, G., Parisi, L., Ridolfi, A., Stefani, G., 1995. Fifteen years of reinjection in the Larderello-Valle Secolo area: analysis of the production data. Proceedings, World Geothermal Congress, pp. 1997−2000.

Capuno, V.T., Sta. Maria, R., Stark, M.A., Minguez, E.B., 2010. Mak-Ban geothermal field, Philippines: 30 years of commercial operation. World Geothermal Congress paper 0649.

Carey, B., 2000. Wairakei 40 plus years of generation. Proceedings, World Geothermal Congress, pp. 3145−3149.

Cataldi et al., 1999: Cataldi, R., Hodgson, S.F., Lund, J.W. (1999). Stories from a Heated Earth. Geothermal Resources Council and International Geothermal Association, pp. 569.

Cathles, L.M., 1977. An analysis of the cooling of intrusive by groundwater convection which includes boiling. Econ. Geol. v72, 804−825.

Celati, R., Squarci, P., Stefani, G.C., Taffi, L., 1977. Analysis of water levels in Larderello region geothermal wells for reconstruction of reservoir pressure trend. Geothermics v6, 183−198.

Chen, D., Wyborn, D., 2009. Habanero field tests in the Cooper Basin, Australia: a proof-of-concept for EGS. Transactions, Geothermal Resources Council v33, 157−164.

Cinco-Ley, H., Samaniego, V.F., 1981. Transient pressure analysis for fractured wells. J. Pet. Tch. v33 (9), 1749−1766.

Clemente, W.C., Villadolid-Abrigo, F.E.L., 1993. The Bulalo geothermal field, Philippines: reservoir characteristics and response to production. Geothermics v22 (5/6), 381−394.

Clotworthy, A.W., 2000. Reinjection into low temperature loss zones. Proceedings, World Geothermal Congress paper 0090.

Clotworthy, A.W., Carey, B.S., Bacon, L.G., 1989. Initial response to production and reinjection at the Ohaaki geothermal field, New Zealand. Transactions, Geothermal Resources Council v13, 505−509.

Clotworthy, A.W., Lawless, J., Ussher, G., 2010. What is the end point for geothermal developments: modelling depletion of geothermal fields. World Geothermal Congress paper 2239.

Clotworthy, A.W., Lovelock, B., Carey, B., 1995. Operational history of the Ohaaki geothermal field, New Zealand. Proceedings, World Geothermal Congress, 1797−1802.

Colebrook, C.F. (February 1939.). Turbulent flow in pipes, with particular reference to the transition region between smooth and rough pipe laws. Journal of the Institution of Civil Engineers (London).

Combs, J., Garg, S.K., 2000. Discharge capability and geothermal reservoir assessment using data from slim holes. Proceedings, World Geothermal Congress, 1065−1070.

Cooper, G.T., Beardsmmore, G.R., 2010. Engineered geothermal systems in the Australian context − resource definition in conductive systems. World Geothermal Congress paper 3114.

Corey, A.T., 1972. Mechanics of heterogeneous fluids in porous media. Wat. Resour. Pub, Ft Collins, Colorado.

Craft, B.C., Hawkins, M.F., 1959. Applied reservoir engineering. Prentice-Hall Inc, Englewood Cliffs NJ. 1959.

Cumming, W., 2009. Geothermal resource conceptual models using surface exploration data. Proceedings, 34th Workshop on Geothermal Reservoir Engineering, Stanford University.

D'Amore, F., Truesdell, A.H., 1979. Models for steam chemistry at Larderello and The Gesyers. Proceedings, 4th Workshop on Geothermal Reservoir Engineering, Stanford University, pp. 283−297.

de Anda, L.F., Septien, J.I., Elizondo, J.R., 1964. Geothermal energy in Mexico. Proc. UN Conf. New Sources of Energy, Sol. Energy, Wind Power Geotherm. Energy 1961 v3, Pap G/77, 149−164.

Dench, N.D., 1980 "Interpretation of fluid pressure measurements in geothermal wells" Proc., 2nd NZ Geothermal Workshop, Auckland University.

Dezhi, B., Jiurong, L., Keyan, Z., 2005. The new national standard for geological exploration of geothermal resources in China. Proceedings, World Geothermal Congress paper 2605.

DiPippo, R., 1997. High-efficiency geothermal plant designs. Transactions, Geothermal Resources Council v21, 393−398.

Donaldson, I.G., 1962. Temperature gradients in the upper layer of the Earth's crust due to convective water flows. J. Geoph. Res. 67, 3449−3459.

Donaldson, I.G., Grant M.A., Bixley P.F., 1981. Non-static reservoirs, the natural state of the geothermal reservoir. SPE10314, 1981.

Dowdle, W.L., Cobb, W.M., 1975. Static formation temperature from logs—an empirical method. J. Pet. Tech. 1975, 1326−1330.

Dwikorianto, T., Abidin, Z., Kamah, Y., Sunaryo, D., Hasibuan, A., Prayoto, 2005. Tracer injection evaluation in Kamojang geothermal field, West Java, Indonesia. Proceedings, World Geothermal Congress paper 1927.

Dykstra, H., Adams, R.H., 1978. Momotombo geothermal reservoir. Proceedings, 3rd Workshop on geothermal reservoir engineering, Stanford University, pp. 96−106.

Earlougher, R.C., 1977. Advances in well test analysis. Soc. Pet. Eng. Dallas.

Economides, M.J., Fehlberg, E.L., 1979. Two short-time buildup test analyses for Shell's Geysers well D-6, a year apart. Proceedings 4th Workshop on Geothermal Reservoir Engineering, Stanford University, pp. 91−97.

Einarsson, T., 1942. Ueber das Wesen der heissen Quellen Islands. Soc. Sci. Isl., Reykjavik.

Elders, W.A., Fridleifsson, G.O., 2005. The Iceland deep drilling project—scientific opportunities. Proceedings, World Geothermal Congress paper 0626.

Ellis, A.J., Mahon, W.A.J., 1977. Chemistry and Geothermal Systems. Academic Press.

Elmi, D., Axelsson, G., 2009. Application of a transient wellbore simulator to wells HE-06 and HE-20 in the Hellisheide geothermal system, SW-Iceland. Proceedings, 34th Workshop on geothermal reservoir engineering, Stanford University.

Enedy, K., 1991. The role of decline curve analysis at The Geysers. Transactions, Geothermal Resources Council Special Report 17, 197—203.

Enedy, S., 1987. Applying flowrate type curves to Geysers steam wells. Proceedings, 12th Workshop on Geothermal Reservoir Engineering, Stanford University, pp. 29—36.

Epperson, I.J., 1983. Beowawe acid stimulation. Transactions, Geothermal Resources Council v7, 409—411.

Fajardo, V.R., Malate, R.C.M., 2005. Estimating the improvement of Tanawon Production wells for acid treatment, Tanawon sector, BacMan geothermal production field, Philippines. Proceedings, World Geothermal Congress paper 1144.

Flores-Armenta, M., Tovar-Aguado, R., 2008. Thermal fracturing of well H-40, Los Humeros geothermal field. Transactions, Geothermal Resources Council v32, 445—448.

Flores-Armenta, M., 2010. Evaluation of acid treatments in Mexican geothermal fields. World Geothermal Congress paper 2420.

Flores-Armenta, M., Barajas, E.N.M., Torres-Rodriguez, M.A., 2006. Productivity analysis and acid treatment of well AZ-9AD at the Los Azufres geothermal field, Mexico. Transactions, Geothermal Resources Council v30, 791—795.

Flores-Armenta, M., Davies, D., Couples, G., Palsson, B., 2005. Stimulation of geothermal wells, can we afford it? Proceedings, World Geothermal Congress paper 1028.

Fukuda, D., Asanuma, M., Hishi, Y., Kotanaka, K., 2005. The first two-phase tracer tests at the Matsukawa vapor-dominated geothermal field, Northeast Japan. Proceedings, World Geothermal Congress paper 1204.

Ganefianto, N., Stimac, J., Azwar, L.S., Pasikki, R., Parini, M., Shidartha, E., et al., 2010. Optimizing production at Salak geothermal field, Indonesia, through injection management. World Geothermal Congress paper 2415.

Garg, S.K., Combs, J., 2010. Appropriate use of USGS volumetric "heat in place" method and Monte Carlo calculations Proc., 34th Workshop on Geothermal Reservoir Engineering, Stanford University.

Garg, S.K., Nakanishi, S., 2000a. Pressure interference tests at the Oguni geothermal field, Northern Kyushu, Japan. Proceedings, World Geothermal Congress paper 0315.

Garg, S.K., Nakanishi, S., 2000b. Pressure interference tests in a fractured geothermal reservoir. In Dynamics of fluids in fractured rock. Geophysical Monograph 122, American Geophysical Union.

Garg, S.K., Pritchett, J.W., 1981. Buildup analysis for two-phase geothermal reservoirs. Transactions, Geothermal Resources Council v5, 287—290.

Garg, S.K., Combs, J., Kodama, M., Gokou, K., 1998. Analysis of production injection data from slim holes and large-diameter wells at the Kirishima geothermal field, Japan. Proceedings, 23rd Workshop on geothermal reservoir Engineering, Stanford University, pp. 64—76.

Garg, S.K., Pritchett, J.W., Combs, J., 2010. Exploring for hidden geothermal systems. World Geothermal Congress paper 1132.

Garg, S.K., Pritchett, J.W., Wannamaker, P.E., Combs, J., 2007. Use of electrical surveys for geothermal reservoir characterisation: Beowawe geothermal field. Transactions, Geothermal Resources Council v31, 341—352.

Genter, A., Evans, K., Cuenot, N., Baticci, F., Dorbath, L., Graff, J., et al., 2009. The EGS Soultz project (France) From reservoir development to electricity production. Transactions, Geothermal Resources Council v33, 389−394.

Geophysical Research Corporation, 6540 East Apache, PO Box 15968, Tulsa, Oklahoma 74112, USA. Amerada RPG-3 Gauge Operator's Manual.

Georgsson, L.S., 1981. A resistivity survey on the plate boundaries in western Reykjanes peninsula, Iceland. Transactions, Geothermal Resources Council v5, 75−78.

Giggenbach, W., 1980. Geothermal gas equilibria. Geochim Cosmochim Acta v44, 2021−2032.

Glanz, J., 2010. Basel project ended. GRC Bulletin v39 (1), 20.

Glover, R.B., Hunt, T., Severne, C.M., 2000. Impacts of development on a natural thermal feature and their mitigation − Ohaaki Pool, New Zealand. Geothermics v29 (4−5), 509−523.

Glover, R.B., Hunt, T.M. Severne, C.M., 1996. Ohaaki Ngawha, Ohaaki Pool. Proc., 18th New Zealand Geothermal Workshop, pp. 77−84.

Gok, I.M., Sarak, H., Onur, M., Serpen, U., Satman, A., 2005. Numerical modelling of the Balcova-Narlidere geothermal field. Proceedings, World Geothermal Congress paper 1132.

Gonzalez, R.C., Alcober, E.H., Siega, F.L., Saw, V.S., Maxino, D.A., Ogena, M.S., et al., 2005. Field management strategies for the 700MW greater Tongonan geothermal field, Leyte, Philippines. Proceedings, World Geothermal Congress paper 241.

Goyal, K.P., Pingol, A.S., 2007. Geysers performance update through 2006. Transactions, Geothermal Resources Council v31, 435−439.

Granados, E., Henneberger, R., Klein, C., Sanyal, S., de Pone, C.B., Forjaz, V., 2000. Development of injection capacity for the expansion of the Ribeira Grande geothermal project, São Miguel, Açores, Portugal. Proceedings, World Geothermal Congress, 3065−3070 [0798].

Grant, M.A., 1979. Fluid state at Wairakei. Proc. 1st NZ geothermal workshop, Auckland University, pp. 79−84.

Grant, M.A., 1981. Ngawha geothermal hydrology. in The Ngawha geothermal area. DSIR geothermal report no.7.

Grant, M.A., 1982. The measurement of permeability by injection tests. Proceedings, 8th workshop on geothermal reservoir engineering, Stanford University, pp. 111−114.

Grant, M.A., 2000. Geothermal resource proving criteria. Proceedings, World Geothermal Congress, 2581−2584.

Grant, M.A., 2009a. Mathematical modelling of Wairakei geothermal field. ANZIAM J. 50(2009), pp. 426−434.

Grant, M.A., 2009b. Optimisation of drilling acceptance criteria. Geothermics 39, 247−253.

Grant, M.A., Bixley, P.F., 1995. An improved algorithm for spinner profile analysis. Proc., NZ Geothermal Workshop, Auckland University.

Grant, M.A., Sorey, M.L., 1979. The compressibility and hydraulic diffusivity of a water-steam flow. Water Resour. Res. v15 (3), 684−686.

Grant, M.A., Wilson, D., 2007. Interference testing at Kawerau 2006−2007. Proc. 29th NZ Geothermal Workshop.

Grant, M.A., Bixley, P.F., Donaldson, I.G., 1983. Internal flows in geothermal wells: their identification and effect on the wellbore temperature and pressure profiles. Soc. Pet. Eng. J. v23 (1), 168−176.

Grant, M.A., Bixley, P.F., Wilson, D.W., 2006. Spinner data analysis to estimate wellbore size and fluid velocity. Proc 28th NZ Geothermal Workshop.

Grant, M.A., Donaldson, I.G., Bixley, P.F., 1982a. Geothermal reservoir engineering. Academic Press, New York.

Grant, M.A., James, R., Bixley, P.F., 1982b. A modified gas correction for the lip-pressure method, Proc 8th Workshop Geothermal Reservoir Engineering, Stanford University, pp. 133–136.

Griggs, J., 2005. A reevaluation of geopressured-geothermal aquifers as an energy source. Proceedings, 30th Workshop on Geothermal Reservoir Engineering, Stanford University, pp. 501–509.

Grindley, G.W., 1966. Geological structure of hydrothermal fields in the Taupo Volcanic Zone, New Zealand. Bull. Volc. v29 (1).

Gringarten, A.C., Ramey, H.J. Jr., 1976. Effect of high-volume vertical fractures on geothermal steam well behavior. 2nd UN Symposium on the development and use of geothermal resources, US Government Printing Office, v2, 1759–1771.

Gudmundsson, J.S., Hauksson, T., 1985. Tracer survey in Svartsengi field 1984. Transactions, Geothermal Resources Council v9, 307–315.

Gudmundsson, J.S., Thorhallson, S., 1986. The Svartsengi reservoir in Iceland. Geothermics v15, 3–15.

Gudmundsson, J.S., 1983. Injection testing in 1982 at the Svartsengi high-temperature field in Iceland. Transactions, Geothermal Resources Council v6, 423–428.

Gudmundsson, J.S., Olsen, G., Thorhallson, S., 1985. Svartsengi field production data and depletion analysis. Proceedings, 10th Workshop on Geothermal Reservoir Engineering, Stanford University, pp. 45–51.

Gunn, C., Freeston, D., 1991. An integrated steady-state wellbore simulation and analysis package. Proc. 13th New Zealand Geothermal Workshop, 161–166.

Hanano, M., Matsuo, G., 1990a. A summary of recent study on the initial state of the Matsukawa geothermal reservoir. Proceedings, 15th Workshop on Geothermal Reservoir Engineering, Stanford University, pp. 39–46.

Hanano, M., Matsuo, G., 1990b. Initial state of the Matsukawa geothermal reservoir: reconstruction of a reservoir pressure profile and its implications. Geothermics v19 (6), 541–560.

Hanano, M., 2004. Contribution of fractures to formation and production of geothermal resources. Renewable & Sustainable Energy Reviews v8, 223–236.

Hanson, J.M., 1979. Tidal pressure response well testing at the Salton Sea geothermal field, California, and Raft River, Idaho. Proceedings, 5th Workshop on geothermal reservoir engineering, Stanford University, pp. 131–137.

Häring, M.O., Schanz, U., Ladner, F., Dyer, B.C., 2008. Characterisation of the Basel 1 enhanced geothermal system. Geothermics 37, 469–495.

Hasan, A.R., Kabir, C.S., 2002. Fluid flow and heat transfer in wellbores. SPE, Richardson, Texas.

Heizler, M.T., Harrison, T.M., 1991. The heating duration and provenance age of rocks in the Salton Sea geothermal field, southern California. J. Volcan. Geoth. Res. v46 (1–2), 73–97.

Hendrickson, R.R., 1975. Preliminary report. Tests on cores from the Wairakei geothermal project, Wairakei, New Zealand. Terra Tek.

Hirtz, P., Lovekin, J., 1995. Tracer Dilution Measurements for Two-Phase Production: Comparative Testing and Operating Experience. World Geothermal Conference.

Hirtz, P., Lovekin, J., Copp, J., Buck, C., Adams, M., 1993. Enthalpy and Mass Flowrate Measurements for Two-phase Geothermal Production by Tracer Dilution Techniques. Proceedings 18th Workshop on Geothermal Reservoir Engineering, Proceedings, Workshop on Geothermal Reservoir Engineering, Stanford University, 1993 SGP-TR-145.

Hitchcock, G.W., Bixley, P.F., 1975. Observations of the Effect of a three-year shutdown at Broadlands geothermal field, New Zealand. Second United Nations conference on the development and use of geothermal resources. US Government Printing Office, 1657–1661.

Hjartarson, A., Axelsson, G., Xu, Y., Production potential assessment of the low-temperature sedimentary geothermal reservoir in Lishuiqiao, Beijing, P.R., of China, based on a 3D numerical simulation study. Proceedings, World Geothermal Congress paper 1142.

Hoang, V., Alamsyah, O., Roberts, J., 2005. Darajat geothermal field expansion performance— a probabilistic forecast. Proceedings, World Geothermal Congress 1153.

Horne, R.N., Szucs, P., 2007. Inferring well-to-well connectivity using nonparametric regression on well histories. Proceedings, 32nd Workshop on Geothermal Reservoir Engineering, Stanford University, pp. 36—43.

Horne, R.N., Satik, C., Mahiya, G., Li, K., Ambusso, W., Tovar, R., et al., 2000. Steam-water relative permeability. Proceedings, World Geothermal Congress, pp. 2609—2615.

Hunt, T.M., Bromley, C.J., 2000. Some environmental changes resulting from development of Ohaaki geothermal field, New Zealand. Proceedings, World Geothermal Congress paper 0045.

Hunt, T.M., 1995. Microgravity measurements at Wairakei geothermal field, New Zealand; a review of 30 years data (1961—1991). Proceedings, World Geothermal Congress, 863—868.

Hutchings, P.G., Wyborn, D., 2006. Hot fractured rock (HFR) geothermal development, Cooper Basin, Australia. 28th NZ Geothermal Workshop, Auckland University.

Ingersoll, L.R., Zobel, O.J., 1913. Mathematical theory of heat conduction. Ginn, Waltham, Massachusetts.

Ishido, T., Pritchett, J.W., 2001. Prediction of magnetic field changes induced by geothermal fluid production and reinjection. Transactions, Geothermal Resources Council v25, 645—649.

ISO 1438/1 (E), Water flow measurement in open channels using weirs and venturi flumes—Part 1: Thin-plate weirs.

ISO 5167-2, Measurement of fluid flow by means of pressure differential devices inserted in circular cross-section conduits running full—Part 2: Orifice plates.

Itoi, R., Matsuzaki, R., Tanaka, T., Kamei, J., 2003. Lumped parameter model analysis for estimating fractions of reinjected water return to produced fluid at Sumikawa, Japan. Transactions, Geothermal Resources Council v27, 387—391.

Iwata, S., Nakano, Y., Granados, E., Butler, S., Robertson-Tait, A., 2002. Mitigation of cyclic production behavior in a geothermal well at the Uenotai geothermal field, Japan. Transactions, Geothermal Resources Council v26, 193—196.

James, C.R., 1966. Measurement of steam-water mixtures discharging at the speed of sound to the atmosphere. New Zealand Engineering 21 (10).

James, C.R., 1965. Powr life of a hydrothermal system. Proc. 2nd Australas. Conf. Hydraul. Fl. Mech. Eng. pp. B211—B234.

Kaplan, U., Nathan, A., da Ponte, C.A.B., 2007. Pico Vermelho geothermal project, Azores, Portugal. Transactions, Geothermal Resources Council v31, 521—524.

Karamarakar, M. and Cheng, P., 1980. A theoretical assessment of James' method for the determination of geothermal bore characteristics. Rep LBL-11498 (GREMP-12), Lawrence Berkeley Laboratory, California.

Kasameyer, P.W., Schroeder, R.C., 1975. Thermal depletion of liquid-dominated geothermal reservoirs with fracture and pore permeability. Proceedings, 1st Workshop on Geothermal Reservoir Engineering, Stanford University, pp. 249—257.

Kasameyer, P.W., Younker, L.W., Hanson, J.M., 1984. Development and application of a hydrothermal model for the Salton Sea geothermal field, California. Geol. Soc. Am. Bull. v95 (10), 1242—1252.

Ketilsson, J., Axelsson, G., Passon, H., Jonsson, M.T., 2008. Production capacity assessment: numerical modelling of geothermal resources, Proceedings, 33rd Workshop on Geothermal Reservoir Engineering, Stanford University, pp. 47—55.

Khan, M.A., Estabrook, R., 2005. New data reduction tools and their application to The Geysers geothermal field. Transactions, Geothermal Resources Council v29, 637—642.

Khan, M.A., Estabrook, R., 2006. New data reduction tools and their application to The Geysers geothermal field. Proceedings, 31st Workshop on Geothermal Reservoir Engineering, Stanford University.

Kiryukhin, A.V., Kopylova, G.N., 2009. iTOUGH2 analysis of the ground water level response to the barometric pressure change (well YZ-5, Kamchatka). Proceedings, 34th workshop on geothermal reservoir engineering, Stanford Univerisity.

Kiryukhin, A.V., 2005. Modeling of the Dachny site Mutnovsky geothermal field (Kamchatka, Russia) in connection with the problem of steam supply for 50MWe power plant. Proceedings, World Geothermal Congress 2005 paper 1126.

Kissling, W.M., Weir, G.J., 2005. The distribution of the geothermal fields in the Taupo Volcanic Zone, New Zealand. Proceedings, World Geothermal Congress 2005 paper 1919.

Kitao, K., Ariki, K., Hatakeyama, K., Wakita, K., 1990. Well stimulation using cold-water injection experiments in the Sumikawa geothermal field, Akita prefecture, Japan. Transactions, Geothermal Resources Council Trans 14, 1219—1224.

Kjaran, S.P., Halldorsson, G.K., Torhallson, S., Eliasson, J., 1979. Reservoir engineering aspects of Svartsengi geothermal area. Transactions, Geothermal Resources Council v3, 337—339.

Kneafsey, T.J., Pruess, K., O'Sullivan, M.J., Bodvarsson, G.S., 2002. Geothermal reservoir simulation to enhance confidence in predictions for nuclear waste disposal. Report LBNL-48124, Lawrence Berkeley National Laboratory, http://www.escholarship.org/uc/item/8hf6q620.

Koenig, J.B., 1991. History of development at The Geysers geothermal field, California. Transactions, Geothermal Resources Council Special Report no 17, 7—18.

Kohl, T., Mégel, T., 2005. Coupled hydro-mechanical modelling of the GPK3 reservoir stimulation at the European EGS site Soultz-sous-Forêts. Proceedings, 30th Workshop on Geothermal Reservoir Engineering, Stanford University.

Kohl, T., Mégel, T., 2007. Predictive modeling of reservoir response to hydraulic stimulations at the European EGS site Soultz-Sous-Forêt. International journal of rock mechanics and mining sciences 2007, vol. 44, no 8, 1118—1131.

Kumamoto, Y., Itoi, R., Tanaka, T., Hazama, Y., 2009. Modeling and numerical analysis of the two-phase geothermal reservoir at Ogiri, Kyushu, Japan. Proceedings, 34th Workshop on geothermal reservoir engineering, Stanford University.

Kuster Subsurface Instruments, PO Box 90909, Long Beach California. KPG Service Manual.

Lachenbruch, A.H., Sass, J.H., Munroe, R.J., Moses Jr., T.H., 1976. Geothermal setting and simple heat conduction models for the Long Valley Caldera. J. Geoph. Res. v81, 769—784.

Ladner, F., Häring, M.O., 2009. Hydraulic characteristics of the Basel 1 enhanced geothermal system. Transactions, Geothermal Resources Council v33, 199—203.

Layman, E., Soemarinda, S., 2003. The Patuha vapor-dominated resource West Java Indonesia. Proceedings, 28th Workshop on geothermal reservoir engineering, Stanford University.

Libert, F., Pasikki, R.G., 2010. Identifying important reservoir characteristics using calibrated wellbore hydraulic models in the Salak geothermal field, Indonesia. World Geothermal Congress paper 2285.

Libert, F., Peter, Pasikki, R., Yoshioka, K., Looner, M., 2009. Real-time acid treatment performance analysis of geothermal wells. Proceedings, 34th Workshop on Geothermal Reservoir Engineering, Stanford University.

Lopez, D.L., Matus, A., Castro, M., Magana, M.I., Sullivan, M., 2008. Implications of Temporal Changes in Solute Concentrations for the Mass Balance of the Berlin Geothermal Reservoir. Transactions, Geothermal Resources Council v32, 459—465.

Lopez, S., Bouchot, V., Lakhssassi, M., Calcagno, P., Grappe, B., 2010. Modeling of Bouillante geothermal field (Guadeloupe, French Lesser Antilles). Proceedings, 35th Workshop on Geothermal Reservoir Engineering, Stanford University.

Malate, R.C.M., Acqui, A.A., 2010. Steam production from the expanded two-phase region in the Southern Negros geothermal field, Philippines. World Geothermal Congress paper 2411.

Mannington, W., O'Sullivan, M.J., Bullivant, D., 2004. Computer modelling of the Wairakei-Tauhara geothermal system, New Zealand. Geothermics 33, 401−419.

Marini, L., Cioni, R., 1985. A chloride method for the determination of the enthalpy of steam/water mixtures discharged from geothermal wells. Geothermics V14 (1), 29−34.

Mathews, C.S., Russell, D.G., 1967. Pressure buildup and flow tests in wells. Monogr. no1, Soc. Pet. Eng., Dallas.

Matsunaga, I., Niitsuma, H., Oikawa, Y., 2005b. Review of the HDR development at Hijiori site, Japan. Proceedings, World Geothermal Congress paper 1635.

Matsunaga, I., Yanagisawa, N., Sugita, H., Tao, H., 2005a. Tracer tests for evaluation of flow in a multi-well and dual fracture system at the Hijiori HDR test site. Proceedings, World Geothermal Congress 2005 paper 1619.

Maturgo, O.O., Sanchez, D.R., Barroca, G.B., 2010. Tracer test using naphthalene disulfonates in Southern Negros geothermal production field, Philippines. Proceedings, World Geothermal Congress 2010 paper 2406.

Mégel, T., Kohl, T., Gérard, A., Rybach, L., Hopkirk, R., Downhole pressures derived from wellhead measurements during hydraulic experiments. Proceedings, World Geothermal Congress 2005 paper 1636.

Menzies, A.J. Pham, M., 1995. A Fieldwide Numerical Simulation Model of The Geysers Geothermal Field, California, USA, Proceedings, World Geothermal Congress, pp. 1697−1702.

Menzies, A.J., Pham, M., 1993. Results from a field-wide numerical model of The Geysers geothermal field, California. Transactions, Geothermal Resources Council v17, 259−265.

Menzies, A.J., 1981. Static formation temperature tests an evaluation. EDC-PNOC Geothermal Conference.

Menzies, A.J., Granados, E.E., Puente, H.G., Pierres, L.O., 1995. Modeling discharge require-ments for deep geothermal wells at the Cerro Prieto geothermal field, Mexico. Proceedings, 20th Workshop on geothermal reservoir engineering, Stanford University, pp. 63−69.

Menzies, A.J., Swanson, R.J., Stimac, J.A., 2007. Design issues for deep geothermal wells in the Bulalo geothermal field, Philippines. Transactions, Geothermal Resources Council v31, 251−256.

MIT, 2007. The future of geothermal energy.

Moench, A.F., Atkinson, P.G., 1978. Transient-pressure analysis in geothermal steam reservoirs with an immobile vaporising phase. Geothermics v7 (2−4), 253−264.

Moore, J.N., Allis, R., Renner, J.L., Mildenhall, D., McCulloch, J., 2002. Petrological evidence for boiling to dryness in the Karaha-Telega Bodas geothermal system, Indonesia. Proceedings 27th Workshop on geothermal reservoir engineering, Stanford University, pp. 223−232.

Mori, Y., 1970. Exploitation of Matsukawa geothermal area. UN Symposium on the development and use of geothermal resources. Special Issue, Geothermics v2 pt2 pp. 1150−1156.

Mroczek, E.K., Stewart, M.K., Scott, B.J., 2004. Chemistry of the Rotorua geothermal field Part 3: Hydrology. IGNS Client Report, 2004/178 to Environment Bay of Plenty.

Muffler, L.P.J. 1977. 1978 USGS geothermal resource assessment. Proceedings, 1st Workshop on Geothermal Reservoir Engineering, Stanford University, pp. 3−8.

Muffler, L.P.J., Cataldi, R., 1978. Methods for the regional assessment of geothermal resources. Geothermics v7, 53–89.

Muffler, L.P.J., 1978. Assessment of geothermal resources of the United States – 1978. U.S. Geological Survey, Circular 790, p. 163.

Murray, L.E., Rohrs, D.T., Rossknecht, T.G., 1995. Resource evaluation and development strategy, Awibengkok field. Proceedings, World Geothermal Congress, pp. 1525–1539.

Muskat, M., 1937. The flow of homogeneous fluids through porous media. McGraw-Hill, New York.

Najurieta, H.L., 1980. A theory for pressure transient analysis in naturally fractured reservoirs. J. Pet. Tch. v32 (7), 1241–1250.

Nakanishi, S., Kawano, Y., Tokada, N., Akasaka, C., Yoshida, M., Iwai, N., 1995. A reservoir simulation of the Oguni field, Japan, using MINC type fracture model. Proceedings, World Geothermal Congress, pp. 1721–1726.

Nakanishi, S., Pritchett, J.W., Tosha, T., 2001. Changes in ground surface geophysical signals induced by geothermal exploitation – computational studies based on a numerical reservoir model for the Oguni geothermal field, Japan. Transactions, Geothermal Resources Council v25, 657–664.

Nakao, S., Ishido, T., Takahashi, Y., 2007. Numerical simulation of tracer testing data at the Uenotai geothermal field, Japan. Proceedings, 32nd Workshop on Geothermal Reservoir Engineering, Stanford University, pp. 207–212.

Nathenson, M., 1975. Physical factors determining the fraction of stored energy recoverable from hydrothermal convection systems and conduction-dominated areas. U.S. Geological Survey, Open-file report 75–525, p 50.

Newsom, J., O'Sullivan, M.J., 2001. Modelling of the Ohaaki geothermal system. Proceedings, 26th Stanford Workshop on geothermal reservoir engineering.

Nordquist, G.A., Acuña, J., Stimac, J., 2010. Precision gravity modelling and interpretation at the Salak geothermal field, Indonesia. World Geothermal Congress paper 1378.

Nordquist, G.A., Protacio, J.A., Acuña, J., 2004. Precision gravity monitoring of the Bulalo geothermal field, Philippines: independent checks and constraints on numerical simulation. Geothermics v33, 37–56.

Norton, D., Knight, J., 1977. Transport phenomena in hydrological systems: Cooling plutons. Am. J. Sci. v277, 937–981.

Nygren, A., Ghassemi, A., Cheng, A., 2005. Effects of cold-water injection on fracture aperture and injection pressure. Transactions, Geothermal Resources Council v29, 183–187.

Okandan, E., 1988. Geothermal reservoir engineering. Kluwer Academic, Dordrecht.

Onur, M., 2010. Analysis of well tests in Afyon Ömer-Gecek geothermal field Turkey. World Geothermal Congress paper 2220.

Onur, M., Aksoy, N., Serpen, U., Satman, A., 2005. Analysis of well-tests in Balcova-Narlidere geothermal field, Turkey. Proceedings, World Geothermal Congress 2005 paper 1131.

Onur, M., Cinar, M., Ilk, D., Valko, P.P., Blasingame, T., Hegeman, P.S., 2008. An investigation of recent deconvolution methods for well-test data analysis. Proc., Soc. Pet. Eng. J., (June 2008), 226.

Onur, M., Sarak, H., Türeyen, Ö.I, 2010. Probabilistic resource estimation of stored and recoverable thermal energy for geothermal systems by volumetric methods. World Geothermal Congress paper 2219.

Orizonte, R.G., Amistoso, A.E., Aqui, A.R., 2000. Reservoir management during 15 years of exploitation: Southern Negros geothermal production field. Proceedings, World Geothermal Congress 2000, 2773–2778.

Orizonte, R.G., Amistoso, A.E., Malate, R.C.M., 2003. Projecting the performance of Nasuji-Sogongon production wells with the additional 20 MWE power plant in the Palinpinon-2 area, Southern Negros geothermal production field, Philippines. Transactions, Geothermal Resources Council v27, 765−770.

Orizonte, R.G., Amistoso, A.E., Malate, R.C.M., 2005. Optimising the Palinpinon-2 geothermal reservoir, Southern Negros geothermal production field, Philippines. Proceedings, World Geothermal Congress 2005 paper 1155.

O'Sullivan, M.J., Pruess, K., Lippmann, M.J., 2001. State of the art of geothermal reservoir simulation. Geothermics 30, 395−429.

O'Sullivan, M.J., Yeh, A., Mannington, W.I., 2009. A history of numerical modelling of Wairakei geothermal field. Geothermics 38, 155−168.

Palaktuna, M., Akin, S., Sayik, t., Karahan, Ç., Bakraç, S., 2010. Interpretation of the short term production test of Seferihisar geothermal field. World Geothermal Congress paper 2234.

Pasikki, R.G., Gilmore, T.G., 2006. Coiled tubing acid stimulation; the case of AWI 8-7 production well in Salak geothermal field, Indonesia. Proceedings, 31st Workshop on Geothermal Reservoir Engineering, Stanford University.

Pasikki, R.G., Libert, F., Yoshioka, K., Leonard, R., 2010. Well stimulation techniques applied at the Salak geothermal field. World Geothermal Congress paper 2274.

Pastor, M.S., Fronda, A.D., Lazaro, V.S., Velasquez, N.B., 2010. Resource assessment of Philippine geothermal areas. World Geothermal Congress paper 1616.

Peter, P., Acuña, J., 2010. Implementing mechanistic pressure drop correlations in geothermal wellbore simulators. World Geothermal Congress paper 3234.

Pham, M., Klein, C., Ponte, C., Cabeças, R., Martins, R., Rangel, G., 2010. Production/injection optimization using numerical modelling at Ribeira Grande, São Miguel, Azores, Portugal. World Geothermal Congress paper 2261.

Ponte, C., Cabeças, R., Martins, R., Rangel, G., Pham, M., Klein, C., 2009a. Numerical modelling for resource management at Ribiera Grande, São Miguel, Azores, Portugal. Transactions, Geothermal Resources Council v33, 847−853.

Ponte, C., Cabeças, R., Rangel, G., Martins, R., Kelin, C., et al., 2010. Conceptual modelling and tracer testing at Ribeira Grande, São Miguel, Azores, Portugal. World Geothermal Congress paper 2260.

Ponte, C., Cabeças, R., Rangel, G., Martins, R., Klein, C., Pham, M., 2009. Conceptual modelling and tracer testing at Ribeira Grande, São Miguel, Azores, Portugal. Transactions, Geothermal Resources Council v33, 839−846.

Porras, E.A., Bjornsson, G., 2010. The Momotombo reservoir performance on 27 years of exploitation. World Geothermal Congress paper 0638.

Pritchett, J.W., 1985. WELBOR: a computer program for calculating flow in a producing geothermal well. S-Cubed report SSS-R-85-7283, La Jolla, California.

Pritchett, J.W., 2007. Geothermal reservoir engineering in the United States since the 1980s. Transactions, Geothermal Resources Council v31, 31−38.

Protacio, J.A.P., Golla, G.U., Nordquist, G.A., Acuña, J., San Andres, R.B., 2000. Gravity and elevation changes at the Bulalo geothermal field, Philippines: independent checks and constraints on numerical simulation. Proceedings, 21st New Zealand geothermal workshop, pp. 115−119.

Pruess, K., Bodvarsson, G., Schroeder, R.C., Witherspoon, P.A., Marconini, R., Neri, G., et al., (1979). Simulation of the depletion of two-phase geothermal reservoirs paper SPE7699, presented 54th Annual TechnnicalConference, SPE-AIME

Pruess, K., Narasimham, T.N., 1985. A practical method for modelling fluid and heat flow in fractured porous media. Soc. Pet. Eng. J., Feb, 14−26.

Pruess, K., Oldenburg, C., Moridiis, G., 1999. TOUGH2 user's guide, version 2.0. Report LBNL-43134, Lawrence Berkeley Laboratory.

Quijano, J., 2000. Exergy analysis for the Ahuachapan and Berlin geothermal fields, El Salvador. Proceedings, World Geothermal Congress, pp. 861–866.

Ramey, H.J., Jr. A reservoir engineering study of the Geysers geothermal field. Submitted as evidence, Reich & Reich, petitioners v Commissioner of Internal Revenue, 1969 Tax Court of the United States, 52, T.C. No.74.

Ramos-Candelaria, M.N., Garcia, S.E., Hermoso, D.Z., Bayrante, L.F., Mejorada, A.V., Application of geochemical techniques to deduce the reservoir performance of the Palinpinon geothermal field, Philippines—an update. Transactions, Geothermal Resources Council v21, pp. 231–239.

Reed, M.J., 2007. An investigation of the Dixie Valley geothermal field, Nevada, using temporal moment analysis of tracer tests. Proceedings, 32nd Workshop on Geothermal Reservoir Engineering, Stanford University.

Reyes, A.G., Giggenbach, W.F., Saleras, J.R.M., Salonga, N.D., Vergara, M.C., 1993. Petrology and geochemistry of Alto Peak, a vapor-cored hydrothermal system, Leyte Province, Philippines. Geothermics 22, 479–519.

Reyes, J.L., Li, K., Horne, R.N., 2003. Estimating water saturation at The Geysers based on historical pressure and temperature production data and by direct measurement. Transactions, Geothermal Resources Council v27, 715–726.

RJV (Rotokawa Joint Venture), 2010. Ngatamariki geothermal power station Resource consent applications and Assessment of environmental effects.

Robertson-Tait, A., Klein, C.W., McLarty, L., 2000. Utility of the data gathered from the Fenton Hill project for development of enhanced geothermal systems. Transactions, Geothermal Resources Council v24, 161–167.

Robinson, R., Iyer, H.M., 1979. Evidence from teleseismic P-wave observations for a low velocity body under the Roosevelt Hot Springs geothermal area, Utah. Transactions, Geothermal Resources Council v3, 585–586.

Rodgriguez, R., Aunzo, A., Kote, J., Gumo, S., 2008. Depressurization as a strategy for mining ore bodies within an active geothermal system. Proceedings, 33rd Workshop on geothermal reservoir engineering, Stanford University.

Rose, P., McCulloch, J., Adams, M., Mella, M., 2005. An EGS experiment under low wellhead pressures. Proceedings, 30th Workshop on Geothermal Reservoir Engineering, Stanford University.

Rose, P., Mella, M., Kasteler, C., Johnson, S.D., The estimation of reservoir pore volume from tracer data. Proceedings, 29th Workshop on Geothermal Reservoir Engineering, Stanford University, pp. 330–338.

Rose, P.E., Mella, M., McCullough, J., 2006. A comparison of hydraulic stimulation attempts at the Soultz, France, and Coso, California Engineered geothermal systems. Proceedings, 31st Workshop on Geothermal Reservoir Engineering, Stanford University.

Rosenberg, M., Wallin, E., Bannister, S., Bourguinon, S., Sherburn, S., Jolly, G., et al., Tauhara stage II geothermal project: Geoscience report. GNS Science consultancy report 2010/38.

Roux, B., Sanyal, S.K., Brown, S., 1979. An improved approach to estimating true reservoir temperature from transient temperature data. Proceedings, 5th Workshop on geothermal reservoir engineering, Stanford University, pp. 373–384.

Salveson, J.O., Cooper, A.M., 1979. Exploration and development of the Heber geothermal field, Imperial Valley, California. Transactions, Geothermal Resource Council v3, 605–608.

Sambrano, B.G., Aragon, G.M., Nogara, J.B., 2010. Quantification of injection fluids effects to Mindanao geothermal production field productivity through a series of tracer tests, Philippines. World Geothermal Congress paper 2418.

Sammis, C.G., An, L., Ershaghi, I., 1992. Determining the 3-D fracture structure in The Geysers geothermal reservoir. Proceedings, 17th Workshop on Geothermal Reservoir Engineering, Stanford University, pp. 79—85.

Sanyal, S.K., Butler, S.J., 2005. An analysis of power generation prospects from enhanced geothermal systems. Transactions, Geothermal Resources Council, Vol. 29, 131—137.

Sanyal, S.K., Sarmiento, Z., 2005. Booking geothermal energy reserves. Transactions, Geothermal Resources Council v29, 467—474.

Sanyal, S.K., 2000. Forty Years of Production History from The Geysers Geothermal Field, California. Trans. Geothermal Resources Council, Vol. 24, 2000.

Sanyal, S.K., 2009. Optimization of the economics of electric power from enhanced geothermal systems. Proceedings, 34th Workshop on Geothermal Reservoir Engineering, Stanford University.

Sanyal, S.K., Butler, S.J., Brown, P.J., Goyal, K., Box, T., 2000. An Investigation of Productivity and Pressure Decline Trends in Geothermal Steam Reservoirs. Proc. World Geothermal Congress, 873—877.

Sanyal, S.K., Granados, E.F., Menzies, A.J., 1995. Injection-related problems encountered in geothermal projects and their mitigation: the United States experience. Proceedings, World Geothermal Congress, 2019—2022.

Sanyal, S.K., Henneberger, R.C., Klein, C.W., Decker, R.W., 2002. A methodology for the assessment of geothermal energy reserves associated with volcanic systems. Transactions, Geothermal Resources Council v26, 59—64.

Sanyal, S.K., Klein, C.W., Lovekin, J.W., Henneberger, R.C., 2004. National assessment of U.S. geothermal resources—a perspective. Transactions, Geothermal Resources Council v28, 355—362.

Sanyal, S.K., Klein, C.W., McNitt, J.R., Henneberger, R.C., MacLeod, K., 2008. Assessment of power generation capacity at The Geysers geothermal field, California. Proceedings, 33rd Workshop on Geothermal Reservoir Engineering, Stanford University.

Sanyal, S.K., Klein, C.W., McNitt, J.R., Henneberger, R.C., MacLeod, K., 2007. Assessment of the power generation capacity of the Western Geopower leasehold at The Geysers geothermal field, California. Transactions, Geothermal Resources Council v31, 447—455.

Sanyal, S.K., Menzies, A.J., Brown, P.J., Enedy, K.L., Enedy, S., 1989. A Systematic Approach to Decline Curve Analysis for The Geysers Steam Field, California. Trans. Geothermal Resources Council, v13, October, 1989.

Sanyal, S.K., Menzies, A.J., Brown, P.J., Enedy, K.L., Enedy, S., 1989. A systematic approach to decline curve analysis for The Geysers steam field, California. Transactions, Geothermal Resources Council v13, 415—421.

Sanyal, S.K., Morrow, J.W., Butler, S.J., 2007. Net power capacity of geothermal wells versus reservoir temperature — a practical perspective. Proceedings, 32nd Workshop on Geothermal Reservoir Engineering, Stanford University.

Sarak, H., Korkmaz, D., Onur, M., Satman, A., 2005. Problems in the use of lumped-parameter models for low-temperature geothermal fields. Proceedings, World Geothermal Congress paper 1134.

Sarak, H., Türeyen, Ö.İ., Onur, M., 2009. Assessment of uncertainty in estimation of stored and recoverable heat energy in geothermal reservoirs by volumetric methods. Proceedings, 34th Workshop on geothermal reservoir engineering, Stanford University.

Sarmiento, Z.F., Björnsson, G., 2007. Reliability of early modelling studies for high-temperature reservoirs in Iceland and The Philippines. Proceedings, 32nd Workshop on Geothermal Reservoir Engineering, Stanford University.

Sarmiento, Z.F., 1986. Waste water injection at Tongonan geothermal field: results and implications. Geothermics 15, 295−308.

Satman, A., Serpen, U., Onur, M., Aksoy, N., 2005. A study on the production and reservoir performance of Balcova-Narlidere geothermal field. Proceedings, World Geothermal paper 1133.

Sausse, J., Dezayes, C., Genter, A., Bisset, A., 2008. Characterization of fracture connectivity and fluid flow pathways derived from geological interpretation and 3D modelling of the deep seated EGS reservoir of Soultz (France). 33rd Workshop on Geothermal Reservoir Engineering, Proceedings, Workshop on Geothermal Reservoir Engineering, Stanford University University.

Seastres, J.S., Salonga, N.D., Saw, V.S., Clotworthy, A.W., 1996. Hydrology of the greater Tongonan geothermal system, Philippines and its implications to field exploitation. Transactions, Geothermal Resources Council v20, 713−720 [has p-z].

Seibt, P., Kabus, F., Hoth, P., 2005. The Neustadt-Glewe geothermal power plant − practical experience in the reinjection of cooled thermal waters into sandstone aquifers. Proceedings, World Geothermal Congress paper 1209.

Shook, G.M., 1998. Prediction of reservoir pore volume from conservative tracer tests. Transactions, Geothermal Resources Council v22, 477−480.

Shook, G.M., 2003. A simple fast method of estimating reservoir geometry from tracer tests. Transactions, Geothermal Resources Council v27, 407−411.

Shook, G.M., 2005. A systematic method for tracer test analysis: an example using Beowawe tracer data. Proceedings, 30th Workshop on Geothermal Reservoir Engineering, Stanford University, pp. 166−171.

Siega, C.H., Saw, V.S., Andrino Jr., R.P., Cañete, G.F., 2005. Well-to-well two-phase injection using a 10in diameter line to initiate well discharge in Mahanagdong geothermal field, Leyte, Philippines. Proceedings, World Geothermal Congress paper 1024.

Silberman, M.L., White, D.E., Keith, T.E.C., Dockter, R.D., 1979. Duration of hydrothermal activity at Steamboat Springs, Nevada, from ages of spatially associated volcanic rocks. Geol. Surv. Prof. Pap. (US) 458-D.

Sore, M.L., 2000. Geothermal development and changes in surficial features: examples from the western United States. Proceedings, World Geothermal Congress paper 0149.

Sorey, M.L., Fradkin, L. Ju., 1979. Validation and comparison of different models of the Wairakei geothermal reservoir. Proc., 5th Workshop on geothermal reservoir engineering, Stanford University, pp. 215−220.

Sorey, M.L., 1980. Numerical code comparison project − a necessary step towards confidence in geothermal reservoir simulators. Proceedings, 5th Workshop on Geothermal Reservoir Engineering, Stanford University, pp. 253−257.

Sta. Maria, R.B., Villadolid-Abrigo, M.F., Sussman, D., Mogen, P.G., 1995. Development strategy for the Bulalo geothermal field, Philippines. World Geothermal Congress, 1803−1805.

Stark, M.A., Box Jr., W.T., Beall, J.J., Goyal, K.P., Pingol, A.S., 2005. The Santa-Rosa − Geysers recharge project, Geysers geothermal field, California, USA. Transactions, Geothermal Resources Council v29, 145−150.

Stefansson, V., Steingrimsson, B., 1980a. Geothermal logging. National Energy Authority, Iceland.

Stefansson, V., Steingrimsson, B., 1980b. Production characteristics of wells tapping two-phase reservoirs at Krafla & Namafjall. Proceedings, 6th Workshop on Geothermal Reservoir Engineering, Stanford University, pp. 49−59.

Stefansson, V., 1997. Geothermal reinjection experience. Geothermics 26, 99—139.

Stefansson, V., 2005. World geothermal assessment. Proceedings, World Geothermal Congress paper 0001.

Steingrimsson, B., Bodvarsson, G.S., Gunnlaugsson, E., Gislason, G., Sigurdsson, O., 2000. Modelling studies of the Nesjavellir geothermal field, Iceland. Proceedings, World Geothermal Congress, pp. 2899—2904.

Stimac, J., Nordquist, G., Suminar, A., Sirad-Azwar, L., 2008. An overview of the Awibengkok geothermal system, Indonesia. Geothermics 37, 300—331.

Straus, J.M., Schubert, G., 1977. Thermal convection of water in a porous medium: effects of temperature- and pressure-dependent thermodynamic and transport properties. J. Geoph. Res. v82 (2), 325—333.

Strobel, C.J., 1976. Field case studies of pressure buildup behavior in Geysers steam wells. Proceedings, 2nd Workshop on Geothermal Reservoir Engineering, Stanford University, pp. 143—149.

Strobel, C.J., 1993. Bulalo field, Philippines: reservoir modelling for prediction of limits to sustainable generation. Proceedings, 18th Workshop on geothermal reservoir engineering, Stanford University, pp. 5—10.

Sugrobov, V.M., 1970. Evaluation of operational reserves of high-temperature waters. United Nations Symposium on the Development and use of Geothermal Resources. Spec. Issue, Geothermics v2, 1256—1260.

Tenma, N., Yamaguchi, T., Tezuka, K., Karasawa, H., 2000. A study of the pressure-flow response of the Hijiori reservoir at the Hijiori HDR test site. Proceedings, World Geothermal Congress paper 0828.

Tenma, N., Yamaguchi, T., Tezuka, K., Oikawa, Y., 2001. Estimation of productivity from the shallow reservoir using pressure monitoring data in SKG-2 at the Hijiori HDR test site. Transactions, Geothermal Resources Council v25, 199—202.

Tenma, N., Yamaguchi, T., Tezuka, K., Kawasaki, K., Zyvoloski, G., 2002. Prductivity changes in the multi-reservoir system at the Hijiori HDR test site during the long-term circulation test. Transactions, Geothermal Resources Council v26, 261—266.

Tezuka, K., Ohsaki, Y., Miyairi, M., Takahashi, Y., 2003. High temperature (400°C) fluid-density logging tool and application in resolving an unstable production mechanism of the well in Uenotai geothermal field in Japan. Transactions, Geothermal Resources Council v27, 737—741.

Thorhallson, S., 1979. Combined generation of heat and electricity from a geothermal brine at Svartsengi in S.W. Iceland. Transactions, Geothermal Resources Council v3, 733—736.

Thorkellson, T., 1910. The hot springs of Iceland. K.Dan. Vidensk. Selsk. Skr.

Thorsteinsson, T., Eliasson, J., 1970. Geohydrology of the the Laugarnes hydrothermal system in Reykjavik, Iceland. United Nations Symposium on the Development and use of Geothermal Resources. Spec. Issue, Geothermics v2, 1191—1204.

Todaka, N., Akasaka, C., Xu, T., Pruess, K., 2005. Reactive geothermal transport simulations to study incomplete neutralization of acid fluid using Multiple Interacting Continua method in Onikobe geothermal field, Japan. Proceedings, World Geothermal Congress paper 0806.

Tokita, H., Haruguchi, K., Kamenosono, H., 2000. Maintaining the rated power output of the Hatchobaru geothermal field through an integrated reservoir management. Proceedings, World Geothermal Congress paper 0381.

Torres, A.C., Lim, W.Q., 2010. Bulalo 99 workover: complex zonal isolation. World Geothermal Congress paper 2141.

Torres-Rodriguez, M.A., Mendoza-Covarrubias, A., Medina-Martinez, M., 2005. An update of the Los Azufres geothermal field, after 21 years of exploitation. Proceedings, World Geothermal Congress paper 0916.

Truesdell, A., Walters, M., Kennedy, M., Lippmann, M., 1993. An integrated model for the origin of The Geysers geothermal field. Transactions, Geothermal Resources Council v17, 273—280.

Truesdell, A.A., Nathenson, M., Frye, G.A., Downhole measurements and fluid chemistry of a Castle Rock steam well, The Geysers, Lake County, California. Geothermics v10 (2), 103—114.

Truesdell, A.H., White, D.E., 1973. Production of superheated steam from vapor-dominated geothermal reservoirs. Geothermics v2 (3—4), 154—173.

Tulinius, H., Correia, H., Sigurdsson, O., 2000. Stimulating a high-enthalpy well by thermal cracking. Proceedings, World Geothermal Congress, 1883—1888.

Türeyen, Ö.İ., Onur, M., Sarak, H., 2009. A generalized nonisothermal lumped-parameter model for liquid-dominated geothermal reservoirs. Proceedings, 34th Workshop on geothermal reservoir engineering, Stanford University.

Türeyen, Ö.İ., Sarak, H., Onur, M., 2007. Assessing uncertainty in future pressure changes predicted by lumped-parameter models: a field application. Proceedings, 32nd Workshop on Geothermal Reservoir Engineering, Stanford University, pp. 253—262.

Upton, P.S., The wellbore simulator SIMU000. Proceedings, World Geothermal Congress, pp. 2851—2856.

Urbino, M.E.G., Zaide, M.C., Malate, R.C.M., Bueza, E.L., 1986. Structural flowpaths of reinjected returns based on tracer tests — Palinpinon 1, Philippines. 7th NZ Geothermal Workshop, pp. 53—58.

Vallejos-Ruiz, O., 2005. Lumped parameter model of the Miravalles geothermal field, Costa Rica. Proceedings, World Geothermal Congress paper 1120.

Verma, M.P., 2008. Moodychart: an ActiveX component to calculate frictional factor for fluid flow in pipelines. Proceedings, 33rd Workshop on Geothermal Reservoir Engineering, Stanford University.

Villa, I.M., Puxeddu, M., 1994. Geochonology of the Larderello geothermal field: new data and the closure temperature. issue. Contr. Mineral. Petrol., v115 (4) 415—426.

Villacorte, J.D., Malate, R.C.M., Horne, R.N., 2010. Application of nonparametric regression on well histories at Mahiao and Mahanagdong sector of Leyte geothermal production field, Philippines. Proceedings, 34th Workshop on Geothermal Reservoir Engineering, Stanford University.

Villadolid, F.L., 1991. The application of natural tracers in geothermal development: the Bulalo, Philippines experience. Proceedings, 13th New Zealand geothermal workshop, Auckland University, pp. 69—74.

von Knebel, W., 1906. Studien in den Thermengebieten Islands. Naturwiss. Rundsch.

Walters, M.A., Sternfeld, J.N., Haizlip, J.R., Drenick, A.F., Combs, J., 1991. A vapor-dominated high-temperature reservoir at The Geysers California. Transactions, Geothermal Resources Council Special Report no 17, 77—87.

Warren, J.E., Root, P.J., 1963. The behaviour of naturally fractured reservoirs. Soc. Pet. Eng. J., Sept 1963, 245—255.

Wattenbarger, R.A., Ramey Jr., H.J., 1969. Well test interpretation of vertically fractured gas wells. J. Pet. Tech. v21 (5), 625—632.

Weinbrandt, R.M., Ramey, H.J. Jr., Cassé, F.J., 1972. The effect of temperature on relative and absolute permeability of sandstones. paper SPE4142, presented SPE-AIME 47th Annual Fall Meeting, Texas.

White, D.E., 1957. Thermal waters of volcanic origin. Geol. Soc. Am. Bull. 68, 1637—1658.

White, D.E., 1967. Some principles of geyser activity, mainly from Steamboat Springs, Nevada. Am. J. Sci. 265, 641—684.

White, D.E., 1968. Hydrology, activity and heat flow of the Steamboat Springs thermal system, Washoe County, Nevada. Geol. Surv. Prof. Pap. (US) 548-C.

White, D.E., Fournier, R.O., Muffler, L.J.P., Truesdell, A.H., 1975. Physical results of research drilling in thermal areas of Yellowstone National Park, Wyoming US Geol. Surv. Prof. Pap. 982

White, D.E., Muffler, L.J.P., Truesdell, A.H., 1971. Vapor-dominated hydrothermal systems compared with hot-water systems. Econ. Geol. v66 (1), 75—97.

Whiting, R.L., Ramey Jr., J.J., 1969. Application of material and energy balances to geothermal steam production. J. Pet. Tech. v21, 893—900.

Williams, C.F., 2004. Development of revised techniques for assessing geothermal resources. Proceedings, 29th Workshop on Geothermal Reservoir Engineering, Stanford University, pp. 276—280.

Williams, C.F., 2007. Updated methods for estimating recovery factors for geothermal resources. Proceedings, 32nd Workshop on Geothermal Reservoir Engineering, Stanford University.

Williamson, K.H., 1992. Development of a Reservoir Model for The Geysers Geothermal Field. Monograph on The Geysers Geothermal Field, Geothermal Resources Council, Special Report, no. 17, pp. 179—187.

Wilson D.M., Gould J., 1989. A simultaneous temperature and pressure tool. 11th New Zealand Geothermal Workshop, 1989.

Wooding, R.A., 1957. Steady state free convection of liquid in a saturated permeable medium. J. Fl. Mech. 2, 273—285.

Wooding, R.A., 1963. Convection in a saturated porous medium at large Rayleigh number or Peclet number. J. Fl. Mech. 15, Pt 4, 527—544.

Wooding, R.A., 1981. Aquifer models of pressure drawdown in the Wairakei-Tauhara geothermal region. Wat. Resour. Res. v17 (1), 83—92.

Wright, M.C., Beall, J.J., 2007. Deep cooling response to injection in the Southeast Geysers. Transactions, Geothermal Resources Council v32, 457—461.

Wyborn, D., de Graaf, L., Hann, S., 2005. Enhanced geothermal development in the Cooper Basin area, South Australia. Transactions, Geothermal Resources Council v29, 151—156.

Yanagisawa, N., Rose, P., Wyborn, D., 2009. First tracer test at Cooper Basin, Australia HDR reservoir. Transactions, Geothermal Resources Council v33, 281—284.

Yglopaz, D., Austria, J.J., Malate, R.C., Buñing, B., Sta. Ana, F.X., Salera, J.R., et al., 2000. A large-scale well stimulation campaign at Mahanagdong geothermal field (Tongonan), Philippines. Proceedings, World Geothermal Congress, 2303—2307.

Yoshioka, K., Stimac, J., 2010. Geologic and geomechanical reservoir simulation modelling of high pressure injection, West Salak, Indonesia. World Geothermal Congress paper 2286.

Zais, E.J., Bodvarsson, G., 1980. Analysis of production decline in geothermal reservoirs. Rep LBL-11215 (GREMP-10), Lawrence Berkeley Laboratory, Berkeley, California.

Zúñiga, S.C., 2010. Design and results of an increasing permeability test carried out in the first deep well drilled in the Borinquen geothermal field, Costa Rica. World Geothermal Congress paper 2211.

Index

Printed in the United States
By Bookmasters